The Landforms of the Humid Tropics, Forests and Savannas

Geographies for Advanced Study

Edited by Professor Stanley H. Beaver, M.A., F.R.G.S.

The Landforms of the Humid Tropics, Forests and Savannas

J. Tricart

Director, Centre de Géographie Appliquée,
University of Strasbourg

Translated by

Conrad J. Kiewiet de Jonge
Professor of Geography
California State University, San Diego

LONGMAN

LONGMAN GROUP LIMITED
London

*Associated companies, branches and representatives
throughout the world*

English translation © Longman Group Limited 1972

Le modelé des régions chaudes, foréts et savanes
first published by Société d'Édition d'Enseignment Supérieur, Paris, 1965

English edition first published 1972

ISBN 0 582 48157 0

**Dedicated to Benita and Anna Maria Kiewiet de Jonge
for their love and forbearance**

ᶜ

*Set in Monophoto Baskerville
and printed in Great Britain by William Clowes & Sons, Limited
London, Beccles and Colchester*

Contents

Contents

List of illustrations

List of tables

Introduction

The land masses of the low latitudes are characterised by particular land-forms, some of which drew the attention of scientific explorers of the past century. These landforms, among others, include the well-known inselbergs that were first named and described by German authors who visited Cameroon and southern Africa. Since then our knowledge of the tropics has increased considerably, especially that of the inhospitable forested zone where observations are difficult. Sugarloaves and monolithic domes have also become incorporated in the scientific literature. A more detailed knowledge of laterites is leading to an abandonment of this equivocal term (that for some means soils and for others indurated crusts) in favour of more precise words such as latosols and cuirasses.[1] The demands of economic development in sparsely inhabited regions have had as a consequence the multiplication of pedologic studies and a beginning of ecologic observations. Tropical soils have been intensely investigated during the last twenty years, and our knowledge of them has changed considerably.

Studies concerning the morphogenic processes of the tropics are unfortunately not as advanced, and certain pedologists have been handicapped in their work by the lack of geomorphological knowledge, a shortcoming that they have tried their best to fill. Cuirasses, so important from both the agricultural and the mining points of view (some are mineral deposits), are closely related to the geomorphic evolution. This is only one specific example of the dependence of a number of pedologic phenomena on the morphogenic processes. The study of tropical geomorphology, like other chapters of climatic geomorphology, has been delayed by the uniformist ideas of 'normal erosion'. Differences between the landforms of the forested tropics or the tropical savannas and the landforms of the humid temperate zone were less evident to early investigators than the differences they observed between the 'normal' relief of the temperate zone and the glaciated relief or the dune covered wastelands of the desert. Consequently the tropics were systematically neglected: no importance was attached to differences in the long-profiles of midlatitude and tropical rivers, to the fact that the

[1] A term long used in the French literature and which is equally well suited to the English language. Its use is highly recommended in order to avoid confusion (see Chapter 1) (KdeJ).

latter are characterised by a high frequency of rapids alternating with almost currentless reaches. The landforms themselves were equally neglected. As to the inselberg concept, introduced by German authors, it was only belatedly and timidly adopted abroad.

To these unfavourable intellectual circumstances resulting from an erroneous methodological orientation were added difficulties inherent in the tropical environment. Interest in the tropics remained limited up to the Second World War. Few regions had been the object of systematic economic development, and those that had, like Malaya, Java, Ceylon, and Hawaii were exploited in a rather pragmatic fashion. No geomorphological studies had been undertaken. Elsewhere, unhealthiness, lack of communications, difficulties of observation in regions covered by vegetation have all helped prevent a beginning of systematic exploration. The only existing studies were those of precursors such as Freise who, though he attributed an excessive rapidity to it, was the first to draw attention to the importance of chemical weathering in the humid tropics. Another precursor of modern trends was Sapper. The ideas of both were partially re-used by de Martonne as a result of a trip to Brazil, but were unfortunately not incorporated in his *Traité de géographie physique.*

It was really only after the Second World War that the systematic geomorphological exploration of the humid tropics was begun. The creation of the Institut Français d'Afrique Noire[1] at Dakar, with resident scientists, has played a very important role in the systematic investigation of the tropics. It made possible, for example, such works as Rougerie's monograph on the forested Ivory Coast (1960). The rapid progress in sanitary techniques after the war enabled an effective fight against endemic diseases (the enemy of field work). The development of air transport made sojourns of several months possible, while the construction of roads not only made travel easy but also provided the indispensable geological cross-sections in regions where rocks are deeply buried by regolith; the availability of air photographs and sometimes of accurate maps also contributed by finally enabling the man in the field to locate himself and to supplement his field observations, always few in number in a forest environment. Lastly, the effort made in the systematic economic development of certain parts of South America and especially Africa have simultaneously created favourable conditions of investigation at the very moment when a revolution in the fundamental concepts of geomorphology (the concept of climatic geomorphology) directed scientific men to the specific problems of these regions.

Nevertheless, the humid tropics are still imperfectly known and the level of our knowledge about them is far from that of the other large climatic zones. Enormous land areas, particularly in South America, still remain practically uninvestigated. Good topographic maps are still exceptional, partly because of the difficulty of making maps in regions where observation is extremely difficult. Even aerial photographs are often unsatisfactory, be-

[1] Now called Institut Fondamental d'Afrique Noire.

cause of atmospheric conditions and the obstruction of vegetation. Field work thus remains arduous and often hazardous, in spite of advancement in tropical medicine. Beyond the roads observation is limited by the lack of wide horizons, want of sufficient rock outcrops, and the slowness of travel. In most cases the data available do not go beyond the precision of a reconnaissance. Moreover, they are most often obtained on accessible itineraries or during the most favourable season, and are consequently partial and perhaps even unreliable. A savanna looks utterly different at the end of the dry and at the end of the wet season. Erroneous ideas can easily arise from observations made systematically during the same season. What kind of picture does the tourist obtain of the Alps when he only sees them in summer?

Finally, the humid tropics present truly scientific difficulties inherent in themselves. Everywhere life is exuberant and assumes great geomorphic importance that in places reaches world maxima. Interactions between biologic, pedogenic, and geomorphic factors assume the greatest complexity. The ground climate in a dense tropical forest differs more from the climate of the free air than in any other biogeographic environment. In savannas brushfires subject the ground to extreme heat: temperatures may rise to 600°C (1 112°F) or even 800°C (1 472°F) for a few minutes. About many such phenomena we have as yet very little exact and quantitative knowledge. Microclimatological measurements, in spite of all their ecologic importance, still remain exceptional. The study of brushfires, which is so important, is still in its initial stage. The study of soil microbiology has hardly begun; but even so, it may be surmised that micro-organisms play a very important role in the migration of iron oxides and in the concentration of aluminium oxides.

Our knowledge of the morphogenic characteristics of the humid tropics therefore remains insufficient, in spite of certain pioneer works (e.g. Rougerie's important monograph) that are far too few in number. We are still in a preliminary phase in which systematic exploration is only beginning. But having become aware of the essential characteristics we can now pose the problems. Some answers are already apparent, though we are short of precise data, particularly quantitative data. We perceive certain factors while others remain elusive due to insufficient observations. We sense the interplay of certain phenomena but are unable to go into details. For example, discontinuous overland flow seems much more important in the forests of southern Ivory Coast than in the forests of the state of Bahia in Brazil. Why? In Ivory Coast the litter is never abundant and the water runs underneath it, as Rougerie has observed. In Bahia, on the contrary, the litter is thick and the topsoil is composed of a spongy humic horizon, both of which retard overland flow. The explanation seems to lie in the manner of decomposition of the organic matter, but what is the role of the type of litter, of micro-organisms and of microclimate? To what degree can these observations be generalised? Because we are ignorant of the processes that cause such differences, and lack microclimatic data, it is difficult to answer these questions, which are of great importance to geomorphology. Under such conditions our present state of knowledge is only provisional. Many questions remain unanswered. We hope they will stimulate research and bring new observations, for which there is great need.

In Chapter 1 we define what we mean by the tropics. We then explain their climatic characteristics and analyse the tropical morphogenic processes, after which we delimit and describe their morphoclimatic subdivisions. Chapter 2 analyses the effects of the tropical climates on the azonal fluvial and littoral processes. Chapter 3 deals with the morphogenic system of the forested wet tropics, and Chapter 4 with the morphogenic system of the alternating wet and dry savanna environments. Lastly, Chapter 5 is devoted to the influence of palaeoclimates on the present tropical landscapes.

1
The tropics defined

From the astronomical point of view the tropics are that zone of the earth that lies between the Tropic of Cancer and the Tropic of Capricorn. They are, of course, in general characterised by high temperatures. Because in geomorphology the temperature criterion is more important than the astronomical, the tropics will here be defined from the point of view of temperature and are therefore not coextensive with the intertropical zone. We must exclude from this zone the cooler highlands and the cool, foggy littoral deserts. These regions are cool (or cold) in spite of their latitude, because of the geographic factors of elevation and the configuration of land and sea. If it is true that all the constantly warm regions of the earth are located within the intertropical zone, it is not true that the whole intertropical zone belongs to a uniformly warm morphoclimatic environment.

Uninterrupted heat is, nevertheless, a zonal phenomenon on the globe. This heat entails very important biologic consequences which affect the land-forming processes and create an original zonal morphogenic climate. The zone affected by this heat must be defined. To do so we first review its climatic characteristics, then point out the peculiar morphogenic processes that accompany them, and, lastly, examine its extent and limits.

Climatic characteristics of the tropics

The tropics, then, are first of all defined by thermal criteria. These must be examined first. But temperature also has important repercussions on humidity, another extremely important factor in geomorphology. These repercussions are studied later.

Temperature regimes

The thermal criterion used to define the tropics has a statistical value only. The tropics do not have a monopoly of high temperatures. The temperature maxima of other zones are often even higher. At Montpellier, France, for instance, temperatures higher than 45°C (113°F) have been registered several times, whereas they have never been recorded in Abidjan or in

Douala. Average temperatures of the warmest month are higher in a number of extratropical regions than they are in a large part of the intertropical zone.

Table 1.1. *Average temperatures of the warmest month*

INTERTROPICAL ZONE			EXTRATROPICAL ZONES		
	°C	°F		°C	°F
Konakry	27·8	82·0	Madrid (655 m)	25·1	77·2
Abidjan	28·3	83·0	Athens	26·6	79·8
Dakar	28·9	84·0	Beirut	28·4	83·2
Accra	27·2	81·0	Orléansville	30·2	86·3
Douala	26·7	80·0	Baghdad	34·4	94·0
Tamatave	26·6	79·8	In Salah	37·2	99·0
Bangalore	26·6	79·8	Teheran (900 m)	29·4	85·0
Colombo	28·3	83·0	Shanghai	26·9	80·5
Djakarta	26·5	79·7	Salt Lake City (1 344 m)	25·0	77·0
Port Moresby	27·8	82·0	Chicago	23·8	74·8
San Salvador (700 m)	26·0	78·8	Phoenix	32·2	90·0
San Juan, P.R.	26·7	80·0	Abilene	27·8	82·0
Port of Spain	26·1	79·0	Charleston	27·2	81·0
Catamarca	26·8	80·2	Washington	24·4	76·0
Rio de Janeiro	25·6	78·0	Asunción	26·9	80·5

In the arid zone (western United States, northern Mexico, northern Sahara, the Persian Gulf area) and in certain dry intramontane basins (central Asia, Iran) the average temperature of the warmest month is often higher than that recorded in a large part of the intertropical zone. Even inside the very intertropical zone, averages of 30°C (86°F) for the warmest month are passed only in a limited number of regions; for example, the Sudanese zone of Africa from Kayes to Djibouti, which is part of the unbroken bloc of arid lands with torrid summers of the Sahara; the south of Arabia, which prolongs the deserts with torrid summers of the Persian Gulf area; and, in the southern hemisphere, the Kalahari and the northwest of Australia, both of which extend somewhat beyond the tropic. Regions in which the warmest month has a temperature of over 30°C (i.e. regions with 'torrid' summers) are slightly more extensive in extratropical latitudes than between the tropics. They are marginal in the intertropical zone, being located close to the Tropics of Cancer and Capricorn, nowhere reaching equatorial latitudes (between 10° N and 10° S latitude).

The localisation of regions characterised by one or more torrid months (with average temperature above 30°C) is therefore significant. It introduces within the intertropical zone an important differentiation which has other aspects, and to which we will return.

It is not so much the presence of very high temperatures that characterises the tropics as it is the absence of cool temperatures. Few are the regions in the intertropical zone outside the highlands and foggy coastal deserts where the mean temperature of the coolest month falls below 20°C (68°F), a temperature which corresponds to the warmest month in a large part of the temperate zone. The only intertropical regions where the average temperature of the coolest month falls below 20°C are situated along the Tropic of Cancer: from the coast of Mauritania to the Red Sea, from the west coast of Arabia to the mouth of the Indus; and, in southeast Asia, in the north of Laos, Tonkin, and the extreme south of China. The drop in temperature, however, is small. Furthermore, most of these intertropical regions with cool months are located within the dry zone. The most important exception is in monsoon Asia, from northeast Burma to Formosa. Hong Kong, for example, has an average temperature of 15°C (59°F) in its coolest month.

With the exception in southeast Asia and the encroachment of the dry zone, the intertropical zone is therefore characterised by the absence of cool months whose mean temperature is below 20°C; and the existence of two thermal regions: one characterised by the presence of torrid months (average above 30°C), the other by their absence. The uninterruptedness of high temperatures during all months of the year is the fundamental characteristic of the tropics. No other climatic zone has this characteristic. Beyond the tropics one or more months of the year are cool even in regions where the warmest month is torrid as in the north of the Sahara, Arabia, India, the Persian Gulf area, or central Australia.

The following additional characteristics are reflected in the temperature regime of the tropics:

(*a*) A small annual temperature range which is always less than 10°C (18°F). But here, as with the averages of the warmest month, it is necessary to distinguish two varieties which, as a matter of fact, coincide with the two thermal regimes mentioned above. A zone closer to the equator has an average range of less than 5°C (9°F), whereas extensive areas near the Tropics of Cancer and Capricorn have ranges between 5° and 10°C (9° and 18°F).

As indicated in Table 1.2, the stations whose average range is over 5°C (9°F) are in one of the following groups:

Stations with at least one torrid month (Kayes, Bobo-Dioulasso, Bombay, Rangoon, Darwin).
Stations close to the tropic in whose vicinity the thermal seasons are more clearly defined (Rio de Janeiro, Key-West, Havana, Tamatave, Hong Kong, Asunción).
Stations located in monsoonal climates (Bombay, Madras, Rangoon, Hong Kong).

These stations are typical of different varieties of intertropical climates some of which are transitional with subtropical climates.

3

Table 1.2. *Temperature ranges in the tropics*

HIGHER THAN 5°C (9°F)	AVERAGES Min.	Max.	RANGE	LOWER THAN 5°C (9°F)	AVERAGES Min.	Max.	RANGE
Dakar	21·7	28·9	7·2	Konakry	26·1	27·8	1·7
	71·0	*84·0*	*13·0*		*79·0*	*82·0*	*3·0*
Kayes	27·2	35·6	8·4	Beyla	22·8	25·6	2·8
	81·0	*96·0*	*15·0*		*73·0*	*78·0*	*5·0*
Bobo-Dioulasso	25·0	31·1	6·1	Abidjan	26·1	28·3	2·2
	77·0	*88·0*	*11·0*		*79·0*	*83·0*	*4·0*
Tamatave	20·5	26·6	6·1	Douala	23·9	26·7	2·8
	69·0	*80·0*	*11·0*		*75·0*	*80·0*	*5·0*
Bombay	23·9	30·0	6·1	Colombo	26·1	28·3	2·2
	75·0	*86·0*	*11·0*		*79·0*	*83·0*	*4·0*
Madras	24·4	32·2	7·8	Saigon	26·0	29·9	3·9
	76·0	*90·0*	*14·0*		*79·0*	*86·0*	*7·0*
Rangoon	25·0	30·6	5·6	Manilla	24·8	27·0	2·2
	77·0	*87·0*	*10·0*		*76·5*	*80·5*	*4·0*
Darwin	25·0	30·0	5·0	Djakarta	25·5	26·5	1·0
	77·0	*86·0*	*9·0*		*78·0*	*79·5*	*1·5*
Havana	22·0	28·0	6·0	Port Moresby	26·1	27·8	1·7
	71·5	*82·5*	*11·0*		*79·0*	*82·0*	*3·0*
Key-West	20·6	28·9	8·3	San Salvador	22·5	26·0	3·5
	69·0	*84·0*	*15·0*		*72·5*	*79·0*	*6·5*
Rio de Janeiro	20·0	25·6	5·6	Port of Spain	23·9	26·1	2·2
	68·0	*78·0*	*10·0*		*75·0*	*79·0*	*4·0*
Asunción	17·0	26·9	9·9	Manaos	26·7	28·3	1·6
	62·5	*80·5*	*18·0*		*80·0*	*83·0*	*3·0*
Hong Kong	15·0	27·8	12·8	Cuiabá	24·1	27·8	3·7
	59·0	*82·0*	*23·0*		*75·5*	*82·0*	*6·5*

Note. °C are given first; °F in italic figures.

(*b*) The regularity of temperatures although not typical everywhere is normal in most of the tropics. The wide range of temperature so characteristic of the midlatitudes, and which takes away much of the significance of mean temperatures, is absent here or limited to certain climatic types which are already transitional to the subtropics. The cold and hot waves which accompany a cyclonic circulation are absent or very much attenuated. For example, the temperature seldom drops below 18°C (64·4°F) at Abidjan in West Africa, and this only happens during the boreal winter when the trade winds are particularly strong and bring dry air masses from the high pressures over the Sahara. Invasions of austral polar air, on the other hand, occur every now and then in South America: above 1 000 m (3 300 ft) in the interior of Paraná, a little to the south of the Tropic of Capricorn, they may cause nocturnal frost; even in Rio de Janeiro they can lower the temperature to 15°C (59°F) and even less. Florida, and to a lesser degree Cuba, is in a

4

similar position as regards boreal polar air. Florida also experiences occasional frosts. In southeast Asia the monsoonal regime provokes an important lowering of the mean temperature of the coolest month as far south as the tropic, as is evident from Hong Kong data. Such cool air invasions, even without the accompaniment of frost, play an important ecologic role. Some tropical plants cannot endure them: *Begoniaceae*, for example, die when the temperature falls below 6–8°C (43–46°F) as the circulation of the sap is impeded. Among cultivated plants the same happens to cassava. The appearance of average minima below approximately 10°C (50°F) constitutes an important ecologic limit. Even if the forest does not change in physiognomy, its floristic composition changes. Some typically intertropical species are entirely eliminated, and the passage toward a subtropical forest is made, as is the case in southern Brazil. Such a change roughly corresponds to mean temperatures of 18°C (64·4°F) for the coolest month or, in short, to the appearance of *cool months*.

(*c*) The daily temperature range is higher than the annual temperature range. Intertropical regions whose annual temperature range is generally lower than 10°C (18°F) have a daily range which, on the contrary, generally exceeds this value. The only regions where the daily range is less than 8–10°C (14–18°F) are islands (West Indies, Hawaii). Even in humid interior regions where the annual range is very low, below 3–4°C (5–7°F), the average daily range exceeds 8–10°C (14–18°F) (Congo Basin, Amazonia). It reaches approximately 15°C (17°C) in semiarid regions such as the Sudan and certain parts of India. Troll has compared daily and annual temperature ranges and made a map of the areas of the world where the daily range exceeds the annual range. This map clearly shows that such areas occur exclusively in the intertropical zone. Except for certain islands (West Indies, Hawaii) of oceanic climate, the northeast of the Deccan influenced by the monsoon, and a narrow fringe close to the Tropic of Cancer in North Vietnam, Arabia, and the Sudan Republic, the entire intertropical zone is included within the area where the daily range exceeds the annual range. Inversely, this area hardly extends beyond the tropics $(23\frac{1}{2}°)$ except for an offshoot in South America along the Andes. This thermal regime is of great geomorphic importance, as such short period temperature variations are easily checked by the screens that constitute the dense forests. Temperature variations affect the air much more than the ground in forested regions. Furthermore, they cannot penetrate the soil to any great depth due to lack of time. In temperate lands variations of temperature between winter and summer are recorded to a depth of some 15 m (50 ft). In the intertropical zone daily temperature oscillations do not reach down further than a few cm, at the most 10 or 15 cm (4 or 6 in). Further down, in the neighbourhood of 1 to 2 m (40 to 80 in), the soil temperature remains constant and 2° to 4°C (4° to 7°F) above the yearly average. It is therefore out of the question to attribute exfoliation and the parting of slabs several decimetres thick (a foot or so) on inselbergs and monolithic domes to thermal action.

5

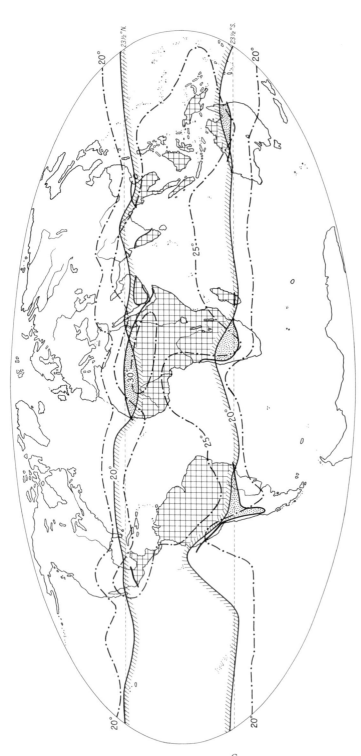

FIG. 1.1. Thermal delimitation of the tropics (data extracted from C. P. Péguy).

The mean annual isotherms of 20°C (68°F), 25°C (77°F), and 30°C (86°F) are shown by lines of alternating dashes and dots; the 20°C (68°F) January (northern hemisphere) and July (southern hemisphere) isotherms, corresponding to the coldest months, are shown by two solid hachured lines.

Areas where the mean daily temperature range exceeds the mean annual temperature range are delimited by a heavy line indicated by a grid of vertical, by horizontal lines when not a single month has a mean temperature below 20°C (68°F), and by dots when this is not the case.

The thermal regime of intertropical regions has therefore a marked originality which affects almost the whole zone. It has a profound unity despite variety and diversity. And this unity (enduring high temperatures, predominance of the daily range over the annual range) contains aspects which have important morphogenic consequences. However, certain differences which occur in this zone in connection with temperature are also encountered in a more pronounced way in connection with the water budget.

The nature and regime of the rainfall

Although the rainfall has common characteristics throughout the tropics, there are great differences which are at the origin of the subdivisions of this climatic zone.

COMMON CHARACTERISTICS

The common characteristics of the rainfall are due both to high temperatures and the zonal characteristics of the atmospheric circulation.

In contrast to the midlatitudes rain is not produced by cyclonic circulation and frontal effects. It results primarily from convection which causes air masses to rise to elevations of 6 000 to 10 000 and even to 15 000 m (20 000, 30 000, 50 000 ft) where the water vapour then condenses. The droplets which result serve as nuclei of condensation during their fall and so increase in size. Cumulonimbus are typical clouds of the tropics. They sometimes form during squalls when local air masses of different humidity are affected by unequal upward movements. Such convectional rains are much affected by unequal relief particularly where a mountain creates a forced ascent and causes a rapid increase in rainfall. This explains the marked contrasts in rainfall which occur on the windward and leeward sides not only of mountains, but even of hills.

Because tropical rains are produced by convection they are generally accompanied by lightning and thunder. These thunderstorms have two distinct characteristics: they are generally localised, which causes considerable differences in the weather records over a short period of time between neighbouring stations; and they are generally intense, especially where there are orographic influences. Here again the intertropical regions do not hold world records; regions such as the Cévennes, the eastern Pyrenees, Liguria, and Japan have occasionally recorded heavier downpours than many an intertropical station. Overall, however (data are far from being obtained systematically), it seems that the tropics have a particularly high frequency of torrential rains, which also contribute a higher proportion than elsewhere to the total annual rainfall. In regions with a dry season the first downpours of the rainy season are usually produced by electrical storms and are particularly violent. Such storms are of great geomorphic importance. The measurements of Barat (1957) have demonstrated that they are characterised by large drops for which, unfortunately, we lack comparative data

7

in other climatic zones. When no obstacle is interposed splash erosion is particularly violent.

The high temperature is partly responsible for the nature of the rainfall. It acts in two ways: on the one hand it inhibits, except at very high elevations, the formation of raindrops from ice crystals, as occurs in medium high clouds in cooler regions; and on the other hand it permits the presence of relatively large quantities of water vapour per cubic metre of air, as the dew point increases rapidly with temperature. Air masses that are cooled during convection therefore release considerable quantities of water through condensation. The result is a high total rainfall in the form of heavy showers or, more often, as intense downpours, wherever convection takes place. If convection cannot take place, aridity is pronounced, as in the littoral deserts where a thermal stratification composed of a cooled air mass in contact with a cold ocean current is overlain by warmer air. The inversion of temperature effectively blocks convection, which excludes rain: only widespread fogs are possible. Outside this extreme case rainfall is also prevented by the existence of a regular and constant horizontal atmospheric circulation which homogenises the air mass, thus precluding differential heating and convection. For this reason regions subjected to the trade winds are dry. The displacement of the trade wind belts causes the fluctuation of the zone of convection which corresponds approximately to the thermal equator. Such has been the traditional and still valid explanation of the nature of the rainfall in the intertropical zone.

Because the potential specific humidity increases rapidly with temperature, warm air has a very high water vapour capacity. Also, an equal variation of temperature produces a much greater volume of condensation between, for example, $25°$ and $30°C$ ($77°$ and $86°F$) than between $5°$ and $10°C$ ($41°$ and $50°F$). When the relative humidity is high, close to 100 per cent, condensations caused by moving air masses are much more copious in a warm than in a cool atmosphere. This explains the fact that the summer rains of midlatitude thunderstorms closely resemble the rains of the tropics. When the requisite conditions are realised, intertropical showers are copious. Daily rainfalls of a few centimetres shed in only a few hours are the rule. Even in dry regions such as the semiarid Sahelian zone of Africa or the subarid zones of the interior of Australia, torrential rains (more than 30 mm: $1·2$ in in 24 hours) are important and contribute a relatively high fraction of the annual total. The generation of rain by convection tends to produce an 'all or nothing' situation so frequent in nature. Either it does not rain because the raindrops evaporate before reaching the ground, or it rains heavily because the larger raindrops escape evaporation and instead rapidly grow in size through coalescence with others during their fall.

Evaporation also increases appreciably with temperature because of the rapid rise in the water vapour capacity of the air. In order to pass from a relative humidity of 80 to 100 per cent at a temperature of $25°C$ ($77°F$) much more water is needed than is required to pass the same conditions at

8

a temperature of 5°C (41°F). Therefore the air is capable of absorbing a great quantity of water vapour as soon as it is heated, but without increasing its relative humidity. Evaporation thus has the potential of being very intense; the ground dries up rapidly after a shower, especially when the relative humidity is not very high. The same is true of the leaves and branches of plants; after the rain an important fraction of the precipitation retained by the aerial organs of the vegetation returns directly to the atmosphere through evaporation. It has been calculated that in the rainforest of the Congo Basin about two-thirds of the precipitation originates in moisture provided by local evaporation and one-third only by extraneous water vapour, principally from the oceans through migrating air masses. A kind of permanent evaporation–condensation regime therefore tends to be established in warm and sufficiently humid regions under the effect of convectional air.

Because evaporation is closely linked to high daily temperature oscillations one should not use the norms of temperate lands to understand the aridity or humidity of a region, a season, or a particular month. A month that receives, for example, 20 mm (0·8 in) of rain in the tropics is a dry month, in fact a *very dry* month. De Martonne's (1926) aridity index (P/T + 10) (12P/T + 10 for monthly averages), conceived in 1923 for humid regions, especially tropical ones, is utterly unusable because it underestimates the importance of temperature. Other authors have proposed a comparison of the precipitation to the mean temperature multiplied by a coefficient. For example, Gaussen (1952) uses P/2T, whereas Birot (1959) proposes P/4T.[1] The latter bases himself on the fact that the column of water evaporated per month in a moderately humid atmosphere and from a soil which is denied desiccation is approximately 100 mm (4 in) for a temperature of 25°C (77°F). This value is obtained from actual measurements made in the wet tropics. Rains of more than 100 mm per month are necessary under such conditions for the water to infiltrate the soil and to feed the ground water reserves. If the average monthly temperature drops to 20°C (68°F), which happens during certain months in the neighbourhood of the Tropics of Cancer and Capricorn, the same approximate formula gives 80 mm (3·2 in) to delimit humid months. We agree with Birot that *dry* months in the intertropical zone are those whose ratio P/4T is less than one. Whereas *very dry* months are those in which this proportion drops below one-half. This criterion may help define a few basic types of tropical climates.

TYPES OF TROPICAL CLIMATE

The proportion of dry months and their distribution throughout the year varies considerably from place to place in the intertropical zone. At certain

[1] In these three formulas P = average precipitation in mm, T = mean temperature in degrees Centigrade. In de Martonne's formula a month is semiarid if the aridity index is below 20, arid if below 5. For Gaussen a dry month is when P < 2T, but he has later modified this (for résumé see Péguy, *Précis de climatologie*, p. 250 (KdeJ)).

stations the whole year is characterised by dry months, even by very dry months. These stations, of course, are located in the dry zone (for example, Lima, Timbuktu). Outside the dry zone there are a fairly large number of climatic types.

Peguy (1961) groups certain climates under the name *irregular equatorial* regime; its type station is Jaluit in the Marshall Islands. The annual rainfall is high: 4 492 mm (180 in). Not a single month receives less than 100 mm (4 in), the minimum being 291 mm (12 in), a very high value. Under such conditions there is plenty of water in the regolith in all seasons and the drainage is good, which is important in the weathering process. Unfortunately we do not know of any study which throws light on the consequences of such a climate on geomorphology. This type, however, is rather limited in extent and principally realised on islands.

The *pure equatorial* regime has been defined long ago as being character-ised by a succession of four seasons during the course of the year: rainy seasons more or less at the time of the equinoxes, a short dry season in August in the northern hemisphere, and a long dry season corresponding to the boreal winter. In a typical station such as Colombo only three months re-ceive less than 100 mm of rain (4 in): August with 97 mm in the short 'dry' season (it can hardly be considered a dry month), and, in the midst of the long dry season, January with 82 mm (3·3 in) and February with 48 mm (2 in). In such a climate, with an annual total of 2 242 mm (90 in), plants benefit almost continuously from abundant water in the soil. It is only in February that it dries up and a deficiency develops, but this is too indistinct and too short to play a limiting role and to exclude certain plant species. The microclimate peculiar to the underbrush of the rainforest indeed con-siderably attenuates the atmospheric drought when it is not more pro-nounced than this, as we will see later. Still, a decrease in rainfall of this order of magnitude is enough to cause a decrease in stream flow, a lowering of the watertable in alluvial plains and on interfluves, and thus may cause the precipitation of certain solutions whose concentrations rise close to saturation.

The *transitional equatorial* regime includes two varieties, according to Pe-guy. One of them, whose type station is Kinshasa, is characterised by the persistence of a still well-marked secondary minimum which receives less than one-twelfth of the annual total: February with 89 mm (3·6 in) is a dry month. The long dry season is more pronounced with three very dry months (June, July, and August) and one dry month (September). The rigour and duration of the long dry season are, for us, the essential characteristics. This type is also found widely represented in Ivory Coast. A station like Abidjan, which receives a total of 1934 mm (77 in) (compare 1 412 mm: 56 in for Kinshasa), has a well marked short dry season: August 50 mm (2 in), Sept. 69 mm (2·7 in); and a long dry season equally well marked: Dec. 71 mm (2·8 in), Jan. 34 mm (1·4 in), and Feb. 47 mm (1·9 in). All transitions exist between this type and the pure equatorial type. For example, Tabou (in southwestern Ivory Coast) with a total of 2 215 mm (86 in) has as poorly marked a short dry season in August, 90 mm (3·6 in), as Colombo; but also a long dry season of three months: Jan. 42 mm (1·7 in), Feb. 54 mm (2·2 in), and March 81 mm (3·2 in). Fifty miles to the north, at Grabo, the short dry season becomes more marked: 61 mm (2·4 in) in July; and the long one less marked: 83 mm (3·3 in) in Jan., and 88 mm (3·5 in) in Feb. Whereas at Sassandra both dry seasons are important; the short dry season (three months below 100 mm—4 in): 23 mm (0·9 in) in August, 42 mm (1·7 in) in Sept., and 93 mm (3·7 in) in Oct.; and the long dry season (four months): 78 mm (3·1 in) in Dec., 21 mm (0·8 in) in Jan. (which is very little), 26 mm (1 in) in Feb., and 75 mm (3 in) in March. The total, it is true, falls to 1 426 mm (57 in) as against 2 327 mm (93 in) at Grabo. It is equal, however, to the total of Kinshasa where the long dry season, which is even drier, is also shorter (three rainless months), and where the short rainy season is not as wet and shorter. From the ecological point of view it seems that the most important fact is the existence of more than two very dry consecutive months. They create a shortage of water which is reflected in the vegetation by the elimination of certain demanding plants which cannot survive. From the hydrological point of view small *marigots*[1] completely dry up in the Sassandra area during the long dry season. The water table swings widely, and soil desiccation assumes important proportions.

The other transitional equatorial regime is represented by Libreville, where the short dry season is only marked by a small breach in a single maximum. The principal dry season is proportionally extended, and there is close resemblance to the tropical (wet–dry) type. February, which represents the short 'dry' season, receives 226 mm (9 in) and is therefore a humid month. From the ecological point of view there is only one wet season alternating with one dry season. The dry season, however, is longer than that of pure equatorial climates. It includes June with 7 mm (0·3 in), July with 3 mm (0·1 in), August with 17 mm (0·7 in), and Sept. with 98 mm (almost 4 in), which is at the limit. The annual total of 2 410 mm (96 in) is high.

When the principal dry season does not exceed three months and the total annual rainfall remains sufficiently high—above 1 200–1 500 mm (48–60 in)—a forest vegetation can maintain itself. However, it can no longer be an evergreen rainforest, demanding in water and incapable of withstanding a three months drought, but must be a semi-evergreen seasonal forest adapted to a fairly considerable seasonal water shortage.

The *tropical* regime is characterised by the alternation of a single wet and a single dry season. Libreville is transitional between this and the equatorial regime. As with the latter, Peguy distinguishes several varieties of tropical regimes.

The *moderately contrasted tropical* regime, of which Djakarta is the type station, is characterised by moderate seasonal pluviometric contrasts. No month receives less than 15 mm (0·6 in) nor more than 500 mm (20 in). The

[1] For definition see pp. 60 and 68.

dry season lasts only four months, only one of which is very dry. June receives
91 mm (3·6 in), July 66 mm (2·6 in), August 34 mm (1·4 in), and September
67 mm (2·7 in). Analogous climates are found in Africa; for example, Man,
in west central Ivory Coast, has an annual total of 1 777 mm (71 in), and the
dry season includes the months of November with 62 mm (2·5 in), December with 23 mm (0·9 in), January with 17 mm (0·7 in), and February with
57 mm (2·3 in). Deciduous seasonal forests are adapted to this type of climate.
It has a large distribution in the southern hemisphere and extends beyond
the tropics (23½° lat.), for instance, in the area of Rio de Janeiro.

The *tropical regime with long rainy season* is characterised by a more accentuated dry season. Some months have totals of less than 15 mm (0·6 in), which
is also the case of some transitional equatorial climates as that of Kinshasa
(with three months without rain). The type station proposed by Peguy is
Majunga on the west coast of Madagascar. The average annual rainfall is
1 563 mm (63 in). The dry season begins in April with 77 mm (3·1 in) and
continues in May with 4 mm (0·16 in), June (0), July (0), August with 3 mm
(0·12 in), Sept. with 4 mm (0·16 in), and Oct. with 31 mm (1·2 in). As a
matter of fact, in this station the dry season in the ecological sense of the
term is longer than the wet season, and therefore it is a bad choice for a type
station. In Madras, on the other hand, five months receive more than
100 mm (4 in), and July, at the beginning of the wet season, 98 mm. Konakry
is more typical although the dry season is very short as there are seven
months with totals over 100 mm (4 in): Dec. receives 10 mm (0·4 in), Jan.
and Feb. 0 mm, March 3 mm (0·12 in), and April 37 mm (1·5 in). Touba,
60 miles north of Man, Ivory Coast, represents a transition to this type. A
secondary minimum, although poorly marked, persists in June which receives 139 mm (5·5 in) as against 149 mm (5·9 in) in May. But a dry season
of five months extends from November, which receives 37 mm (1·5 in), to
May, which receives 81 mm (3·2 in), with totals of 18 mm (0·7 in) each for
December, January, and February. The year receives 1 337 mm (53 in).
These climates are on the limit of the forest. With high annual totals of over
1 500 mm (60 in), as at Konakry, deciduous seasonal forests will still grow.
When the figures drop below this value, as at Touba, such forests find it hard
to survive and give way to wooded savannas which, in many cases, seem to
have developed as the result of anthropic degradation.

The *tropical regime with short rainy season* is characterised both by smaller
pluviometric totals, generally less than 1 500 mm (60 in) and as low as 1 200
mm (48 in), and a rainy season limited to five months at the most. As the
dry season lengthens it also becomes more pronounced. Peguy assigns Kayes,
Mali, as the characteristic station of this variety. As a matter of fact, it is a
marginal station as its average annual rainfall is only 736 mm (29 in). In
West Africa most authors place the limit between the Sudanese zone, characterised by savanna, and the Sahelian zone, the domain of thorn shrubs
and desert grasses, at an average rainfall of 750 mm (30 in), a limit which is
both ecologically and geomorphologically valid, and beyond which lies the

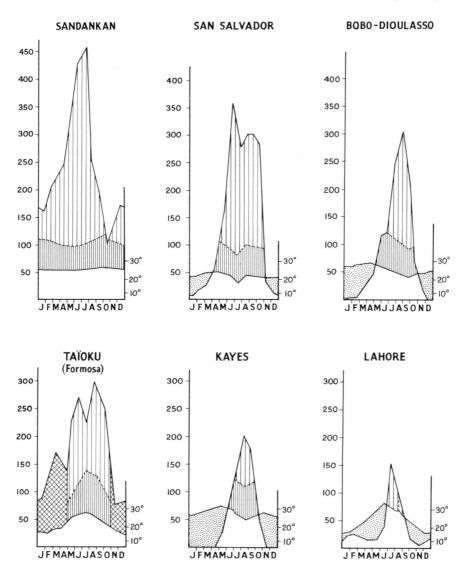

FIG. 1.2. Climatic diagrams (after Péguy, 1961)
        Vertical shading:
          Dense: humid months
          Open: very humid months
        Cross-hatching: 'temperate' months (mean temp. below 20°C:68°F)
          Dots: dry months
Note the conditions favourable to weathering in a subtropical location such as that of
Taïoku, Formosa.

dry zone. At Kayes the dry season lasts eight months from October (48 mm) (1·9 in) to May (15 mm) (0·6 in), with only 7 mm (0·28 in) in November, 4 mm (0·16 in) in December, 0 mm in January, February, and March, and 1 mm in April. In monsoon Asia a similar regime exists with sometimes much higher totals, as at Bombay (1 804 mm : 72 in) where the dry season also lasts eight months from October to May. The monthly totals of this season differ but little from those of Kayes, Mali. Only the rains of the wet season are more copious. This climate does not allow forest growth outside of flood plains where forests endure the dry season thanks to shallow and abundant ground water (gallery forests). The vegetation reflects a very marked seasonal aspect either as herbaceous formations (savannas) or as shrubby formations which dry up completely (*campo cerrado* of Brazil).

Therefore, whereas the tropics as a whole are delimited by thermal criteria, the degree of moisture determines the subdivisions. Before going any further into the details regarding the limits and subdivisions of the tropics, we will examine the characteristic morphogenic processes of the humid tropical zone.

Morphogenic processes characteristic of the humid tropics

Geomorphic effects of heat

Heat has a very great influence on a large number of morphogenic processes, especially those which are of a chemical nature. Statistically, Van 't Hof's law states that an increase of 10 °C (18 °F) multiplies the intensity of the reactions by 2·5. However, there is also a biological aspect: living things function at optimum temperatures, which vary according to species, but which are normally between 20° and 25 °C (68° and 77 °F). Heat, therefore, has a great influence on biochemical mechanisms. It not only modifies the biochemical mechanisms that are characteristic of other morphoclimatic zones but causes new ones to appear. We therefore first consider the effects of heat in general, and then analyse the specific mechanisms it sets in motion and define their climatic framework.

The biochemical processes include three successive steps:

1. The creation of organic matter, in the form of plants, from minerals and soil water through photosynthesis.
2. The supply of organic matter on the ground in the form of decomposing dead vegetal debris (humus).
3. The mineralisation of this humus with the help of a complex fauna and flora dominated by microscopic organisms which live off the organic matter and reduce it to mineral residues.

These various phenomena affect the morphogenic processes because plants, by means of their roots, modify the soil water budget, absorb minerals, and thus affect the weathering process. The supply of organic debris,

14

followed by their mineralisation, returns to the soil the mineral elements that were extracted from it, but in a different form and with another mobility. For example, certain savanna grasses contain as much as 5 per cent and even 7 per cent silica. This silica is extracted from the mineral environment by processes that are not well understood. It is later returned to the soil in the form of colloidal silica which is easily entrained.

The growth of vegetation, therefore, produces a genuine exchange between the soil and the lower layers of the atmosphere, an exchange which includes, first of all, the circulation of important quantities of water required by photosynthesis, the respiration and transpiration of plants, and the transfer of mineral matter. These exchanges, of course, operate with an intensity which is a function of temperature. But photosynthesis and respiration, the two biological activities essential to plants, do not have the same optimum temperature.

The optimum temperature for the creation of vegetal matter by photosynthesis is relatively low. It normally lies between 20° and 25°C (68° and 77°F) and approaches 30°C (86°F) only when the atmospheric content of carbon dioxide gas is unusually high. These temperatures are common in the intertropical zone. However, an essential fact must be taken into account: the temperatures which characterise a climate are those of the atmosphere and are measured under shelter, whereas the temperatures which concern plants are real temperatures which are strongly influenced by the local setting.

As a whole, therefore, the intertropical zone offers optimum thermal conditions for photosynthesis. The conditions are realised almost throughout the year and not only during a few summer months as in the temperate zone. This is why, for example, a tropical evergreen rainforest produces three to four times as much vegetal matter per hectare as a midlatitude forest. There is, therefore, a much greater production of litter.

Looked at in detail there are some small but important temperature variations. Temperatures sometimes pass the vital optimum, especially with the appearance of torrid months in the wet–dry tropics. Such months, which are also nearly always dry months, are characterised by a reduction in photosynthesis. Due to the lack of water photosynthesis is almost completely at a standstill. Plant life is slowed down exactly as in winter in the midlatitudes. The wet tropics without torrid temperatures are therefore much more favourable to photosynthesis and the growth of vegetal matter than the alternating wet–dry tropics with too high temperatures and unequally distributed rainfall. These ecologic differences are reflected in the plant formations as we pass from the evergreen rainforest (with optimum conditions of temperature and humidity) to the deciduous seasonal forest and the savanna. In a dense forest where the carbon dioxide content of the air may be twice that of the free atmosphere, the growth of sapling trees and some underbrush is much favoured because the temperatures are almost constantly above 25°C (77°F).

The respiration optimum is reached at much higher temperatures, between 35° and 50°C (95° and 122°F). This explains why the higher animals, such as birds and mammals, which do not employ photosynthesis, have a constant body temperature lying between these values. For plants, therefore, an optimum vitality is realised at temperatures of 20° to 25°C (68° to 77°F). Photosynthesis is then at a maximum for normal, free atmospheric conditions, and respiration–transpiration is not too intense. At higher temperatures photosynthesis decreases while the consumption of carbohydrates, which are decomposed in CO_2 and H_2O through respiration, increases. The increase of respiration simultaneously produces a loss of water which results in a more rapid extraction of water from the soil and probably causes an impoverishment of the sap in mineral matter. If there were no reaction under such conditions, the plant would lose through respiration all the carbohydrates it produces in photosynthesis. Its growth would be checked, and even its reserves would gradually be exhausted.

For this reason the metabolism of tropical plants changes during the most torrid months. Most of them reduce the activities of respiration and transpiration simply by shedding their leaves. Others adopt xerophytic characteristics such as glossy or hairy leaves (the latter impede the circulation of the air), or smaller leaves, in order to protect themselves from the excesses of respiration and transpiration during the hottest hours.

Such morphologic characteristics appear as soon as there is a seasonal shortage of moisture, as in the deciduous seasonal tropical forest. They increase when temperatures well above the optimum for photosynthesis are added to it; the resulting vegetation then is savanna or campo cerrado. If the humidity drops even lower there are further adaptations producing the xerophytic thorny formations of the dry regions of the earth.

The influence of temperature, combined with that of humidity, is reflected in the decomposition (mineralisation) of organic matter so important in pedogenesis, weathering and the elementary forms of surface drainage.

The dead organic matter has two geomorphic aspects:

1. The accumulation of organic matter on the ground, where it plays a very important role in lessening splash erosion, impeding overland flow, and absorbing water as a sponge, which in the case of humus may amount to three or four times its dry weight.
2. The decomposition of organic matter, which produces substances that are very active in the weathering of the subsurface.

The accumulation of organic matter is the result of an algebraic sum: the rate of conveyance of the vegetal debris and the rate of its decomposition. In the tropics the rate of decomposition is always high, so that in spite of an abundant supply little humus remains on the ground, often less than in temperate forests. Humus can only accumulate when powerful limiting factors impede the action of the micro-organisms which decompose it; the lack of oxygen is one of them. Very poorly drained soils tend to be peaty and anaerobic, which allows the accumulation, notably in marshes, of organic matter which only partly decomposes, resulting in peaty soils. Bottomlands flooded during a good part of the year, especially in the wet tropics, offer good examples. In savanna regions, where the seasons are much more contrasted, real peat is the exception, and bottomlands are only characterised by blackish or greyish humic soils. In the wet tropics the organic matter of poorly drained or very acid soils is less prone to decomposition: bacteria are hampered by pHs lower than 5·5. These soils are often humic at the surface, especially the soils which form on very acid, pure quartz sands. A layer of litter and a humic horizon a few centimetres thick may even develop because of high acidity. Such soils, which are transitional to tropical podzols, are common on the old Dunkirkian or Eemian[1] barrier beaches of Ivory Coast or eastern Brazil. They also occur on fixed dunes. Mohr and Van Baren (1954) report that on homoclinal ridges in Indonesia the faceslopes with good drainage are often characterised by oxidising conditions which cause the rapid disappearance of the organic matter and result in the formation of yellow or reddish soils, while the backslopes with poor drainage display grey humic soils with podzolic tendency. In hilly regions the poorly drained soils of the basal slopes become progressively richer in organic matter toward the humid bottomlands. The well drained hilltops, on the contrary, are characterised by an oxidising environment and soils that are coloured by iron hydroxides. Such catenas are found in savannas as well as in forests, but they are better developed in the latter because of a more abundant supply of organic matter.

Apart from poorly drained bottomlands and very acidic soils on quartz sands, the humid tropics have a characteristically poorly developed humic horizon. In general it is better developed under rainforests than under deciduous seasonal forests, a fact which plays an important role in the discontinuity of overland flow.

The decomposition of organic matter takes place in different ways according to temperature and, of course, also the pH and the abundance of oxygen which control the living conditions of micro-organisms. Whereas the proportion of fulvic to humic acid is about equal in the soils of the temperate zone, fulvic acid, according to Bachelier (1960), clearly predominates in ferrallitic soils. This is because different microbic floras transform the organic matter in the intertropical and temperate zones. The species of bacteria which destroy prehumic and humic acids are very active at temperatures of 27° to 28°C (80·6° to 82·4°F) and become completely inactive at temperatures of 16° to 20°C (60·8° to 68°F). Whereas fulvic acids are destroyed during the mineralisation of humus in temperate regions, in the humid tropics they are not destroyed but accumulate. Unfortunately we still know very little about the reaction of these fulvic acids on the minerals of the regolith and the rocks. As a whole, they seem to be more active than humic and prehumic acids. According to Mohr and Van Baren (1954):

[1] See footnotes pp. 87 and 90.

(*a*) bacteria have a high thermal optimum which is above 30°C (86°F) and especially above 35°C (95°F); at such temperatures they play an important role in the dissociation of carbohydrates;

(*b*) fungi, on the contrary, have a lower temperature optimum, not too different from the higher plants, which is generally comprised between 18° and 25°C (64·4° and 77°F); they also prefer an acid environment with a pH between 5·5 and 3·5;

(*c*) moulds need much oxygen but do not tolerate temperatures higher than 30° to 35°C (86° to 95°F).

The divers conditions obtaining in tropical soils are thus quite unequally favourable to different categories of micro-organisms.

A forest soil at a depth of a few decimetres (a foot or so) has a wellnigh constant temperature, generally a few degrees higher than the mean annual temperature of the air, that is, somewhere between 25° and 30°C (77° and 86°F). Whereas bacteria and moulds abound at this depth, fungi prefer the uppermost horizon where they find their optimum temperature at night. But fungi destroy much less organic matter than bacteria and moulds. They transform and fix the organic matter in their organism, many species incorporating about one-third of their weight as against one per cent only for bacteria. The superficial conditions that are more favourable to fungi thus allow the preservation of a little humus, especially when it is highly acid. This preservation occurs mostly in rather humid but not excessively warm environments: in short, in rainforests.

In savannas ground temperatures are much higher than under forest because of the existence of torrid months and because of the poor thermal protection provided by the vegetation. Temperatures of 40° to 50°C (104° to 122°F) are frequently encountered on a poorly protected soil, and temperatures as high as 70° to 75°C (158° to 167°F) have been recorded on bare black soils in India. Whereas this heat is unfavourable to fungi, drought is unfavourable to moulds. Only certain bacteria prosper under such conditions; they rapidly destroy the organic matter, and there is practically no humus or litter except in floodable bottomlands where the absence of oxygen hampers decomposition. During the rainy season bacteria find an environment very favourable to their wellbeing in the saturated soil. Even though our knowledge is still scanty, it does seem that bacteria play an important role in the fixation of iron. One might thus explain certain mechanisms which enable the mobilisation during the rainy season of the iron incorporated in certain species of bacteria, and released in the dry season when the desiccation of the soil causes the massive death of the micro-organisms. But this is still only an hypothesis.

Under deciduous seasonal forests greater drought and higher temperatures are detrimental to fungi and moulds during part of the year, whereas during the rainy season they benefit from conditions which differ but little from those of the rainforest. It is an intermediate situation. This might explain the scarcity of humus, which is rapidly decomposed by bacteria lack-

18

ing any sustained competition from fungi whose role as fixing agents is minor. It is, in any case, a fact of personal observation that deciduous seasonal forests generally have only little humus and that their soils in most cases reveal a mere layer of woody litter. They are therefore much less protected against overland flow than the soils of the rainforests.

The predominant influence of temperature, combined with that of humidity, creates unique conditions of supply and of decomposition of the humus in the humid tropics. These conditions correspond to the three great plant formation-types found there: the rainforest, the deciduous seasonal forest, and the savanna or campo cerrado. Humus, it should be noted, plays a very important geomorphic role:

(*a*) It is one of the cements of soil aggregates and thus determines their resistance to splash erosion. The humic soils of the rainforest are therefore not as much affected by splash erosion as the soils, poor in humus, of the deciduous seasonal forest and, *a fortiori*, of the savanna.

(*b*) Humus increases the superficial porosity of the soil and therefore facilitates infiltration at the expense of overland flow. But this mechanism, of course, only functions when rains are not excessive.

(*c*) Lastly, through its decomposition humus allows the formation of soluble iron humates. It therefore plays a major role in the liberation of iron oxides, which are the most mobile products of weathering of the humid tropical zone.

Rock weathering in the humid tropics

One of the characteristics of the humid tropics is the scarcity of rock outcrops. In forests it is necessary to go to stream rapids to find an *in situ* ledge. Even in savannas large expanses reveal only weathered rock formations with, locally, some residual boulders (tors), which may be more or less displaced.

Sharply contrasting with these vast surfaces without outcrops are isolated, often entirely rocky hills that have no vegetation and are dotted here and there by only a few xerophytic plants. Nowhere else in the world is there such a sharp contrast between bare monolithic hills and smooth deeply weathered lowlands. This contrast is at the origin of the terms inselberg, monolithic dome and sugarloaf. Inselbergs are rocky residual hills or uplands that rise in isolation above rather uniform plains. Monolithic domes (sometimes called bornhardts) and sugarloaves are usually more restricted in area and stand in striking contrast above neighbouring, deeply weathered convex hills or ridges (called *meias laranjas*—half oranges—in Brazil). These contrasts are particularly striking in regions of crystalline rocks, especially granites and gneisses, which in other climates do not produce a thick and widespread regolith. Monolithic domes and sugarloaves are sometimes covered with equally original microforms such as flutings, lapiés, and solution hollows although the rocks have nothing in common with limestones but may be syenites, diabases, basalts, or even sandstones.

The microforms are alike throughout the humid tropics. They are only details yet reflect differences in the rate of evolution which are better known qualitatively than quantitatively. Like intense and deep weathering they are a zonal phenomenon that has an important dynamic geomorphic significance.

We examine first the general aspects of rock weathering in the tropics, before turning to the specific effect of the wet and the alternating wet–dry tropics on the weathering processes.

The slightness of mechanical weathering and the microforms on outcropping rocks

Rocky hills, which are more or less common throughout the intertropical zone, offer ideal conditions for the study of mechanical weathering. Vegetation on them is scarce and often reduced to but a few xerophytic epiphytes which cling to almost perfectly clean rock surfaces. Sometimes fissures retain a little weathered material where shrubs and even trees may grow, largely re-using, as the trees of the rainforest, their own detritus which remains in place. In this case fissures gradually widen, and a zonal type of weathering will eventually ensue. We will disregard this intermediate case and consider only monolithic domes that are bare or have but a few scattered epiphytes.

Fig. 1.3. Monolithic domes north of Petrópolis, Brazil (drawing by D. Tricart after photo by J. Tricart)

The nuclei of massive gneiss with curved joints stand in relief in the form of monolithic domes with large, smooth, and fluted flanks which rise above ridges sculptured by landslides (scars in the foreground) in weathered and jointed gneiss. Example of differential erosion.

The microforms on such monolithic domes are the following:

Flutings and lapiés   These are always oriented down slope; the lapiés originate from the deepening and ramification of the flutings which are the initial forms. Their development varies according to the type of rock. It is obviously on limestone that lapiés are best developed. They are characterised by sharp crests, pinnacles, pillars, and spires. Giant lapiés, with several metres of local relief, may, however, even form on basic igneous rocks such as the syenites of the Agulhas Negras (Itatiaia), Brazil. These giant lapiés are found at an elevation where night frost occurs every cold season. But lapiés of the same type, though less broad, may be observed on diabases

FIG. 1.4. Lapiés on quartzites near Palmeiras, State of Bahia, Brazil
Tropical climate transitional between wet–dry and semiarid: 750 mm (30 in)
of rain. A dark ferruginous varnish covers the rock surface. Note how bedding
planes stand out because of more intensive weathering in these areas. Solution
cavities and sharp ridges analogous to lapiés are developed by corrosion on
the top surface.

which produce bare outcrops on hilltops of only 50 to 100 m elevation (160 to 330 ft) at Monrovia, Liberia. Here all intervention of frost-weathering is, of course, excluded. On acidic crystalline rocks, such as granites and gneisses, no lapiés occur but only flutings up to 1 m (3 ft) deep occasionally interrupted by small breaks and solution hollows. On sandstones and quartzites the influence of bedding and of differences in consolidation becomes preponderant, and lapiés are generally chequered. Rock steps and capping rocks tend to form overhanging scarps. In spite of differences in lithology such microforms are very common and occur in a great diversity of rocks, whereas they are restricted to limestones and certain sandstones in other morphoclimatic zones.

FIG. 1.5. Fluted monolithic dome near Petrópolis, Brazil (drawing by D. Tricart after photo by J. Tricart)
Curved jointing produces overhangs and makes possible the growth of trees and shrubs; nevertheless, sufficient stability allows flutings to form.

Solution hollows (bowls or potholes)   These do not form in limestones and are infrequent in sandstones and quartzites but common in crystalline rocks, especially acidic (granites and gneisses) although they also occur in syenites, for example. Such hollows are small closed depressions a few decimetres deep, generally 10 to 20 cm (4 to 8 in), with a diameter close to one metre, that may reach two m (6 ft) but rarely more. Such hollows, of course, only develop on rather level surfaces. As soon as the slope passes 5° to 10°, they give way to lapiés and flutings. They are therefore mostly found on top of flat monolithic domes. Often plant debris mixed with corrosion residues accumulates in them. They are then transformed into tiny alluvial basins, true natural flower pots in which plants make their home. Further evolution then changes their features, and they become more and more like rock surfaces buried below regolith. In Surinam Bakker, who has studied them, calls them *oriçangas* when they are occupied by vegetation, and *pingen* when they are not. The latter can only exist at the base of rock surfaces periodically washed by rain water, causing them to overflow and be cleaned out, thwarting accumulation of detritus which would convert them into oriçangas and assure their eventual disappearance. Bottoms are mostly fairly smooth, even flat. The rim then overhangs a continuous nick a few centimetres above the bottom.

FIG. 1.6. Lapiés on granite near Singapore (drawing by
D. Tricart after photo by Tschan Hsi Lin, 1961)
Height of the block about 8 m (25 ft). The vertical flutings
correspond to the flow of rainwater. The joints cutting
diagonally through them have been widened into furrows.

Exfoliation scars   When these are clear and distinct the preceding microforms are absent or poorly developed. Only vague flutings, at the most, are
then encountered. There is thus a clear antinomy between the two groups
of forms. Exfoliation generally affects slabs at least half a metre thick (20 in)
and sometimes several metres. It often happens, as for example on certain
exfoliation domes in the area of Rio de Janeiro and Teresopolis, Brazil, that
caps of non-exfoliated rock persist on top of a dome, with a few trees growing on ancient joints, whereas the same rock sheet has exfoliated on the
flanks of the dome, creating a chaos of blocks at the base. The peeling of exfoliation sheets leaves behind angular scars which reflect the joint pattern.
Exfoliation is sudden, catastrophic, whereas the formation of flutings, lapiés,

23

and solution hollows is slow and gradual. This explains the antinomy between the two groups of microforms.

Occasionally attempts have been made to explain exfoliation scars by
mechanical actions. Most often invoked is thermoclastism, or fragmentation
due to temperature variations without the intervention of frost. Although
granites are relatively dilatable, this explanation is unfounded. Slabs are
much too thick. Even taking into account the temperature variations of bare
rock surfaces, which are much larger than those recorded at a weather station, and the possibility of sudden cooling when a cloudburst with raindrops having a temperature of about 20°C (68°F) drops a shower on rocks
which the sun may have heated to 50°C (122°F) and even 55°C (131°F), the
process proves incapable of explaining the formation of slabs which are in
all cases thicker than 30 to 50 cm (1 to 2 ft), often 2 to 3 m (6 to 10 ft), or even
more—10 m (33 ft) on certain domes of the Serra do Mar. Such temperature variations do not penetrate the rock very deeply at all, only a few millimetres, especially when they are sudden. They spall splinters and chips, not
slabs, and occur mostly in dry climates. That this phenomenon is never observed in the humid tropics is probably due to the smaller temperature

FIG. 1.7. Exfoliation dome near Petrópolis, Brazil (drawing by D. Tricart after
photo by J. Tricart)

Clearly visible are the curved joints which, once widened, make possible the
exfoliation through the detachment of slabs close to 10 m (30 ft) thick; the shrubs
give some idea of the scale. Lower down, an isolated scar marks the eventual
loosening of a slab several decimetres thick (about a foot). Flutings are visible
on the surfaces cleared by exfoliation.

range, the less frequent sudden changes of temperature and, also, to the protection of a film of lichen and algae which make the rock surfaces dangerously slippery when wet.

Exfoliation is narrowly guided by joints. The sliding planes extend into the mass of the monolithic dome as joints that are at first widened by corrosion but eventually grade into tight joints which mine blasts reveal in quar-

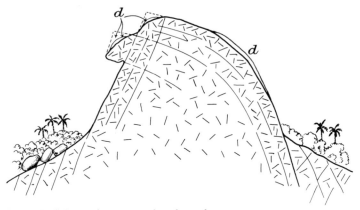

FIG. 1.8. Schematic cross-section through a sugarloaf (according to P. Birot, 1968)
At *d*, exfoliation slabs in process of detachment. The general form is guided by joints. The sugarloaf corresponds to a less jointed mass in the midst of more densely jointed rocks.

ries. In many places such as the Guianas, the Serra do Mar, northern Cameroon, and central Ivory Coast, the most perfect exfoliation domes are composed of compact masses of intrusive rocks broken by concentrically curved joints that guide exfoliation and determine the form of the dome. The problem relating to exfoliation therefore corresponds only to that of the widening of the joints. Examination of the joint surfaces clearly reveals corrosive action. The mechanism thus has certain analogies with the one that produces the two other types of microforms: flutings and solution hollows.

Bakker (1957) made highly interesting and precise observations and measurements of chemical weathering on monolithic domes in Surinam. The results, it seems safe to say, may be generalised to the wet tropics because the area studied by Bakker is characterised by an annual rainfall of 2 300 mm (92 in) and a dense rainforest. The lichens and algae which cover the domes produce a slightly basic micro-environment at the rock surface. The water gathered in solution hollows at the base of the domes was found to have a pH of 7·6 to 8·2. This water probably causes some silica to be dissolved and precipitated almost simultaneously as a thin protective film on the rock surface. In this way the granite would be waterproofed and capable of rapid drying after rain, thereby retarding the weathering process. In the sun the temperature of the rock surface might reach 60° to 70°C (140° to 158°F). A peculiar kind of weathering therefore occurs on monolithic domes. It produces montmorillonite, which is characteristic of well drained and non-acidic environments, and illite. Kaolinite also forms, but in relatively small proportions for a wet tropical climate. Analyses have revealed an average of 30 per cent montmorillonite and 25 per cent illite in the clays of the solution hollows at the base of a monolithic dome; 100 m (330 ft) further away kaolinite is again predominant in the ordinary regolith and accounts for 85 per cent of the clays.

Such montmorillonite weathering is related to a unique local environment and may have important consequences. Indeed, when montmorillonite is formed from certain feldspars small accumulations of this clay become enclosed between the non-weathered or slightly weathered crystals in the still non-dissociated rock. They are subject to frequent variations in humidity that cause important volume changes as the shrinking coefficient of montmorillonite is one of the highest of the clays. The resulting mechanical effects undoubtedly produce a granular disintegration. Fluting, lapiés, and especially solution hollows could thus be explained. A period of moistening is needed for water to penetrate an almost impermeable rock and cause weathering and swelling of the accumulations of montmorillonite produced from certain weathered crystals. It is in small depressions that the mechanism functions best. Undulations in the microrelief, where moisture persists, thus tend to be subject to a certain amount of granular disintegration. They will have a tendency to increase in size. On slopes they are transformed into flutings whose deepening is also helped by a very weak mechanical erosion caused by quartz grains dragged along by the running water. Slightly deeper areas in the flutings betray a weak zone where a chain reaction is set in motion (the deeper areas become moist faster so that weathering and disintegration are accelerated, which increases the depth, and so on). Solution hollows form under identical conditions but on lesser slopes. The fraction of rain water that infiltrates into joints corrodes the joint surfaces and acts in the same fashion. Changes in the volume of montmorillonite facilitate the opening of the joints, therefore the continuation of the weathering, until they are wide open and gravity induces the sliding that causes the exfoliation. Monique Michel, in an unpublished work, reports that she has observed minute quartz grains, a few tens of microns in diameter, in the microscopic joints of crystalline rocks but now detached from the joint walls and jamming a wide open joint. In this way clay with a certain water content can accumulate and fill the joint. If the water content increases the clay swells and acts as a wedge. Such a mechanism may play a role in the initial stage of joint widening and of granular disintegration.

The measurements of Bakker coincide with certain results of Pouquet (1956) in Guinea. The latter investigator has measured rain water with pHs as high as 9 at the beginning of squalls. After a half hour of rain pHs fell to 8.5–6.8. On bare monolithic domes, with an almost complete absence of organic matter, nothing modifies the pH of the rain water, and so it keeps its alkalinity as it enters into contact with the rock. The result is the formation of montmorillonite and the dissolution of some silica. In the rainforest, on the contrary, the pH of rain water changes as it passes through the foliage, trickling down leaves and trunks and, probably to an even greater degree, as it contacts the ground. Infiltration of the water into the soil, where the liberation of carbon dioxide through the decomposition of organic matter is considerable, must still further decrease the pH value. Indeed Pouquet has measured pHs of 4·8 to 5·5 in wells and in a subterranean stream. Under the free atmosphere the water of surface streams again has a less acidic pH, comprised between 5·9 and 6·7, because exchanges with the atmosphere limit the carbon dioxide content to much lower values than those found in soils. Bare rock surfaces, therefore, find themselves in very particular weathering conditions. Even so, they are mostly sculptured by chemical processes, which explains the similarities of their microforms with those observed in the karsts of the midlatitudes. The alkalinity of rainwater favours the immediate corrosion of acidic rocks, the only biotic intervention being that of algae and lichens, about which very little is known. The formation of lapiés on basic rocks such as diabases, which is uncommon, seems to have taken place under the regolith in an acidic environment. The lapiés microrelief would only have become exposed as an outcrop at a later date as a result of erosion. This is corroborated in the area of Monrovia, Liberia, both by field observation and the nature of the weathered products filling the joints.

A last common characteristic of corrosion processes on bare rock outcrops is their slowness. The persistence of solution hollows on the gentle slopes of monolithic domes indicates a meagre supply of debris. Indeed extremely violent rain is necessary to make such hollows overflow with sufficient intensity to eject the enclosed materials. But under the most humid climates the power of colonisation by plants is very great, and a hollow in which some loose alluvium has accumulated is quickly occupied by plants and subsequently drawn into irreversible evolution which transforms it into an oriçanga. Bakker believes that when this happens the silica dissolved on the surface of the neighbouring rock is precipitated into the oriçanga, causing its eventual induration, the final result of which will be a micro-inversion of relief. The often highly developed flutings hardly modify the dome's shape which is in most cases controlled by structural influences; it is thus possible to reconstruct the initial form of the hill (Barbeau and Gèze, 1957). Such monolithic domes usually overlook surfaces or erosion that date back to the Tertiary and so are 10 to 20 million years old, sometimes even more. They are truly the result of sculpture, not of demolition, and the rate of denudation calculated by Freise (1938) in Brazil seems highly exaggerated. The only rapid mechanism is exfoliation which is due to the intervention of gravity in the removal of the slabs. Moderate chemical action affecting a joint plane may cause the exfoliation of many cubic metres of rock because

of a rupture in the balance of forces. But it is a genuine resonance phenomenon which must be correctly interpreted. This explains why lapiés and
flutings are absent where exfoliation takes place: they do not have the time
to form.

GENERAL CHARACTERISTICS OF SUBSURFACE ROCK WEATHERING

The nature of subsurface rock weathering is very different from what it is
under the atmosphere. Below a depth of $\frac{1}{2}$ to 1 m (2 to 3 ft) temperatures remain practically constant, with a value slightly above that of the mean
annual temperature of the free air, probably because of the absorption of
radiations by the ground surface and the liberation of heat from the decomposition of organic matter. At least under forest the soil atmosphere contains a high proportion of carbon dioxide: up to 15 per cent and even
higher. Water which filters through it can therefore dissolve a certain quantity and thus acquire an acid reaction. Where it is not neutralised by alkaline rocks, subterranean water is acid. In wells in the Fouta Djallon, around
Labé, Pouquet (1956) found pHs of 4·8 to 5·5. In forested Ivory Coast
Rougerie (1960) measured 4·1 in alluvium, 4·2 to 5·3 on granites for ground
water at different depths, 5 on schists, 5·3 to 5·7 on shales, 5 to 6 on Tertiary
sands, and between 6 and 6·4 on basic igneous rocks, which is a value close
to neutral. In general the pH varies vertically along the same profile, especially on acidic rocks. On granite, for example, the pH is lowest near the surface (4·3 at the surface, 4·2 at a slight depth) and increases with depth to
reach a maximum of 5·3 at -25 m (-82 ft). The pH is highest at the base
of the regolith. These observations were made under forest. Comparative
data are unfortunately not available for savannas.

Unchanging subsurface temperatures and slightly acidic waters are two
general characteristics which strongly influence weathering in the humid
tropics. They cause certain mechanisms which play an important geomorphic role; these mechanisms are kaolinisation and the migration of iron
oxides which both are dependent on water. It is therefore necessary to begin
with a study of the role of water in the regolith.

The role of water is important because it is at the origin of all the reactions
causing weathering. In the humid tropics many important reactions are
slow and require an intimate contact between the water and the mineral
environment. The effects of waterlogging are particularly significant, for
example in the decomposition of granite. When water percolates rapidly
the weathering process is slow and not very intense. To bring about a complete 'rotting' of the rock, a prolonged soaking is necessary which, alone,
enables water to penetrate between crystals or into fissures. Feldspars can
then kaolinise and dark minerals liberate oxides which are invariably hydrated. The regolith, which always contains a certain proportion of clay
endowed with a certain water capacity, plays a role of compression and
thus makes an important contribution in speeding up the weathering pro-

cess of the underlying rocks. Under rainforests, where the soil never dries, weathering processes thus encounter optimum conditions. Nevertheless, in very porous formations, such as alluvial sands and gravels incised a few metres by rivers and where drainage is excellent, moisture is not fully retained; this can strongly impede weathering even under forest and in a very wet climate receiving 2 000mm (80in) of rain or more. We have observed such a case on the Andean piedmont flanking Lake Maracaibo between La Fría and El Vigía. Fans containing feldspathic and micaceous sands and small pebbles of granite, gneiss and micaschist deposited during the last glacial marine regression (pre-Flandrian regression) are still practically fresh. Only a few traces of iron oxide migration are found in them. Under similar conditions bouldery mudflows reveal rotten crystalline pebbles. Waterlogging is therefore an essential condition of advanced weathering. Even in a very wet tropical environment it may be impeded by excellent drainage of the regolith. The more so in the alternating wet–dry tropics where rain is much more irregular. In depressions where water gathers and persists for a long time, intense weathering goes on during a great part of the year, even the whole year. It then proceeds in a manner little different from that of the wet tropics. On the other hand, on top of hills, especially where the regolith is sandy and permeable, as is the case on granites and certain gneisses, waterlogging is only realised sporadically. Weathering is impeded and slow. If spheroidal, it produces corestones at minor depths. The removal of the finer fractions of the regolith, especially by overland flow, often exposes the corestones at the surface, creating the characteristic landforms present in many savannas: ridge tops covered by tors, and poorly drained lowlands of thick regolith sometimes pockmarked by shallow, closed depressions.

At the rock contact water penetrates in two different ways. On the one hand there is a very slow penetration in the very mass of the rock as a result of its microporosity, that is by infiltration between the crystals and into the voids not filled by the cement. On the other hand there is a more rapid penetration due to macroporosity caused by fissures and joints that produce discontinuities in the rocky mass. Of course some fissures or joints may be very tight, especially in igneous rocks, making the penetration of water extremely difficult as that which takes place between crystals. Eventually, however, they are opened up. In order for the water to penetrate between crystals or along fissures much time is needed, for it proceeds by very slow capillary movement. Permanent or in any case seasonal waterlogging is necessary. The effect of compression of the overlying regolith is indispensable. Its absence on rocky hills explains their peculiar way of weathering, combining widening of joints to direct superficial corrosion that is most intense in places where water persists longest, resulting in solution hollows and flutings. The rate of deep weathering, independently of the petrographic nature of the rocks, depends in large part on the water regime at the rock surface, which in turn depends on the climate and on the physical properties of the regolith. A very argillaceous regolith, as is found for example on shales or on basic rocks, lets water infiltrate very slowly but has a high waterholding capacity favouring a rather permanent waterlogging. With due allowance the situation may be compared to that which obtains under a slowly melting snow cover. Exactly as in this case, conditions are favourable to the formation of lapiés on limestone. But in a tropical climate a very argillaceous regolith may form underground *cryptolapiés* at the base of the regolith on basic rocks such as basalts and diabases. On the contrary,

FIG. 1.9. Cryptolapiés at Mamba Point, Monrovia, Liberia
These forms are developed at the contact of the diabase bedrock and a mottled subsoil. The subsoil has been washed away and the cryptolapiés exposed. This has increased the relief of the cryptolapiés.

in a sandy regolith, as that found on granites, water circulation is faster because of greater permeability. If joints exist, they drain the subsurface water, causing an intense ground water circulation, as reported by Leinz and Vieira de Carvalho (1957) in the neighbourhood of São Paulo, Brazil. Weathering then proceeds rapidly along the joints and slowly at the direct contact of the intervening blocks where water penetrates only as a result of microporosity. We will see that this mechanism explains the type-weathering profiles of jointed acidic rocks. If the same rocks are jointless, or if there are only closed joints more or less parallel to the ground surface, microporosity alone functions, and the progression of weathering in the rock mass is much slower. Significant differential weathering may result from it.

The migration of iron oxides in the soils of the humid tropics is a very important biochemical process. Its mechanism, however, is still insufficiently understood. In the initial stage of rock weathering there is first the dissolution of the basic elements K, Ca, Mg, and Na which decreases the acidity of the water or even makes it slightly alkaline. It enables iron to remain in colloidal suspension in a hydrated form. As the weathering proceeds rust stains may appear around the biotites. But this iron does not migrate very far as it is flocculated by the basic elements. To be removed the water must contain carbon dioxide or, preferably, humic acids. The carbon dioxide enables the formation of a soluble ferrous bicarbonate, which is exported. The mechanism, however, only operates in a reducing anaerobic environment. Ferrous bicarbonate is easily oxidised and then precipitated in the form of ferric iron. To remove the iron in the form of dissolved ferrous bicarbonate, the soil must be deprived of oxygen, which means air. This obtains in waterlogged soils on condition that the water contains enough carbon dioxide in solution. The mechanism therefore operates especially in seasonally flooded marshes where it forms part of the process of gleying.[1] It also seems to function under forest in certain layers of the regolith, particularly in the mottled clays of latosols where the iron may be removed during the wet season when the soil is waterlogged. In the dry season the air penetrates the fissures, especially the desiccation cracks, causing yellow and reddish colorations in these places due to the more or less dehydrated iron hydroxides.

Humic acids are capable of dispersing not only the ferrous iron but even the ferric iron, and thus allowing its migration. Such ferrohumic complexes are relatively stable, enabling a distant exportation.

The behaviour of iron is therefore not the same along the whole weathering profile. At its base, where rock weathering begins, it is one of the first elements to be liberated by the dark minerals, notably biotite. But it cannot migrate very far in a basic or even a slightly acidic environment. It hardly moves, staining the neighbouring minerals. In parts of the profile subject to seasonal waterlogging, it migrates only a little in the form of a ferrous carbonate; mottling may occur on condition that the environment is not alkaline, because in that case the iron remains fixed. When humic acids intervene and drainage is good, distant migration is possible. This is why, generally, especially on acidic rocks, soil profiles show effects of podzolisation immediately below the humic horizon. These effects produce a yellow or yellow-ochre horizon, with some iron, on acidic rocks such as granites, gneisses, or even micaschists, or, more rarely, a whitish or grey horizon, with very little iron, as on sands and sandstones. Below this pallid zone are the mottled clays of typical latosols developed on acidic crystalline rocks.

Long distance transport is also favoured by certain bacteria which surround themselves with a film of iron oxide. The conditions of their development have unfortunately not been studied.

All things considered, the evacuation of free sesquioxides is favoured, according to d'Hoore (1954), by high temperatures, reducing conditions and the presence of organic (especially oxalic and tartaric) acids. All these conditions are realised to the highest degree in the humid tropics. Although the

[1] or gleization.

role of iron migration is important in other climatic zones, as in the process of podzolisation in the humid temperate zone and even in an attenuated form in the tundras, it is in the humid tropics that its geomorphic importance is greatest. Indeed, here, the sesquioxides migrating in the water are later concentrated in the form of concretions and then cuirasses which have a definite and original effect on landforms.

Conditions moreover differ materially from the wet tropics to the alternating wet–dry tropics. Under forest, as Rougerie reminds us, abundant moisture causes a swift leaching of the alkaline elements which causes rapid acidification very soon after the beginning of the weathering process. The soil in its deeper horizons is very often waterlogged. All these conditions favour the departure of sesquioxides, especially in the presence of organic acids. It is only higher up in the more aerated horizons that part of the iron oxides persist, in poorly hydrated forms, imparting reddish colours under the superficial, podzolised yellowish clays. In savanna regions the supply of organic matter and of humic acids is less abundant. The soil dries up seasonally. Iron migration takes place in the rainy season, but in the dry season there is oxidation and precipitation of the hydroxides. Evaporation at the capillary fringe above the water table helps the process considerably. The leaching of iron is therefore less, and is restricted to particularly favourable sites, whereas its precipitation is much more common and occurs in the form of concretions and, in extreme cases, of cuirasses.

Once more we recognise the subdivision of the humid tropics into wet and wet–dry as it turned out to be in the study of the climatic factors.

The formation of clays is also particularly active in the humid tropics. It completely modifies the geomorphic behaviour of certain rocks such as granites, gneisses, and micaschists that weather into a very clayey regolith which is the prime locus of erosional phenomena. Throughout the humid tropics kaolinite is abundant and generally predominant on acidic rocks. It also occurs in various proportions on other rocks. The kaolinisation of acidic rocks is, however, a typical zonal characteristic.

In acidic rocks the first phase of weathering affects the alkaline minerals with a K, Ca, Na, or Mg base. It produces a dissolution of the bases that neutralises the acidity of subterranean water and thus causes the formation of montmorillonite, as Erhart (1935, 1938) and later Leneuf (1959) have shown. This explains why there is formation of montmorillonite clay on monolithic domes that do not have soils but where this first phase can take place. Later, the remaining part of the weathering process proceeds in an acidic environment due to the rapid leaching of the bases, and kaolinite is then formed. In fact, montmorillonite is unstable in a sufficiently humid climate which assures good soil drainage and leaching of bases; it only occurs in a fugitive manner and in small quantities, at the very base of the regolith. Most clays are kaolinites; reworked, they even form the alluvium of fluvial deposits and of coastal marshes. Montmorillonite is stable only in climates too dry to enable a good leaching of bases and where, therefore, acidifica-

32

tion of the soil profile is hampered. Here is a sure criterion to delimit the humid tropics from the semiarid zone. In the Caatinga regions of Brazil or in the Australian bush, films of salt accumulate between the crystalline hills, indicating an insufficiency of salt leaching. Kaolinite cannot form in such circumstances.

The initial montmorillonite phase of the weathering of crystalline rocks may perhaps explain the opening of tight joints and the frequent bursting of quartz crystals along joints, as we have explained above in relation to exfoliation domes. Montmorillonite is indeed a clay whose variations of volume as a function of moisture are the greatest. Ordinary seasonal variations of moisture in non-waterlogged soils would be enough to cause an important mechanical action on a microscopic scale, resulting in a widening of joints and thus facilitating the penetration of water and the realisation of later phases of weathering.

It is perhaps during this initial weathering phase that a certain amount of dissolution of silica takes place, as silica solutions are weakest in an acidic environment and colloidal suspensions are stable only when the environment is alkaline. Acidification of the regolith after the initial montmorillonite phase would thus explain the coatings of siliceous plush which often occur on quartz grains in tropical regolith. The silica seems to be dissolved or put into suspension during the initial phase and precipitated during the acidification of the environment: when, for example, an increase in moisture brings more acidic waters from the higher part of the soil profile. In the dry tropics where there is a lesser leaching of bases, the dissolution of silica should be easier and its migrations farther. One might thus explain the silications of semihumid and subhumid zones where silica is often precipitated together with iron in certain types of cuirasses, and in marshy calcareous deposits where it is more abundant. (Millot *et al.*, 1959.)

The dissolution of ions in rock minerals through hydrolysis is accelerated by the presence of carbon dioxide in the water. Depending on the concentration of the carbonic gas and the ions concerned, the rate of dissolution is increased twofold to fivefold. Not only is the crystalline mass dissociated but part of it, in turn, is dissolved into colloidal solutions. Whereas in the case of the alkaline feldspars K and Na are liberated and entrained in solution, the silica and alumina go into colloidal suspension and are combined to form kaolin provided the environment is both sufficiently humid and acid; under optimum conditions the new compound even crystallises into regular hexagons of kaolinite. The colloidal suspension of alumina probably does not have the time to become isolated, as it is immediately recombined into kaolinite, according to Mohr (1944). As to the calcic feldspars, according to Rougerie (1960), they loose all their Ca only in the presence of intense leaching by acidic waters. As long as some Ca remains, silica may be exported. On the other hand, alumina can migrate as long as there is acidification by humic acids. Muscovite, a potassic mica, reacts more or less as the alkaline feldspars. In basic minerals, like amphiboles and pyroxenes, hydrolysis is swift and causes the liberation of iron. If acidification is not too intense silica can be exported together with the bases, enabling the isolation of alumina which, in the most favourable circumstances, may crystallise, taking the form of *gibbsite*.

In the wet tropics the abundance of ground water and the decomposition of large quantities of organic matter supplied by a luxuriant vegetation causes an intense leaching of the bases. The environment becomes progressively more acidic. Under such conditions minerals containing iron are vigorously attacked by hydrolysis, and iron hydroxides, which are easily leached, are formed, silica, which is not very mobile and remains almost completely in place, recombines (with alumina) to form kaolinite, and clays migrate through the soil profile as they are carried away toward greater depths and dispersed by acidity, resulting in a certain amount of podzolisation.

In this way is formed a sandy argillaceous regolith several metres thick, with an upper yellowish or ochre-grey podzolised horizon (the pallid zone), often 2 m (6 ft) thick, and a reddish lower horizon richer in clay and in poorly hydrated iron hydroxides (goethite, $Fe_2O_3 . H_2O$). Such a regolith is typical of the rainforest.

This evolution is, of course, normal on acidic rocks where an acidic environment is more quickly realised after the elimination of a limited quantity of bases. On basic rocks the leaching of bases is more difficult; the environment remains alkaline for a relatively long time, enabling the leaching of more silica. Acidification is also retarded because the calcium ions cause the coagulation of the humus, which reduces the supply of humic acids at deeper levels. More iron therefore remains, causing a darker, red-brown and purplish subsoil. If drainage is poor alumina may become isolated, and on level marshy surfaces the end result may be the formation of a ferruginous bauxite. The elimination of the iron from the ore, resulting in commercial bauxites, can only occur during a later phase of resumed erosion and intense leaching. High concentrations of alumina will not occur if the drainage of the regolith is good but, instead, clays less rich in silica than kaolinite. The isolation of alumina is an exceptional phenomenon, and the formation of bauxitic cuirasses is always very localised.

As to ferruginous cuirasses, their formation is impossible in a perhumid environment: the leaching of iron is much too intense. The hydroxides remain in a diffuse state, adsorbed by the clays which they stain. Moreover, one should not be misled by the often bright colours of these clays; very little iron is needed to cause them.

The wet tropics, characterised by intense soil leaching, are adverse to the formation of cuirasses, except for some bauxitic cuirasses which form under exceptional conditions. Such a humid environment, above all, causes a deep weathering of the rocks and produces an important amount of clay. The resulting regolith not only profoundly modifies the way in which the rocks are attacked by the agents of the weather, but also has very different mechanical properties from the parent rocks themselves.

In the alternating wet–dry tropics the leaching process is not the same. The regolith is no longer saturated with water but periodically dries up more or less completely for a certain period of time. There is oxidation, and the mobility of iron is reduced. Liberated in one place, it is precipitated in another at a short distance. Concretions appear, and where the iron is more abundant, they may become welded together into a cuirass.

The weathering phases of the bedrock are also partly different. The initial phase, that of the hydrolysis of K, Ca, Na, and Mg, remains the same. But it becomes discontinuous in time as it can only take place when the base of the regolith is saturated, which means that the process is slowed down. The capacity of the water to penetrate into the fresh rock now becomes more important, resulting in a more irregular weathering front than in the wet tropics. The regolith is therefore characterised by a great variation in thickness, with pockets alternating with underground pinnacles which sometimes pierce the ground surface in the form of tors.

The last phase, which is acidic in the wet tropics, is very different in the wet–dry tropics where it is subject to seasonal variations. The leaching of bases is not as intense. There probably is, as Rougerie (1960) has suggested, an acidic reaction during the wet season when the regolith is full of water and the products of decomposition of the organic matter are washed down to lower depths, and a basic (or at least a less acidic) reaction during the dry season. The humic complexes are then no longer leached, and the less abundant water tends to become saturated in soluble bases. The formation of kaolinite is impeded and now occurs together with illite, which is a result of less advanced weathering, and with montmorillonite, which has not been transformed into kaolinite. The local conditions, as in all transitional zones, naturally play a much greater role than in the wet tropics.

Less clay is therefore being produced although the variety is greater, and cuirasses form more readily depending, of course, on the supply of iron hydroxides, especially those provided by lateral migration.

On basic rocks the cuirasses may also be bauxitic. The leaching of bases is now impeded by the dry season, excluding the acidification of the regolith. Under such conditions silica migrates more easily and is exported, which favours the isolation of alumina and the formation of gibbsite $(Al(OH)_3)$. Rougerie (1960) thinks that bauxitic cuirasses form more readily in an alternating wet–dry tropical climate than in a wet tropical climate.

CLASSIFICATION AND TERMINOLOGY OF TROPICAL SOILS AND CUIRASSES

As Rougerie (1960) has pointed out, the concepts concerning the genesis of tropical soils and cuirasses have considerably changed during recent years. Detailed studies undertaken in certain specific regions (Netherlands East Indies, West Indies) shortly before the Second World War have not been followed up by more research until the 1950s. These studies show that a number of general concepts, which were deduced from theoretical considerations rather than based on actual observation, were far from

35

corresponding to reality. A large part of our present vocabulary was created to serve these concepts. Now, in spite of progress most of it is still in use, and even new terms have been added. The present terminology is therefore unsatisfactory as it is composed not only of old terms whose meaning has changed although unequally according to authors, but of new terms some of which have a double meaning.

The central term is that of *laterite* introduced by Buchanan (1807, vol. 2, p. 441) and based on observations made in India. Derived from the Latin *later* (brick), it designates, etymologically, a coherent, red, and porous formation with which walls can be built. In certain cases it hardens subaerially through evaporation and precipitation of the iron oxides. It is consequently, in fact, a *cuirass*, which contains variable proportions of iron and aluminium oxides. 'Laterite' thus has become the equivalent of tropical soil. Buchanan's definition is still found in Lacroix (1913) in one of the first mineralogical studies devoted to cuirasses. For his part, Harrassovitz (1930), in a premature synthesis made on the eve of the first detailed studies which are at the base of our present concepts, put forth an inexact scheme which became generally accepted. He maintained that clays were unstable in a tropical environment, that the silica and the alumina became dissociated, thus enabling the formation of laterites with an essentially alumina and iron base. These inexact concepts were tied to an overestimation of the rate and intensity of tropical weathering, as found, for example, in Freise (1938) who attributed fantastic and untenable values to the corrosion of Brazilian monolithic domes: almost instant dissolution! Later we will see in what consequences such romantic exaggerations can result.

Under Harrassovitz's influence laterite became the symbol of the tropics. Every tropical soil was a laterite or was destined to become one shortly. Some authors have even suggested, although they have not affirmed it, that all that was necessary to cause the appearance of laterite within a few years was to fell the forest. Differences have later appeared in the interpretation of the term. While the British adhered to Buchanan's original definition, other authors, French as well as Americans, extended the word to all the zonal tropical soils. Laterisation therefore became synonymous to tropical pedogenesis. The French, for their part, created two expressions to differentiate two types of products: *lateritic clays* for the non-consolidated regolith, including the kaolinitic regolith of the wet tropics; and *lateritic cuirasses* for the consolidated regolith, whether it is aluminiferous (bauxitic cuirasses) or ferruginous (ferruginous cuirasses).

The American Kellogg (1941) created the expression *latosol* to designate both as a result of a certain amount of wear of the term laterite. The latosols are therefore the zonal intertropical soils.

At the present time it has become practically impossible, in France as well as in America and Britain, to use the term 'laterite' as it is no longer clearly understood how to interpret it; a number of British authors now use the French meaning, while others keep on using the traditional one. The only solution is to abandon the term altogether; here we will speak of latosols in a very general sense, of ferruginous or bauxitic cuirasses when the chemical nature of the cuirasses has been determined, otherwise they will be designated as lateritic, ferrallitic (see below), or simply as cuirasses. Much confusion would be avoided if the word 'cuirass' were to be internationally adopted.

The present terminology of the French school of pedology, synthesised by Duchaufour (1960, 1965) is as follows:

The *ferrallitic soils* form a large group of soils characterised by total hydrolysis in a neutral or slightly alkaline environment resulting in the leaching of iron and in the isolation of alumina. They correspond, therefore, to the lateritic soils of former French authors. *Typical ferrallitic soils* are characterised by a ratio of the argillaceous fraction SiO_2/Al_2O_3 that is lower than 1·7, which can only occur on rocks poor in silica. *Weakly ferrallitic soils* are characterised by a low content of alumina and abundant kaolinite; the ratio is then close to 2. In poorly drained areas where the leaching of bases is impeded or in bottomlands where organic matter decomposes slowly, montmorillonite, characteristic of *black tropical soils*, is formed. At higher elevations the slower decomposition of the organic matter impedes the rubefaction of iron oxides. Soils are no longer reddish but brown or brown-black. A more important layer of humus then forms the superficial horizons. These soils are *brown ferrallitic*. They typify the lowest highland zone; for example, certain mossy forests above 800 m (2 600 ft) in the Venezuelan Andes. As to the cuirasses, they are designated by the expression *ferrallitic cuirasses*.

The ratio SiO_2/Al_2O_3 has been criticised by Rougerie (1960), who remarks that many soil analyses in Ivory Coast, in an environment that ought to produce typical ferrallitic soils, are characterised by ratios higher than 2. The reason is the great abundance of kaolinite. Moreover, the calculation of the ratio only takes into consideration combined silica, whereas free silica, at least in part newly formed, often seems to exist in the form of a fine sand or of a plush on quartz grains. Free alumina may also be present. For this reason it seems to be better and more in accordance with the facts to speak of *kaolinitic weathering*, which is characteristic of an acidic environment with incomplete hydrolysis, the process of ferrallitisation functioning especially well on rocks poor in silica.

Special characteristics of differential erosion

The mechanisms which we have just analysed function in different ways on different rocks. They are therefore the cause of differential erosion in regions of varied lithology. This differential erosion takes on two aspects:

1. A quantitative aspect, which is due to different rates of erosion on different kinds of rock. After a certain amount of time the more resistant rocks will systematically produce the higher reliefs, whereas the more easily disintegrated rocks will correspond to the depressions.

2. A qualitative aspect, which is reflected in the particular outlines of the eroded landforms: steepness of slopes, density of drainage, shape of valleys (wide or narrow floor, entrenchment).

In extreme cases, which are particularly well developed in dry regions, the interfaces of different rock strata are eroded into stripped surfaces, and the different facies of the superimposed strata are clearly reflected in the morphology of the resulting slopes, whether dipslopes or faceslopes. Such distinct reliefs produced by differential erosion are exceptional in the humid tropics because of the importance of chemical weathering which, in general, causes the bedrock to be buried beneath several metres of regolith. This residual mantle produces a 'softening' of the landforms. This 'softening' effect may be compared to a thick snow cover which softens the outlines of the landscape it buries.

The 'softening' effect of the regolith of course varies with the morphoclimatic subdivisions of the tropics, depending on the intensity of the weathering process. The thicker the regolith, the greater the burying effect. Only general differences of a statistical nature remain (drainage density, form and steepness of slopes, and relative height of the interfluves, which depends on the two preceding elements). One may speak of an *erosion facies*. In Lower Ivory Coast Rougerie (1960) has masterfully analysed the erosion facies differences of granites and micaschists.

The thickness of the regolith depends on three factors:

1. The climate, in which temperature and humidity simultaneously intervene, the wet tropical climate being the more favourable to weathering, as we have already indicated.
2. The type of rock, which contains the whole problem of differential erosion.
3. The intensity of erosion. In the wet tropics forested slopes may exceed $30°$ to $45°$, depending on the type of rock. The regolith then becomes unstable and is periodically swept away by landslides. It is therefore thinner, occasionally exposing the bedrock.

Let us examine the influence of this last factor. Landslides especially affect a large number of recent fold mountains subject to current tectonic movements in Indonesia, Vietnam, Central America, and the northern Andes. Some definitely structural landforms may be observed under Venezuela's tropical rainforest, which receives between 1 400 and 5 000 to 6 000 mm (56 and 200 to 250 in) of rain. They are always composed of alternating quartzites and shales (or slates). Quartzites or quartzitic sandstones, as we will see later, are in fact the rocks which are most resistant to weathering and are never covered by more than a thin regolith about equal in thickness to that found in temperate lands. Shales, on the contrary, rapidly weather into clays commonly having a thickness of several metres. On steep slopes they are continuously affected by creep and periodically by earthflows. In this way the contacts between quartzites and shales may be exposed, especially if the dip is considerable. Hogbacks with perfect dipslopes thus form under dense forests at the foot of the Andes. But the dip must be about 45°. When it is only 5° to 10°, for example, well developed cuestas appear. They have more or less distinct escarpments (depending on the location) if a caprock of quartzite overlies mudstones that are prone to landsliding. Each slide, which originates with the help of a small perched ground water level at the base of the quartzite, brings down a segment of the caprock, creating a small scarp which is eventually blunted by scree. If the slides are repeated often enough a sharp cliff persists. This commonly happens on slopes whose steepness is maintained by the sapping action of a stream at its base.

The climate intervenes mainly through the medium of the water available for the weathering process, as temperatures are, by definition, everywhere high. Here again we recognise the importance of the subdivision of the tropics primarily according to the degree of humidity, which in this case depends on the conditions of permeation existing at the weathering front.

In the wet tropics where the soil never dries, weathering is a continuous process. Weathering conditions are optimum if the capillary porosity of the regolith is high enough to permit the seepage of a rather important quantity of water that will permeate the contact with the underlying rocks and that is sufficiently renewed to contain only weak concentrations of dissolved products. Hydrolysis is then at an optimum. The water penetrates into microscopic openings such as the fissures of crystals and the interstices between crystals. A weathering front thus progresses downward from the large and permeable joints and the base of the regolith. All the rocks are affected but at unequal rates which depend on the petrochemical composition and the density of the joints. A high density, of course, multiplies the surfaces that are being attacked. If, on the contrary, the regolith is not very permeable but rather argillaceous, then the amount of water that reaches the unweathered rocks is smaller, no matter how high the rainfall. Saturation is not always assured, and solutions tend to be more concentrated, which, of course, retards hydrolysis. Moreover, differences in the degree of permeation on a small scale assume a greater significance and set off chain reactions: any well permeated rock area decays faster and is transformed into a cavity that concentrates the water blocked by the plane separating the regolith from the unweathered rock. The removal of material may also cause some settling that increases permeability and therefore the water supply. Such irregularities become progressively more important and develop into subterranean lapiés or cryptolapiés. They are absent on rocks with a very permeable regolith such as granite or gneiss. They are, however, very well developed on basalts, diabases or limestones that yield very argillaceous and not very permeable waste products. Under such conditions weathering is also slower than in the former case, and the regolith is thinner.

In the alternating wet–dry tropics the weathering process is intermittent except in topographic depressions where subsurface water persists throughout the year. But even in this case the non-removal of the water supply produces a high concentration of solutions that slows down the weathering process or even stops it altogether when certain elements are not precipitated (iron, for example). The average rate of weathering is thereby decreased. Moreover, the mechanisms that require a thorough permeation are impeded as the waterlogging of the base of the regolith is not always realised. Most of the subsurface water is drained by joints rather than by microfissures in the unweathered rock at the contact of the regolith. In short, lacking intensity, weathering is limited to the easiest prey. Large rock fractures have greater influence than in wetter climates that produce continuous waterlogging. Topography also plays a more important role. Ridges and hills are characterised by rapid infiltration mainly along joints, whereas elsewhere hydrolysis remains embryonic. On crystalline rocks weathering is mainly spheroidal. The regolith is thin and irregular. It is also more intensely washed by overland flow, especially if the soils are incompletely weathered and therefore somewhat sandy; this and a poor structure make them easy prey to splash erosion. Once again there is a chain reaction: more overland flow produces less weathering, and less weathering produces less clay; the soils are therefore sandy and have a poor structure, which again augments overland flow; the augmented overland flow in turn increases erosion but decreases compression and, therefore, the amount of permeation. For this reasons tors and castle koppies often appear on interfluves. They are all the more frequent as drought increases and chemical weathering is reduced. In this way one passes through a series of transitions from the wooded savannas to the semiarid zone and finally the arid zone where every hill is rocky. The topographic depressions, on the contrary, offer much more favourable conditions to intense weathering because of an abundant supply of water during nearly the whole of the rainy season and sometimes even longer in the case of floodable bottomlands. Waterlogging does not only take place along joints, but weathering becomes general as in the wet tropics, its rate only is reduced. In this way pockets of deep regolith frequently form under topographic depressions, especially if deep drainage is assured as, for example, in highly jointed rocks. This is the origin of certain closed depressions in crystalline environments such as are found in the state of Bahia, Brazil, which receives about 1 000 mm (40 in) of rainfall yearly.

Even taking into account these topographic and climatic influences, differential erosion is a very complex phenomenon in the humid tropics. The factors that control differential erosion are:

(a) The petrographic composition of the rocks, which determines the nature and the clayiness of the regolith. The regolith, in turn, has repercussions on the conditions of water infiltration that determine the degree of permeation at the contact of the regolith and the underlying rocks, with all the consequences we have just described on the kind of weathering itself and the density of the drainage net; and on the processes of erosion that operate on the regolith (such as sheet-erosion, creep and earthflows) and sculpture the landforms, determining, among other things, their form and degree of slope, that in turn affect the thickness of the more or less effectively denuded regolith.

A complicated system of interactions therefore works towards a physiographic equilibrium that depends at once on climate, vegetation and lithology. This equilibrium is simultaneously characterised by a certain water budget, a certain balance in the evolution of the regolith (thickening at the base due to progressive weathering, and superficial truncation due to sheet-erosion), a certain organisation of the surface drainage, and a certain type of slope (degree of slope and kind of profile). Each lithologic environment will have its own kind of physiographic equilibrium under given morpho-climatic conditions that are stable for a sufficiently long period of time.

(*b*) The joint spacing, which determines underground drainage and is so important in the weathering process.

We here again encounter the difference between macro- and micro-porosity that, no matter what the petrographic composition of the rocks, brings about a weathering front. This front is either irregular or jagged in rocks with large fractures, revealing deep pockets in highly fractured zones, *cryptopinnacles* in more widely fractured areas, and spheroidal weathering in places where joints are regularly spaced at about one metre (3ft); or it is regular and continuous with a rapid transition from the unweathered rock to a rather evolved regolith in areas of massive rocks (that is, a non-jointed intrusive mass or the top of a thick massive bed without joints).

An alternating wet–dry tropical climate, where chemical weathering is still intense although discontinuous in time, favours, as we have seen, irregular weathering fronts. But even in a wet tropical climate the difference between the two types of fronts shows up clearly. Similar rocks, whether acidic or basic, rich or poor in quartz, may reveal one or the other type within a short distance. But the irregular fronts progress more rapidly than the regular fronts, the rocks disintegrate more deeply, making them apt to slide or prone to stream erosion. Areas of regular weathering fronts therefore tend to lag behind and thus to produce residual reliefs. Many inselbergs of the wet–dry tropics and nearly all the monolithic domes of the wet tropics have this origin. Petrographic composition and joint spacing, therefore, combine to determine the rate of erosion.

One could, on a physiographic basis, regroup all the well consolidated rocks, the only ones whose weathering plays a decisive geomorphic role, into five major classes:

1. *The acidic igneous rocks* with a high quartz content. Their quartz minerals are highly resistant to weathering and seem to corrode only in so far as iron hydroxides become affixed to them. The leaching of the iron oxide, owing to organic acids, causes the liberation of a little silica and the appearance of microscopic corrosion forms. This mechanism seems partly responsible for the fragmentation of the quartz of latosols.

Acidic igneous terrains are always characterised by a regolith with a high proportion of quartz sand. This sand increases the permeability of the regolith and allows it to develop to a very great thickness, up to 50 or even 100 m (160–330 ft). Granites generally produce more quartz sand than do gneisses, so that they have a more typically developed regolith. The subsurface water then circulates easily, especially if the climate is not too humid because, in spite of a decreased production of clay, there is then an increased precipitation of iron hydroxides that flocculate the clays and make them more permeable.

The differences in the rate of weathering that depend on the type of weathering front (regular or irregular) are highest in acidic igneus rocks. Compact rock masses tend to be transformed into important residual reliefs even in the wet tropics. If they are cleared of their regolith, a law of 'all or nothing' operates, and the monolithic domes deprived of their mantle evolve very slowly, accentuating their characteristics of residual reliefs.

2. *The basic igneous rocks*, rich in ferromagnesian minerals and poor in quartz. They produce much iron and clay; the regolith is generally dark, very argillaceous, and has a slow rate of infiltration. The regolith is always very much thinner than on acidic igneous rocks, usually not more than a few metres thick. It absorbs a smaller quantity of water, which favours overland flow and results in a less regular drainage. The dry seasons are thereby emphasised, which is also reflected by the vegetation and the weathering processes. Cryptolapiés are common. The influence of the type of weathering front on the rate of weathering seems smaller on the whole than on acidic igneous rocks. Cuirassing phenomena play a very important role, in part favoured by the high water-holding capacity and therefore capillary action of the regolith. Basic igneous rocks are the locus of bauxitic as well as ferruginous cuirasses especially where the dry season is sufficiently long. They play an important role in differential erosion and frequently cause inversions of relief. Cuirasses always originate in local depressions where dissolved substances become concentrated, thus indurating valley bottoms and piedmonts; later, by differential erosion, these are transformed into tablelands.

3. *The shales, mudstones, and schists* (especially micaschists and sericitic schists). They have, from the physiographic point of view, many characteristics in common with the basic igneous rocks. They also produce a very argillaceous, not very permeable, and generally thin regolith. Their permeability increases as the result of the flocculation of the clays by iron hydroxides and the formation of concretions if the rocks produce a high enough proportion of iron in a not too humid climate. These rocks break up easily at the base of the regolith, generally under the effect of hydration, and a mash of clay with usually small stony fragments assures the transition between the regolith and the unweathered rock. This zone is more permeable, which favours weathering and sometimes sliding. The role of joints is insignificant. Cuirasses may form, especially on micaschists or on very sericitic schists.

4. *All siliceous rocks.* They are the most difficult to weather; their resistance depends on the degree of cementation and the presence of impurities. In the wet–dry tropics the most resistant siliceous rocks are ferruginous quartzites, wellnigh indestructible, and on which cuirasses may develop. They form the highest reliefs of West Africa, such as Mount Nimba, or of the Ferriferous Quadrilateral of Brazil.

Siliceous quartzites, too, are very difficult to disintegrate, but in the presence of organic matter traces of iron oxide, nearly always present, enable some decementation that ends in the liberation of a little sand. But the mechanism is very slow and the resultant sandy soils have a poor structure and are very thin, often not more than a few decimetres (a foot or so). Given enough time they finally become loamy, which implies a progressive fragmentation of the quartz grains, a process that has not been studied up to now. The slowness of these processes and the very limited removal of dissolved substances maintain the very high resistance of these quartzites throughout the humid tropics. Decementation seems to be a little more intense in the most humid regions that alone are capable of maintaining a rainforest on these rocks. Perhaps the decomposition of the litter, basic in certain stages, is a necessary condition for a minor dissolution of the silica, as is demonstrated when the weathered products of the rock are examined. But in all cases such siliceous quartzites always form residual reliefs and are the only ones to display clear structural landforms even in the wet tropics.

Sandstones disintegrate more easily than quartzites because they are not as well cemented, especially if the cement contains clay, iron oxides, or, of course, calcium carbonate. The same is true if feldspars or micas are found among the grains. The weathered products that result are sands that progressively evolve into more or less argillaceous loams. Because they are very permeable they may be quite thick, especially if the sandstones are poorly cemented and rich in easily weathered impurities (minerals and argillaceous and ferruginous cements). The weathering front depends on the degree of cementation of the rock. In poorly consolidated sandstones the front is indistinct and stony as in shales. In compact sandstones it is more regular. Only such sandstones, especially when interbedded with shales, can produce clear structural landforms.

5. *The limestones.* These represent a special case that will be the object of a separate study. The tropical environment offers specific conditions to the development of karst topography. Carbon dioxide, the basic cause of its formation, is present in great quantities in the soils but is less soluble in water with each increase in temperature. There is therefore intensive dissolution at the base of the regolith, which is favourable to the development of cryptolapiés. Dissolution sets free great quantities of impurities to bury outcrops and fill fractures, permitting a substantial and persistent subaerial drainage. The ground water, on the other hand, is relatively poor in CO_2 and therefore not very corrosive. The formation of chimneys and open pits

is thereby impeded. Subterranean forms are less well developed than surface forms, and caves develop mainly at the contact of flood plains where there is more water circulation and consequently more intense corrosion.

Having reviewed the specific physiographic characteristics of the humid tropics, we will now examine the areal extent of the humid tropics and its subdivisions.

Areal extent and subdivisions of the humid tropics

At the beginning of this chapter we defined the tropics from the point of view of temperature (average monthly temperatures always above 20°C: 68°F). Now we must define the tropics from the geomorphological point of view, which is more difficult. Where the tropics are bordered by the dry zone there is no major problem in delimitation. This margin also implies that the tropics, according to our definition, are humid. But this implication may be confusing, and therefore we will speak rather of the humid tropics as, of course, there are also dry tropics. On the east side of the continents, however, the humid tropics merge with the humid subtropics, creating a problem of delimitation, as the thermal criterion is not enough from the geomorphological point of view. Here certain regions that lie outside of the humid tropics (as we have defined them) have nevertheless typical humid tropical landforms. Hong Kong is a good example. Moreover, important differences appear within the humid tropics in plant formations as well as in the nature of the landforming processes.

The limits of the humid tropics and the problem of the humid subtropics

The problem of delimiting the humid tropics is twofold. In certain areas, as in Africa north of the Equator, the humid tropics are separated from the humid midlatitudes by a wide zone of arid and semiarid lands. The contrast is clear. On the other hand, in Atlantic Brazil and in Vietnam and the south of China temperatures drop while the forest continues uninterrupted. Although it displays gradual changes the forest extends well beyond the tropics. There is here a delicate problem of delimitation.

Limits with the dry zone

As we have shown at the beginning of this chapter, precise climatic criteria can be used to separate the humid tropical zone from the dry zone.

The zone of transition between the humid tropics and the dry zone, *sensu lato* (the semiarid tropics, *sensu stricto*) is simultaneously characterised by: (*a*) an important increase in temperature range, which favours aridity. 'Torrid' months, above 30°C (86°F), appear while others drop below 20°C (68°F). But, at the same time, the daily range remains considerable. As in all climates with a high degree of insolation, temperatures measured under shelter are very different from actual temperatures at the ground, the only ones that count from the geomorphological point of view. Near Timbuktu, in the south of the Sahara, the actual temperature on certain days may drop to only a few degrees above 0°C (32°F). Daily variations of 20 to 25°C (36 to 45°F) under shelter are not exceptional. The ground, however, is subject to variations of 30° to 40°C (54° to 72°F) during the cool season, and sometimes even more. Microexfoliation (flaking) by chips only a few millimetres thick prevails over other mechanisms of physical weathering on outcropping rocks. Furthermore, an intense evaporation, which is due to great heat as much as to dryness of the air, produces desert varnish.

(*b*) a decrease of precipitation. The air is now dry most of the time. The total annual rainfall decreases as the rainy season becomes shorter, and the rain is increasingly in the form of isolated showers separated by sunny periods with high evaporation. Moisture is fleeting. Infiltration, outside certain privileged environments such as sheets of sand, decreases even faster than precipitation as much on account of rain falling on a soil hardened by drought as due to the very character of the soils themselves, being thin, discontinuous, and often very young. Apart from floodplains the soils are moistened only sporadically. Weathering is impeded. Alternations of wetting and drying permit migrations of salt immediately below the rock surface, producing desert varnish and other coatings.

The balance between the various morphogenic factors changes. While mechanical weathering becomes more important, chemical weathering becomes less so. There is, of course, no sharp limit between the two as changes are gradual. There is, however, a more rapid change in the neighbourhood of the 750 mm (30 in) isohyet. In West Africa it is marked by:

the appearance of desert varnish;
the disappearance of red soils in Senegal, close to the 800 mm (32 in) isohyet (palaeosols excluded, of course);
the transition from savanna to a formation of grasses, euphorbia, and acacias known as the Sahelian zone;
the gradual disappearance of kaolinite in the regolith at the expense of a greater proportion of montmorillonite.

In other areas also, a critical annual rainfall not higher than 750 to 800 mm (30 to 32 in) seems to correspond to a change in plant formations. In southern Africa exactly the same changes are found as in West Africa. In Brazil the xerophytic *caatinga* formation also begins with a rainfall of this order, whereas the more humid areas are covered with *agreste*, a shrubby formation with a few isolated trees, a few cacti of a limited number of species, and only a few thorn shrubs. Agreste is a variety of campo cerrado peculiar to the Northeast. This campo cerrado formation is very extensive in the interior of Brazil and is the approximate equivalent of the West Africa savanna. In Australia, too, it is near the 700 mm (28 in) isohyet that

45

the more humid vegetation of the north, with Indonesian affinities, disappears to make room for xerophytic formations rich in endemic plants. Similar changes occur in the north of Peru with the end of the coastal desert, and in Venezuela along the dry Caribbean fringe.

The Brazilian caatinga and the African Sahel are therefore included with the dry regions, whereas the Sudanese savannas and the campos cerrados and their equivalents are included with the humid tropics. The transition between these two large groups of plant formation-types corresponds approximately to a mean annual rainfall of 700 to 800 mm (28 to 32 in). The delimitation of the humid tropics is as follows on the various continents:

In Middle America it excludes the north of Yucatán and, of course, the semiarid highlands of Mexico.

In South America it excludes, in addition to the mountains, the whole western littoral desert almost up to the Ecuadorian border, the dry Caribbean fringe of Venezuela, and the interior of the Brazilian Northeast.

In Africa the transition is approximately along the 15th or 16th parallel north latitude, somewhat to the north of Cape Verde, through Kayes, to the south of Mopti and Lake Chad, to Sennar and the foot of the Ethiopian Highlands.

In southern Africa are excluded the coastal fringes of Angola, influenced by the Benguela Current, a junior replica of the Humboldt Current; the Kalahari Basin south of the Cubango River and the Okovanggo Swamp; the extreme south of Madagascar and the coastal fringe near Tulear.

In Asia a few areas of southwest Arabia receive more than 700 mm (28 in) of rain, but they are all mountainous. In India only a small area in Cutch close to the Tropic of Cancer is out of bounds.

In Australia a large part of the tropical north receives more than 700 mm (28 in) of rain. This humid zone begins at the Fitzroy River, includes the Kimberley Hills, the area north of Daly Waters, the fringe around the Gulf of Carpentaria, and the York Peninsula.

The margins of the humid tropics are all clearly defined in these areas.

THE PROBLEM OF THE HUMID SUBTROPICS

Where the humid tropics do not border the dry zone, the problem of its limits is more complex. The rainfall criterion, so important a determinant in vegetation and geomorphology, is now wanting. Differences in temperature, for their part, are less sharp. In a humid environment the daily temperature range gradually decreases, whereas the annual range increases. Furthermore, towards either Tropic the same daily range gradually becomes smaller than the annual range.

The plant formations, however, do not change notably. In the north of Vietnam, the south of Atlantic Brazil, or along the Gulf of Mexico, an observer who is not a botanist can leave the humid tropics without noticing any significant change in the plant cover. The physiognomy of the landscape remains essentially unchanged. The most important differences occur in the floristic composition. Certain species which cannot tolerate the lower temperatures disappear. Others, characteristic of transitional forests, such as the Araucarias of Brazil, make their appearance.

Geomorphic changes are just as meagre, perhaps even less evident. Rocks continue to be thoroughly weathered and weathering profiles do not show any notable change. At the most there is perhaps more organic matter on the ground and in the soil. It would reflect a somewhat slower mineralisation as the seasonal lowering of the temperature favours the activities of fungi, which temporarily fix the humus, above that of bacteria, which quickly decompose it. But within the humid tropics themselves there are even greater differences in the geography of humus. As to the landforms, they remain essentially the same. The crystalline landforms in the vicinity of Santos, to the south of the Tropic of Capricorn, are exactly like those of the neighbourhood of Rio de Janeiro which is in the intertropical zone. One recognises the same knolls periodically cleared by slides triggered off by particularly copious and protracted rains. Even further south, around Porto Alegre, the weathering profiles are still the same. The only difference is that they are generally somewhat thinner and contain a higher proportion of illite. The landforms are less subdued by the regolith, and the hillsides vary more in profile. But these are only details although we are already quite far removed from the Tropic of Capricorn (about 30° south latitude).

There are, then, adjacent to the humid tropics forested regions that look very much like the tropical rainforest and in which many tropical species are still represented. The regolith and landforms differ only in details that were acquired extremely slowly and that, due to lack of precise studies, are often difficult to disclose. We will call these regions the *humid subtropics*. They are indeed located in the immediate vicinity of the Tropics of Cancer and Capricorn, mainly poleward of them, but sometimes actually equatorwards, as in the south of China.

The geomorphic similarities between the humid tropics and subtropics only reflect a climatic resemblance. Differences are in fact minor. Let us take, for example, the case of Hong Kong. The average temperature of the coldest month drops to 15°C (59°F), but the warmest month reaches an average of 27·8°C (82°F). It never freezes, thereby eliminating one of the most important ecological limiting factors. To be sure, average temperatures of 15°C (59°F) considerably decrease biological activity, especially that of bacteria. But they do not interrupt them. It seems that it is essentially the overall speed of the weathering process that is affected, not its nature. The weathered products studied in Hong Kong by Ruxton and Berry (1957) will in fact serve us as a type, as we have been able to verify their general occurrence in the humid tropics. It seems that there are only seasonal variations of thermal origin in the intensity of the weathering process. They are comparable to those of hydric origin within the humid tropics, as occur, for example, under deciduous seasonal forest.

47

The water budget of humid subtropical stations varies more than the temperature regime. Some stations, like Hong Kong, have a dry season, corresponding to the winter monsoon, therefore with lower temperatures. The rainy season is also the warm season, and conditions similar to those of a wet tropical climate are realised. The weathering which then takes place is similar to that which occurs in the wet tropics. In the winter it is slowed down because of the double effect of drought and of lower temperatures, which, however, reduce the effects of drought. The end result is identical to that which obtains in a wet tropical climate. Only the average rate of speed of weathering is reduced. The difference would therefore be quantitative rather than qualitative. But as systematic measurements are lacking we cannot as yet evaluate these purely quantitative differences.

Other stations have a more equally distributed rainfall; for instance, Porto Alegre, Brazil; no month receives less than 75 mm (3 in). Taking into account the temperature, there is therefore no dry month. Weathering is perennial, and its rate only depends on temperature. Temperatures fluctuate between monthly averages of 13°C (55·4°F) and 25·5°C (77·9°F). As a whole, temperature conditions are therefore slightly less favourable than in Hong Kong, but three months have averages above 24°C (75·2°F), which differ but little from typical wet tropical stations. This being so, it is easy to understand why the regolith differs but little from that of the constantly

Table 1.3. *Climatic characteristics of humid subtropical stations*

	HONG KONG		PORTO ALEGRE	
TEMPERATURE	°C	°F	°C	°F
Maximum	27·8	82·0	25·5	78·0
Minimum	15·5	60·0	12·5	54·5
RAINFALL	mm	in	mm	in
January	25	1·0	101	4·0
February	33	1·3	85	3·3
March	83	3·3	92	3·6
April	138	5·4	105	4·1
May	316	12·4	98	3·9
June	413	16·3	120	4·7
July	403	15·9	106	4·2
August	375	14·8	152	6·0
September	318	12·5	111	4·4
October	133	5·2	77	3·0
November	29	1·1	76	3·0
December	23	0·9	106	4·2
Total	2 291	90·0	1 232	48·5

humid tropics, it being only slightly less evolved. Marginal conditions are approached.

The coincidence of high temperatures and high rainfall during part of the year in monsoonal climates considerably favours the weathering process and brings about far to the north, all the way to Korea, a certain likeness with the tropics. But beginning in central China, the seasonal occurrence of frost introduces important modifications. Here the subtropics properly so-called come to an end, and the midlatitudes begin. But it is a particular variety of midlatitude environment: that of the monsoon climates or east coast continental locations with hot summers. An analogous situation exists in southeastern United States. In Florida cool winters are only very exceptionally accompanied by frost. In Georgia and the Carolinas, as in southern Japan, the winter cold alternates with a quasitropical summer. Unfortunately we have only very few morphoclimatic data about these transitional regions.

The typical humid subtropical environment is found, therefore, in the south of Atlantic Brazil, in southern China, Formosa, the extreme south of Japan, North Vietnam, the southeastern United States, essentially Florida, the south of Georgia, and the coastal fringe of the Gulf of Mexico. Almost identical conditions are also found on certain islands, such as the Bermudas, or certain humid mediterranean regions without cold winters, such as the coastal fringe of northern Portugal.

Subdivisions of the humid tropics

The distribution of the plant formation-types is once again a valuable guide to climatic geomorphology. Plants reflect the climatic data and also constitute a screen that considerably affects the progress of the initial landforming processes.

Three biochores share the humid tropics: the forest, the bush (such as the campo cerrado), and the savannas. The first occupies a realm without a too marked seasonal drought; the two others, on the contrary, are adapted to it.

THE REALM OF THE FOREST

The forest occupies a realm where there is always enough water in the soil. We may first recall that in West Africa, where the study of the tropical environment has received a great deal of attention, most authors agree that the forest requires an annual rainfall higher than 1 200 to 1 500 mm (48 to 60 in); and a dry season of less than four consecutive months.

Of course, the soil water reserves play an important role. Rougerie reminds us that they may partially compensate for an insufficient annual rainfall or a slightly too protracted dry season. In Ivory Coast the limit forest–savanna is clearly influenced by the type of regolith. The Precambrian Birrimian schists that weather into a rather impermeable argillaceous regolith carry a forest which receives an annual rainfall of as little as 1 200 mm

(48 in), whereas the more permeable regolith found on granites needs at least 1 500 mm (60 in) to carry a forest. This thick regolith holds more water, but it filters through faster and feeds seepages; and as the roots of the trees are short, they are soon out of reach of the falling watertable. In the thinner argillaceous regolith of schists or basic igneous rocks, on the contrary, a lesser quantity of water is effectively retained and infiltrates slowly. More of it remains near the surface after a period of drought, and the trees do not suffer. On the schists of eastern Ivory Coast the forest appears as soon as the 1 100 mm (44 in) isohyet is reached although anthropic influences are as important here as in the rest of West Africa. Furthermore, as we will see when we study climatic changes, the forest has here reconquered areas which were occupied 10 or 15 000 years ago by savannas formed in drier palaeoclimates.

Former French West Africa, whose geomorphology is better known than that of any other humid tropical region, is therefore not the best region in which to study the limit of the forest. In other places, such as in South America, forests grow with less rainfall than in Ivory Coast. For example, in Venezuela shrub formations of campo cerrado are gradually being replaced by a dry forest (deciduous seasonal) in areas where the mean annual rainfall is only 900 to 1 000 mm (36 to 40 in). The same is true on the Atlantic coast of Brazil.

The tropical forest is composed of a number of varieties, which Rougerie (1960) has described for Ivory Coast. In doing this he has discussed the still rather uncertain terminology of ecologists. He distinguishes two basic varieties: one growing under an equatorial type of climate with abundant rain and without a marked dry season, and the other under an alternating wet–dry climate with a long dry season. Furthermore, in areas where the influence of man at the cost of the forest has played a minor role, as in South America, a third variety must be distinguished, that of the dry forest.

The forest of the wet tropics is first of all characterised by its evergreen aspect. For this reason it is often referred to as an *evergreen forest*. Each of its trees, which are never subject to drought, has its own biological rhythm. No unfavourable season imposes a reduction in the metabolism resulting in leaf fall as is the case in the winter of the temperate zone and in the dry season of the wet–dry tropics. The ground surface is therefore continuously provided with falling debris, never by a massive seasonal defoliation. This is one reason why the litter is so thin.

Another important aspect of the forest of the wet tropics is the height of its trees. They compete for light and leave the ground in a quasiperpetual shade that is seldom breached by sunlight. The underbrush is therefore sparse. Apart from the rotting trunks of fallen trees there are few obstacles. The undergrowth is almost limited to the young trees which make haste to reach the higher levels and generally have only very thin, threadlike trunks. There is a big difference between the climate and the microclimate of such a forest.

This evergreen forest is usually called *rainforest* (*forêt pluviale, Regenwald*). The French also speak of *forêt ombrophile* because of the shade it produces. Rougerie, however, has pointed out that rain is not the only hydric factor. Total humidity is more important and includes that which is maintained by the forest cover itself. This has led to the creation of the French expression *forêt hygrophile* (hygrophilous forest) which is more accurate.

The dry forest is affected by seasonal drought which imposes changes in metabolism. There are two basic differences with the rainforest. The first difference is a seasonal rhythm in metabolism which results in leaf fall during the dry season. The great majority of species is affected. But there are always some exceptions even when the dry season is relatively long. For example, on the margin of the semiarid Sahelian zone, an acacia, *Faidherbia albida,* is green during the dry season. Where the dry season is short such exceptions are more numerous so that there is a gradual transition to the evergreen rainforest. This seasonal rhythm produces an irregular supply of organic matter. It enables more sunlight to reach the ground and results in a different physiognomy. The second difference is a great development of the lower forest layers with shrubs, low plants, and sometimes grasses where the forest opens up and takes on a park-like aspect. Passage through it may then become very difficult. In some parts of South America a very dense tier of shrubs dominated by occasional small trees—about 20 m (66 ft) high—rises above a lower storey of dense bushes often intermixed with cacti.

Dry forests are characterised by several different aspects which depend on the degree of humidity. Some dry forests are transitional to the campo cerrado, and from the geomorphological point of view they are similar.

Many terms have been proposed to name these dry forests. We will follow Eyre (1963) and call them *deciduous seasonal forests,* which insists on leaf fall. But we have seen that leaf fall may vary depending on the species. A better criterion is difficult to come by. In French the terms *tropophile* and *mésophile* have been suggested. *Forêt tropophile* implies an intense dry season. Rougerie (1960) thinks that this term cannot be applied to any of the forests of Ivory Coast. Comparison with what we often have observed in South America makes us agree with him. We will therefore reserve this expression (or deciduous seasonal forest) for the driest tropical forests, absent in West Africa because the action of man has destroyed them and replaced them with savannas, if ever they have existed. On the other hand they are well represented in South America where they are transitional between the campo cerrado and the semi-evergreen forest. In Brazil they are sometimes designated by the popular term *cerradão,* which is an augmentative of *cerrado.* The deciduous seasonal forest is therefore, for us, a low open tree formation with a dense layer of shrubs rising above a less dense underbrush. The geomorphic effect of such a forest is similar to that of the campo cerrado.

The French *forêt mésophile* corresponds to Eyre's semi-evergreen seasonal forest. It is characterised by a clearly predominant tree storey but with a seasonal metabolic rhythm and a more developed undergrowth than is found in the evergreen forest. It represents the first step in the transition towards drought, whereas the deciduous seasonal forest represents, on the contrary, an adaptation to a rather protracted drought and constitutes the transition to the campo cerrado. The non-evergreen forests mentioned by Rougerie in Ivory Coast are all of the semi-evergreen variety.

Besides these climactic forests mention should be made of the *secondary forests* which are the result of anthropic degradations. Where humidity is sufficient, a forest destroyed by man will regrow, but it takes a very long time before it has again acquired its original physiognomy. Light-loving species, which are rare in the evergreen forest, are the first to colonise the ground, as the *Musanga smithii (parasolier)* of West Africa. The secondary forest is much denser in its lower storeys than the *primary forest* unaffected by man. Sometimes it looks like coppice. Its geomorphic effect is therefore different.

In this work the realm of the evergreen and semi-evergreen seasonal forests is referred to from the climatic point of view as the *wet tropics*, whereas the realm of the deciduous seasonal forest, the savannas, and the campos cerrados is referred to as the alternating *wet–dry tropics*. Both realms together constituting the *humid tropics*.

SAVANNAS AND CAMPOS CERRADOS

Regions generally drier than those occupied by forest are covered by two rather different types of vegetation: the savannas and the campos cerrados.

The savannas are characterised by a cover of grasses, often rhizomic, which grow in large tufts during the rainy season. They are 1 or 2 m (3 to 6 ft) high, sometimes even more. Shrubs 2 to 5 m (6 to 16 ft) high, often gnarled, occasionally grouped in groves, are scattered about more or less densely. In some regions of South America, particularly on the Chapada Diamantina, in Bahia, open landscapes resembling savannas and with similar morphogenic processes are covered by widely spaced shrubs 1 to 1·5 m (3 to 5 ft) high. Locally they are called *tabuleiros*. They resemble the *chaparral* and *matorral bajo* of the Spanish-speaking countries. Typical savannas also exist in South America; for example, in the Venezuelan *llanos*. In Brazil they are called *campos limpos*. When some shrubs and trees are present they are called *campos sujos*.

There are, then, several varieties of savannas that are determined by the association of various woody plants with grasses. The latter, however, are the dominant element. The *grassy savannas* (campos limpos) are mainly composed of distinct tussocky grasses between which the soil is bare. One's foot will always fit between the tufts. The *shrubby savannas* are characterised by an intermingling of shrubs with grasses. Shrubs may even form the whole plant cover by themselves (transition to campos cerrados).

Park savannas (campos sujos) are distinguished by a scattering of trees. Such are the savannas of West Africa where lone Borassus palms dominate the grasses. In the centre of Ivory Coast, on the margin of the semi-evergreen forest, large isolated trees form a very open cover over the grasses. In other cases trees are clustered in groves that are particularly dense at their margins and are very difficult to penetrate. Distinct islets of dense forest several tens, even hundreds of metres in diameter (100 to 1 000 ft) may therefore be found in the midst of grassy savannas. A microclimate totally different from that of the open savanna and which resembles that of the semi-evergreen forest reigns in such groves. Because of it they include several species of forest trees. Such groves may be found in the llanos of Venezuela and in some of the littoral savannas of Ivory Coast. Such landscapes may be referred to as *grove savannas*.

The dense plant cover of these groves stands in marked contrast with the open character of the savanna. This open character is found in all kinds of savannas and permits a considerable amount of overland flow. Because of it many groves have progressively been left standing in slight relief a few decimetres (a foot or so) above the surrounding savanna. The watertable is therefore somewhat deeper below the surface, and in some regions which are seasonally marshy this may contribute, after a certain period of time, to the maintenance of the forest. Such microreliefs always underscore the groves of the marshy savannas of Assinie in Ivory Coast.

A savanna therefore has a geomorphic environment which is quite different from that of a forest. The soil is less protected from the elements of the weather. Savanna plants also produce a litter which is quite different from that of the forest. Savannas and forests thus constitute two major subdivisions of the humid tropics.

Savannas are much more extensive in Africa than in South America, and climate is not wholly accountable for these differences. In South America many regions with protracted dry seasons, particularly in Brazil, are not covered by savannas. Instead one passes from the deciduous forest of the agreste to the more or less dense but grassless caatingas of the dry Northeast.

Typical of South America is the campo cerrado. It has equivalents in Australia where it has long been adapted to drought and tolerates a much lower rainfall of approximately 400 to 500 mm (16 to 20 in). The campo cerrado is a dense shrubby formation with a serried coppice-like aspect, difficult to penetrate. Some shrubs are thorny. There is no lower grass storey but only a few shrub saplings and occasionally some lower bushes which may be spiny. Certain trees, belonging to a restricted number of species and always well spaced, overtop the whole. The biological rhythm is as seasonal as that of the savanna. The campo cerrado or *mattoral* (of the Spanish-speaking countries) is grey, dry, forbidding, and bristling during the dry season. It grows leaves and turns green in the rainy season. It provides the ground with scoriaceous plant debris which decompose slowly and produce a permanent but thin litter. Overland movement is more difficult than in savannas.

The great extent of savannas in Africa must be imputed to man. Many of them are pyroseres adapted to recurrent brush fires. They are mainly

53

composed of rhyzomic grasses, which again sprout without difficulty after the first rains, and of shrubs with thick fire-resistant bark in the manner of the cork-oak. Such fires have great geomorphic importance because they prevent the humification of the soil, harden its surface, modify colloids, and favour runoff. It seems that fires have eliminated the African equivalents of the campos cerrados and even the tropical deciduous forests north of the Equator. The contact between forest and savanna is particularly sharp, with frequent juxtaposition of savanna pyroseres and semi-evergreen tropical forests. Even gallery forests can often be explained by fires. Fires do not succeed in destroying plants rich in sap growing in humid places. Gallery forests are at present being formed in many parts of South America where fires have recently become popular for the creation of pastures. In Venezuela deciduous seasonal forests and even semi-evergreen seasonal forests are burned to this effect. They resist along watercourses where humidity is high. Once the pasture is created, it is periodically burned, but the gallery forest persists.

The problem of the boundary between forest and savanna is very complex in Africa, as Rougerie (1960) has pointed out. Three kinds of factors interfere with one another:

1. the present ecologic influences, which are not only climatic but edaphic and geomorphic. Some sites are particularly exposed to fires; others which are more humid favour the growth of more water-loving, less combustible plants;
2. human influences, which, through the medium of fires kindled more or less frequently depend on the density of population, migrations, and certain superstitions (sacred mountains, and sacred woods, which are spared);
3. palaeoclimatic influences, which produced a larger distribution of savannas and campos cerrados during certain drier periods of the Quaternary. In the state of São Paulo, Brazil, Setzer has demonstrated that certain species of cerrado persist in a relict state in the forest which probably engulfed it. The littoral savannas of Ivory Coast are believed to be island remnants which are gradually being swallowed up by a slowly encroaching forest front. They are thought to be relicts from drier palaeoclimates and are now restricted to the most sandy soils.

The limit of the tropical forest is therefore the result of complex phenomena, and all simplistic explanations should be distrusted. The many protrusions and enclaves of this limit can then be better understood.

The humid tropics constitute a highly original physiographic environment. As the dry regions or the glaciated regions, the humid tropics represent a major morphoclimatic subdivision of the earth. Because of it certain polyzonal mechanisms which act in the littoral or in the fluvial domain are profoundly modified in this specific environment. Before studying the zonal geomorphology of the wet and the wet–dry tropics, we will first concern ourselves with the influence of the tropical climates on the polyzonal littoral and fluvial processes.

Bibliographic orientation

General works

We have listed basic references from which to undertake a study of the relief of the humid tropics. Works prior to 1945 have mainly a historical value.

Basic references

BIROT, P. (1959) *Géographie physique générale de la zone intertropicale*, Paris, CDU, 244 p. 15 fig.

BIROT, P. (1968) *The Cycle of Erosion in Different Climates*, trans. from the French edn (1960) by C. Ian Jackson and Keith M. Clayton, University of Calfornia Press, 144 p.
Birot's 1959 publication is the most interesting and contains very useful views on the relationships between climate, vegetation, soil formation, and geomorphology.

MARTONNE, E. DE (1946) 'Géographie zonale: la zone tropicale', *Ann. Géogr.* **55**, 1–18.

ROUGERIE, G. (1960) 'Le façonnement actuel des modelés en Côte d'Ivoire forestière', *Mém. IFAN*, no. 58, 542 p., 134 fig., 92 phot.
This doctoral dissertation widely transcends the local frame and includes very valuable summations concerning various general problems.

SAPPER, K. (1935) *Geomorphologie der feuchten Tropen*, Geogr. Schriften, herausg. von A. Hettner, Teubner, Leipzig, 150 p.
Partly obsolete.

Other references

BORNHARDT, W. (1900) *Zur Oberflächengestaltung und Geologie Deutsch Ostafrikas*, collection Deutsch Ostafrika, Berlin, vol. 7, 595 p., 28 pl., atlas incl. 8 maps.

FREISE, F. W. (1933) 'Brasilianische Zuckerhutberge', *Z. Geomorph.* **8**, 49–66.

FREISE, F. W. (1938) 'Inselberge und Inselberglandschafte in Granit und Gneissgebiete Brasiliens', *Z. Geomorph.*, pp. 137–68.

JESSEN, O. (1936) *Reisen und Forschungen in Angola*, Berlin, Reimer.

KREBS, N. (1942) 'Morphologische Beobachtungen in Central-India and Rajputana', *Z. Ges. Erdk. Berlin*.

MARTONNE, E. DE (1939) 'Sur la formation des pains de sucre au Brésil', *C.R. Acad. Sc.* **208**, 1163–5.

MARTONNE, E. DE (1940) 'Problèmes morphologiques au Brésil tropical atlantique', *Ann. Géogr.* **49**, 1–27, 106–29.

MARTONNE, E. DE and BIROT, P. (1944) 'Sur l'évolution des versants en climat tropical humide', *C.R. Acad. Sc.* **218**, 529–32.

PASSARGE, S. (1924) 'Das Problem der Skulptur Inselberglandschaften', *Petermanns Geogr. Mitt.*

Climate and vegetation

This list contains only a limited choice of particularly useful or interesting works in order not to leave our subject and unnecessarily burden this orientation with too many specialised references about climate and vegetation.

Climate

Besides general works on climatology, such as PEGUY's *Précis de climatologie*, Paris, Masson, 1961 (new edn in the press), data about the humid tropics in particular will be found in:

BARAT, C. (1957) 'Pluviologie et aquidimétrie dans la zone intertropicale', *Mém. IFAN*, no. 49, 80 p., 15 pl.

CARTER, D. B. (1954) *Climates of Africa and India according to Thornthwaite's 1948 classification*, Centerton, N.J., Johns Hopkins Univ. Publ. in climatology, **7**, no. 4, 25 p., 3 fig., 2 pl.

GARNIER, B. J. (1956) 'A method of computing potential evapo-transpiration in West Africa', *Bull. IFAN*, ser. A, **18**, 665–76.

HIERNAUX, C. (1955) 'Sur un nouvel indice d'humidité proposé pour l'Afrique occidentale', *Bull. IFAN*, sér. A, **17**, 1–6.

NICOLAS, J. P. (1956) 'Essais sur deux nouveaux indices climatiques saisonniers pour la zone intertropicale', *Bull. IFAN*, sér. A, **18**, 653–64.

PORTERES, R. (1934) 'Sur un indice de sécheresse dans les régions tropicales forestières. Indices en Côte d'Ivoire', *Bull. Comm. Et. Hist. Sc. AOF*, pp. 417–27.

SCAETTA, H. (1934) *Le Climat écologique de la dorsale Nil-Congo*, Inst. Royal Col. Belge, Sect. Sc. Nat. et Méd., Mém. in 4°, no. 3.

Vegetation and ecology

AUBREVILLE, A. (1938) 'La forêt coloniale; les forêts de l'Afrique occidentale française', *Ann. Acad. Sc. Col.* **9**, 244 p.

AUBREVILLE, A. (1948) 'Ancienneté de la destruction de la courverture forestière primitive de l'Afrique tropicale', *Bull. Agron. Congo Belge*, **40**, no. 2, 1347–52.

AUBREVILLE, A. (1949) *Contribution à la paléohistoire des forêts de l'Afrique tropicale*, Soc. Edit. Géogr. et Col.

BEARD, J. S. (1953) 'The savanna vegetation of northern tropical America', *Ecol. Monogr.* **23**, 149–215.

BIROT, P. (1965) *Les Formations végétales du globe*, SEDES, Paris, 1965, 508 p.

EMBERGER, L., MANGENOT, G., and MIEGE, J. (1950) 'Existence d'associations végétales typiques dans la forêt dense équatoriale', *CR Acad, Sc.* **231**, 640–2.

EYRE, S. R. (1968) *Vegetation and Soils*, 2nd edn, London, Arnold, 328 p.

KITTREDGE, J. (1948) *Forest Influences*, New York, McGraw-Hill, 394 p.
General work more oriented towards the temperate zone but containing some interesting data.

LAMPRECHT, H. (1961) 'Tropenwälder und tropische Waldwirtschaft', *Schweiz. Z. Forstwesen*, **32**, 1–110.
Some interesting basic data.

LEMEE, G. (1961) 'Effets des caractères du sol sur la localisation de la végétation en zones équatoriale et tropicale humide', Unesco, *Humid. Trop. Zone*, Abidjan Symposium, 1959, pp. 25–39.

LEMEE, G. (1967) *Précis de biogéographie*, Paris, Masson, 358 p.

MANGENOT, G. (1951) 'Une formule simple permettant de caractériser les climats de l'Afrique tropicale dans leur rapports avec la végétation', *Rev. Gén. Botanique*, pp. 353–69.

MIEGE, J. (1955) 'Les savanes et forêts claires de Côte d'Ivoire', *Etudes Eburnéennes*, **4**, 62–83.

PITOT, A. (1953) 'Feux sauvages, végétation et sols en A.O.F.', *Bull. IFAN*, sér. A, **15**, 1369–83.

RICHARDS, P. W. (1952) *The Tropical Rainforest*, Cambridge University Press, 450 p., 45 fig., 15 phot. pl.

RICHARDS, P. W. (1961) 'The types of vegetation in the humid tropics in relation to the soil', Unesco, *Humid Trop. Zone, Abidjan Symposium*, Jan. 1959, pp. 15–20.

ROBERTY, G. (1942) *Contribution à l'étude phytogéographique de l'Afrique occidentale*, doct. diss., Geneva, 150 p., 1 separate map.

SCHNELL, R. (1950) *La Forêt dense; introduction à l'étude botanique de la région forestière d'Afrique occidentale*, Paris, Lechevalier, 330 p., 13 fig., 22 phot. pl.

SCHNELL, R. (1960) 'Notes sur la végétation et la flore des plateaux gréseux de la Moyenne Guinée et de leurs abords', *Rev. Gén. Botanique*, **67**, 325–99.

TROCHAIN, J. (1952) 'Les territoires phytogéographiques de l'Afrique Noire Française d'après leur pluviométrie', *Recueil Trav. Sc. Montpellier*, Sér. Botanique, no. 5.

TROLL, C. (1958) 'Zur Physiognomik der Tropengewächse', *Jahresber. Ges. Freunden ... Univ. zu Bonn*, 75 p., 68 phot.

TROLL, C. (1959) 'Der Physiognomik der Gewächse als Ausdruck der ökologischen Lebensbedingungen', *Deutscher Geographentag*, Berlin, pp. 97–122.

WECK, J. (1959) 'Regenwälder, eine vergleichende Studie forstlichen Produktionspotential', *Die Erde*, **90**, 10–37.
Good summation on varietal distribution and their ecologic conditions; presentation of climatic formula.

Weathering and soil formation

This list includes only general works that recognise the originality of the humid tropics. More specialised studies will be listed at the ends of chapters 3 and 4.

BACHELIER, G. (1960) 'Sur l'orientation différente des processus d'humidification dans les sols bruns tempérés et les sols ferrallitiques des régions équatoriales', *Agron. Tropicale*, **15**, no. 3, 320–4.

BERLIER, Y., DABIN, B., and LENEUF, N. (1956) 'Comparaisons physiques, chimiques et microbiologiques entre les sols de forêt et de savane sur les sables tertiaires de la Basse Côte d'Ivoire', *Intern. Congr. Soil Sc.*, Paris, vol. E, pp. 499–502.

BETREMIEUX, R. (1951) 'Etude expérimentale de l'évolution du fer et du manganèse dans le sol', *Ann. Agron.*, sér. A, **2**, no. 3, 193–295.

BRANNER, J. C. (1948) 'Decomposição das rochas do Brasil', *Bol. Geográfico*, Rio de Janeiro, **59**, 1266–1300.

BUNTING, B. T. (1965) *The Geography of Soil*, London, Hutchinson, 213 p.

CORRENS, C. W. (1943) 'Die Stoffwanderungen in der Erdrinde', *Naturwissenschaften*, **21**, no. 3–4.

CRAENE, A. DE and LARUELLE, J. (1955) 'Genèse et altération des latosols tropicaux et équatoriaux', *Bull. Agron. Congo Belge*, **46**, no. 5, pp. 1113–1243.

EDEN, T. (1964) *Elements of Tropical Soil Science*, 2nd edn, London, Macmillan, 164 p.

ERHART, H. (1935, 1938) *Traité de pédologie*, Strasbourg, Edit. Institut pédologique du Bas-Rhin, 2 vols.

HARRASSOVITZ, H. (1930) 'Boden der tropischen Regionen', *Handbuch der Bodenlehre*, Berlin, vol. 2, 362–432.
Especially for historic interest.

HOORE, J. D' (1954) *L'accumulation des sesquioxides libres dans les sols tropicaux*, Publ. Institut National pour l'Etude Agronomique du Congo, sér. scientifique, no. 62, 132 p., 23 fig., 37 phot.

JENNY, H., GESSEL, S. P., and BINGHAM, F. T. (1949) 'Comparative study of decomposition rates of organic matter in temperate and tropical regions', *Soil Sc.* **68**, 419–32.

MILLOT, G. and BONIFAS, M. (1955) 'Transformations isovolumétriques dans les phénomènes de latéritisation et de bauxitisation', *Bull. Serv. Carte Géol. Alsace-Lorraine*, **8**, 1–20.

MOHR, E. C. J. and VAN BAREN, F. A. (1954) *Tropical Soils: a critical study of soil genesis as related to climate, rock and vegetation*, The Hague, van Hoeve, 498 p.
Basic work and numerous selected references, to which we refer.

SCHLAUFELBERGER, P. (1955) 'Eignen sich die Regenfaktoren langs zur exakten Klassifikation der tropischen Böden', *Petermanns Geogr. Mitt.* **99**, 99–106.

SCHOKALSKAYA, S. (1953) *Die Böden Afrikas. Die Bedingungen der Bodenbildung. Die Böden und ihre Klassifikation*, Berlin, Akademie Verlag, 403 p., 18 figs., 1 folded map.

2

Modifications of the polyzonal processes in the humid tropics[1]

The high temperatures of the tropical environment considerably affect the actions of certain polyzonal morphogenic processes whose effects are felt well beyond the area in which they occur. For example, the shore processes are modified because the intensity of solar radiation makes possible the existence of encrusting algae and hexacorals. The latter require plenty of light and a sea temperature between 25° and 30°C (77° and 86°F). It is common knowledge that corals create original forms such as fringing and barrier reefs, which considerably affect the character and evolution of ordinary coasts. A good example of the effect of a humid tropical climate on the morphogenic processes is the predominance of chemical over physical weathering. Because of it the alluvium of tropical rivers is characterised by the near absence of pebbles and falls within the two size ranges of clays and sands. These materials not only influence the nature of the depositional forms but also the kind of wear the bedrock of stream channels is subjected to, which is not more than a light mechanical abrasion. The result is that the long-profiles of tropical rivers are quite different from the long-profiles of midlatitude rivers. Indeed the rivers of the tropics have numerous rapids and falls that alternate with quiet reaches of very low gradient, facts that were long ago noted by explorers.

These characteristics, which are only a few among many, show that if the mechanisms of stream flow or the movements of the sea along the shore always obey the same physical laws and remain intrinsically the same, they nevertheless manage to produce original landforms in the tropics. This is due to the fact that these mechanisms operate in a *distinct* environment. For example, the swell and its rollers on the shore, which are an almost universal phenomenon outside of regions of permanent pack ice (for this reason the swell is not really azonal but polyzonal) produce different littoral forms if the swell breaks freely at the foot of cliffs or if it breaks against the coral structures of a barrier reef. Here the peculiar zonal influence of the tropics is evident. Analogous differences may be observed between the shores of midlatitude estuaries and the mangrove protected littorals of intertropical

[1] For the concept of zonality refer to Tricart and Cailleux: *Introduction to Climatic Geomorphology*, chapter 3 (KdeJ).

58

estuaries, or between the wearing of fluvial rockbars in midlatitude and intertropical streams. In the latter case the laws of hydrodynamics remain the same, but pebbles abrade much more than sand and, of course, much more than clay. The predominantly mechanical fragmentation which occurs between spates in extratropical climates is almost completely non-existent in the tropics, and rapids, which form easily, wear down slowly and incompletely. These differences in the kind of fluvial erosion of course affect and even determine the whole fluvial morphology and therefore play a leading role in the general evolution of the relief.

Such are the kinds of problems that we must presently examine before proceeding to a detailed study of the two morphoclimatic zones of the humid tropics. We first investigate the originality of the fluvial processes as these have more points in common with the whole humid tropical zone and play a larger role in the general evolution of the landforms. After this we consider the originality of the littoral processes, in which differences related to the degree of humidity are more pronounced.

Originality of the fluvial processes

It has been known for a long time that tropical streams have many rapids. Even powerful rivers have imposing rapids close to their mouths in places where under other climates profiles have long been regularised. The most striking example is that of the Congo, one of the largest rivers in the world, which has not been able to notch the edge of the Congo Basin whose waters it drains through its present passage since the Miocene or perhaps even since the Oligocene. Even if this region is less stable than has long been affirmed, it has nevertheless not been affected by any major crustal uplift since the beginning of the Neogene. Low elevations are the proof of it. Comparison with the Loire, which is also forced to incise a crystalline oldland to drain its waters from a sedimentary basin is instructive. This river, which is much smaller than the Congo, has nevertheless succeeded in establishing a profile of equilibrium through the oldland and even in incising its channel well below its present bottom, and all this since the uplift that caused the regression of the Faluns Sea during the Pliocene Epoch.

We can continue our comparisons with the rivers of French Guiana, Ivory Coast, and the Moroccan meseta. The streams of the latter have cut deep valleys with essentially regularised long-profiles through a region which has been slowly uparched during the Plio-Quaternary. In contrast, in Ivory Coast potent rivers carrying much more water than those of Morocco have hardly notched the crystalline oldland affected by an uplift equivalent to that of the Moroccan meseta as a result of movement along the Lagoon Fault, an uplift that is probably somewhat more recent than that of the Moroccan meseta. It is noteworthy that valleys well incised in the Tertiary sands of the *Continental Terminal* suddenly come to an end upstream from their contact with the crystalline oldland. A series of rapids follows, beyond which the rivers flow on the non-incised surface of the crystalline plateau at an altitude of about 100 m (330 ft). An important river like the Sassandra which empties into the ocean without passing through the sandy coastal strip nevertheless drops over rapids at the head of its estuary. Exactly the same may be observed in French Guiana where Quaternary uplifts, if they exist, have been extremely limited.

There is then in the tropics a striking inhibition to regressive erosion, which even affects very large rivers. The only major rivers which escape it are those which debouch in regions of subsidence and which, because of it, have constructed vast alluvial plains in which there is no place for regressive erosion, such as the Amazon, the Ganges, the Mekong, or the Irrawaddy. These rivers have, on the contrary, produced an enormous amount of continuous deposition. They are paradoxical contrasts. How can certain intertropical rivers construct enormous alluvial plains when the incision of their beds often encounters exceptional difficulties? If one adheres to the classical norms of 'normal erosion', there is here a serious contradiction. This contradiction can only be removed by detailed study and by analysing the peculiar conditions in which the fluvial processes of the tropics take place.

Distinctive forms of tropical river channels

The channels of tropical rivers often have characteristic forms that deserve specific names; for instance, the term *marigot* used in Africa of French expression to designate the seasonal streams that dry up during the dry season and are then reduced to a chain of ponds. It is, however, more often because of certain details, and because of the relative frequency of certain forms, that tropical rivers are differentiated from others, especially those of the mid-latitudes. But there are no specific geomorphic phenomena in the fluvial domain as, for example, the coral reefs in the littoral domain.

MOBILE STREAM CHANNELS

Mobile stream channels are cut from friable alluvium, regolith, or rocks whose texture is such that the materials can be moved by hydrodynamic forces, which means that the form of the stream channel is rapidly adjusted to them. Streams with mobile channels are normally developed in alluvial plains, but they also exist in friable rocks such as shales, poorly consolidated sandstones, and conglomerates. In the humid tropics, as opposed to other morphoclimatic zones, streams with mobile channels are exceptionally frequent even on solid rocks because of the importance of weathering.

Valley bottoms in the wet tropics are favourable to weathering because of their constant humidity. The presence of ground water allows water to infiltrate the joints and fissures of the fresh rock and to weather it little by little. Even with poor ground water circulation due to lack of permeability or lack of slope, weathering ends by being very effective in the long run. Various engineering works carried out for the construction of dams and bridges have generally revealed a considerable thickness of regolith below the surface of tropical valleys. In crystalline rocks it is often necessary to go down 10 to 20 m (30 to 60 ft) below the alluvium to encounter the zone of widened joints transitional to the fresh rock. The regolith may even be very much thicker. For instance, at Pampulha near Belo Horizonte, Brazil, in a region of campos cerrados, it reaches some 100 m (300 ft) in a small Tertiary valley not reached by the Quaternary wave of erosion.

In the wet–dry tropics the advantage of the weathering process in valley bottoms over that of the interfluves is also significant. In savannas the regolith is generally thicker under lowlands especially where there are depressions. It is possible that the acidity of the partially anaerobic soils of flood zones plays a role in weathering. Unfortunately we do not know of any detailed study of the regolith of a flood plain. Because cross-sections in such plains are seldom observed, most of our information comes from deep samplings.

Because of intense weathering many streams flowing through regions of solid rocks have mobile channels even if the alluvium only forms a thin veneer, 3 to 4 m thick (9 to 12 ft), on the valley bottom. Such a disposition occurs when two conditions are simultaneously realised: (*a*) an easy penetration of the water into joints and fissures; the ease of penetration is all the more important as the ground water circulation is difficult and slow; and (*b*) a good susceptibility to weathering. For these reasons lithologic differences are usually important along the watercourses of the wet as well as of the wet–dry tropics. It causes an alternation of reaches with mobile channels and of rapids on rockbars, many of which only form minor breaks. The *Institut Géographique National* of France has demonstrated these features by levelling surveys carried out along the long-profiles of a number of French Guiana rivers. We once more note the importance of differential erosion so important in all of the tropics.

Weathering also determines the peculiar texture of the alluvium. The exceptional outcrops of fresh rock, the weakness of the forces of disintegration working upon them, the slowing down of overland flow by the comblike effect of the plant cover all contribute to the fact that the hillsides provide the streams with only a very small amount of coarse debris. The only important coarse fractions come from earthflows that have stripped off coarse material from the lower layers of the subsoil transitional to the fresh rock. Furthermore, such earthflows should reach a riverbed. If they spread on a gentle slope or on the flood plain the material continues to weather before being reworked by the stream.

The origins of the coarser fractions of the river alluvium are therefore as follows:

(*a*) Debris torn from rockbars which lie in the path of the riverbed. These rockbars are always resistant to weathering, like monolithic domes, and therefore only provide a small amount of debris (this case will be examined in more detail later).

(*b*) Products brought by earthflows set in motion in the regolith by heavy rains. Abrupt slopes (over 45°) directly overlooking the river are necessary, which is seldom realised outside of mountains. Such earthflows often contribute mainly fine material and only a few large boulders, which originated in the zone of spheroidal weathering. A larger inflow of rocks requires, as in the case of the crystalline Andes of Venezuela, an intense tectonic fragmentation that prevents the formation of corestones, or, as in the case of the volcanic terrains of El Salvador, an earthflow produced from the clayey residue of poorly cemented pyroclasts that entrains large rock fragments such as blocks, bombs, and cinders.

(*c*) The reworking of old gravelly alluvium, such as the well-known 'underbank gravels'[1] of West African miners, deposited under climatic conditions different from the present one.

The transportation of coarse debris, later transformed into pebbles, to the river channels of lowlands and even of steep hilly regions is therefore exceptional. The same is true even in many mountainous regions, such as the Serra do Mar of Brazil, and only a few recent mountain ranges (Venezuelan Andes, El Salvador) have torrents with irregular mobile channels veneered with pebbles. But even in such cases there are differences with other morphoclimatic zones. The pebbles are much less stable. They are also affected, even in stream beds, by the intense weathering mechanisms characteristic of the humid tropics. Siliceous rocks such as quartz and quartzites are less affected than other rocks. Moreover, landslides and earthflows are more apt to carry fragments of them, especially the debris of quartz veins that survive in the midst of the regolith. The proportion of siliceous pebbles is, therefore, always very high in coarse alluvium: 85 to 100 per cent, whether in French Guiana, Brazil, Colombia (on the lower Magdalena), or West Africa. Crystalline rock fragments, for instance, disappear rapidly. One or two kilometres below a rockbar that is the source of them or below an undermined riverbank producing crystalline gravel, they have practically disappeared. Their progression down the very gentle gradients between rapids is very slow as they move forward only during spates. Weathering therefore proceeds faster than transportation, and there is enough time to reduce gravel to sand between floods. The result is that there is a rapid but gradual proportional downstream increase in the number of siliceous pebbles; less subject to weathering they have a better chance to persist until the next spate when they make a new leap forward. Near Quaimadas, Bahia, Brazil, although in the semiarid tropics (500 to 600 mm: 20 to 24 in of rainfall), the proportion of reworked siliceous pebbles of a terrace increases from 85 to 97 per cent in 2 km. Even so, the siliceous pebbles themselves are not stable in tropical rivers. At low water they become oxidised. Hydroxides penetrate into fissures and even into the crystalline structure of the quartz which they tint. These hydroxides may again be

---

[1] 'Graviers sous berge'.

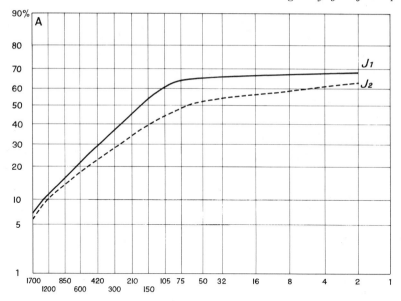

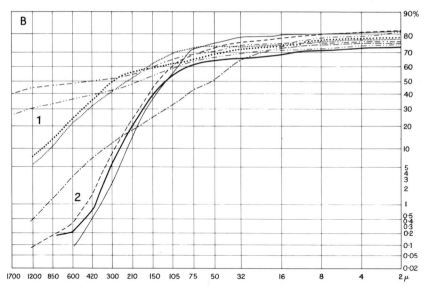

FIG. 2.1. Alluvial grade sizes according to Bakker and Muller (1957)

A. Grade sizes of weathered granites and pegmatites in Surinam. Note the absence of silts (1 to 75 microns).

B. Alluvial deposits of Surinam and Brazil.

Group 1 is composed of terrace materials from the Surinam River. The smaller proportion of coarse sands seems to be due to a beginning of weathering.

Group 2 is composed of the present alluvium of the Surinam River and the Rio São Francisco, Brazil. Note the same absence of silts as in the regolith (A) and the general similitude of the curves; the materials, however, are better sorted in the fluvial deposits.

dissolved, thereby weakening the ferruginised fissures and preparing the pebbles for further fragmentation which eventually reduces them to sand. The slowness of transport on such level reaches and, especially, its very uneven movement, results in a progressive elimination of the pebbles downstream from each outcrop. It seems that the distance over which the reduction takes place is of the order of 1 to 3 km (0·6 to 2 miles) for crystalline pebbles and 10 to 15 km (6 to 9 miles) for quartzes 4 to 6 cm (1·6 to 2·4 in) in diameter.

The alluvium of intertropical rivers with the exception of that found in certain types of mountains is therefore poor in gravel. Viewed in cross-section, gravel is usually limited to a few runs of scattered pebbles in the midst of fine sediments including sands and clays. The insignificance of silts has been demonstrated in numerous analyses by Bakker (1957b) on Surinam

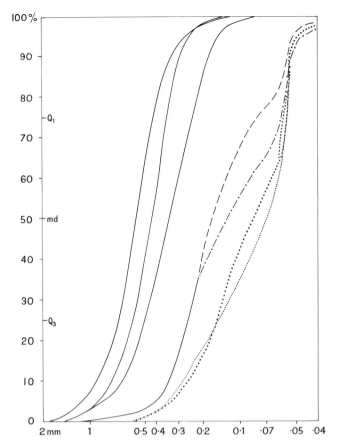

FIG. 2.2 Grade size curves of alluvial sands in the area of Recife, Brazil, according to Ottmann and Ottmann (1960)

To the left, in solid lines: curves of fluvial sediments. To the right, in dashes and dots: curves of estuarine sands, noticeably finer and occasionally bimodal.

rivers. The textural curves are normally bimodal with a mixture of two fractions: sands and clays. This is easily explained. The silts, when they are not the product of reworking, are essentially the result of mechanical actions whose extreme end product they are (like glacial rock flour and products of intense frost-riving). Whereas the clays are derived from the weathered products of non-quartzose minerals, the sands are the residue of quartzes that have resisted weathering. The latter are relatively more abundant in regions composed of rocks rich in quartz such as granites, but less abundant in regions composed, for example, of basalts. Clays contain a high proportion of kaolinite, a typical product of humid tropical weathering. Watertable fluctuations in alluvial plains, lastly, cause the incorporation of precipitated dissolved products, especially hydroxides but also silica, in the alluvial deposits of the flood plain. Such products are, of course, more abundant in areas where swinging water tables are ample and regular, that is, in regions with a marked dry season. They may result in the formation of so-called *groundwater cuirasses*. In a more regularly humid climate iron hydroxides only colour the clays or may even form some concretions. The alluvial deposits of the humid tropics therefore have certain distinct characteristics the knowledge of which may be useful in palaeogeographic reconstructions:

(*a*) There is no clear stratification as each layer abandoned after every flood is quickly disturbed by burrowing animals and plant roots. Only runs of scattered pebbles remain more or less undisturbed.

(*b*) There is poor sorting as a result of the comblike effect of the flood plain vegetation which is composed of forest even in savanna regions (gallery forests). There are two modes: one sandy and the other clayey, not counting occasional pebbles.

(*c*) There is a considerable proportion of kaolinite and iron hydroxide that imparts a red colour to the deposits. Hydroxides are unequally distributed, forming accumulations along discontinuities such as old desiccation cracks, old roots, and sandy laminations that increase the permeability. In the wet–dry tropics they become concentrated in concretions and cuirasses.

(*d*) There is a very high proportion of quartz, sometimes of quartzite and other siliceous rocks such as flint or chert in sands and gravels.

These characteristics influence the development of river channels as well as the fluvial processes themselves.

Poor textural sorting, a high proportion of clays, and even a beginning of cementation by iron hydroxides increase resistance to erosion. Undermined banks do not seriously crumble and retreat slower than in the more sandy deposits of the Sahelian zone or the silty to gravelly alluvium of temperate lands. This rule, of course, does not apply to certain piedmont regions where the proximity of mountains subjected to different morphogenic conditions modifies the nature of the alluvium, giving it a closer resemblance to that of the temperate zone. Even during high floods the mass of materials once

FIG. 2.3. Swamp forest near Dabou, Ivory Coast

Demonstrates the effective comblike effect of the plant cover, causing accumulation of clays in colloidal solution together with non-decomposed organic matter. Extremely level alluviation.

again set in motion in the stream channel always remains relatively modest. Even where a river undermines a low terrace rich in 'underbank gravels', as on the upper Gambia studied by Michel (1959), excessive gravel banks do not form but only modest deposits that are in transit. Gravelly braided channels seldom form in intertropical rivers. Even meanders seem to be less common, except in estuaries, than in the temperate zone.

The river load, on the other hand, is very mobile. During spates clays cover vast distances and suddenly emerge into the sea if the high waters persist long enough and do not drain into swamps. At low water, on the contrary, when the current comes to a near standstill and stagnation develops, clays accumulate in depressions and form sticky coatings with the help of biological activities on pebbles, rocks and the bottom itself. Resistance to erosion is thereby increased. The clays that find their way to swamps are trapped and help build up although very slowly the alluvial deposits of the flood plain. They may accumulate at a great distance from the riverbed, which results in the spreading of a vast but thin veneer of clayey alluvium.

Sand texture varies according to lithology and climatic types. It is finer and less abundant where chemical weathering is more active. It is coarser and mixed with ferruginous pisolites where there is a marked dry season as in the Sudanese zone of West Africa. Fine sands are often moved in suspension on eddies and are deposited as soon as they leave the river channel, trapped as they are by the vegetation. But this happens only during high floods. Lesser floods abandon only clays, which become mixed with the older sands by biological action. In the wet tropics, except in sandstone regions that supply a large quantity of sand, natural levees are generally poorly developed and only slightly more sandy than the back-swamps. Such conditions are unfavourable to the formation of meanders because deposition on the convex banks is insufficient to maintain a harmonious development of the bends. In the wet–dry tropics sands are more abundant, at least in rivers of a certain importance, because of the lesser intensity of chemical weathering and a more violent discharge. Perfect, almost pure sandy natural levees are then formed, with clay deposits in the intermediate back-swamps. This is also the case of the Senegal in the Sahelian zone or of the Peruvian Andean piedmont in a zone of evergreen rainforest. Magnificent meanders and natural levees then develop.

The fluvial processes are also influenced by the vegetation. But it is not the evergreen rainforest which has the greatest effect. In fact, in spite of its density it does not much hamper the undermining of riverbanks for the tree roots are too superficial and the sapping of the banks proceeds at a lower level. These conditions easily cause trees to topple and to form bridges across streams, sometimes creating inextricable congestions which hamper navigation. At low water wood rotting in near stagnant pools probably influences the chemistry of alluvial plains, but in a way still unknown. On small streams that do not produce a breach in the forest canopy and do not have a characteristic bank vegetation composed of light seeking species, the degree of sapping depends on the nature of the regolith, contrary to what happens in the temperate zone where bank thickets are a very effective protection. A high flood multiplies the number of small streams in disorderly fashion and produces toppled trees and obstructions of branches, which rot after the waters have subsided and become stagnant. For these reasons such small streams are poorly graded. On large rivers, forming an opening through the forest, banks are better protected by thickets rich in shrubs which limit the sapping. As a whole, there is less undermining and the channels are generally well graded.

In savanna regions small streams are distinctly different from the streams of other morphoclimatic regions. The flow of water is interrupted during the dry season. There is hardly a transition between absence of flow and flood, which is unfavourable to the establishment of a graded profile. In most cases there is no continuous stream channel. A large part of the valley floor constitutes a floodable bottomland submerged below a few decimetres (1 to 3 ft) of water during the rainy season, where the water drains as best as it can between the vegetation, whether gallery forest or tall grass. In certain

FIG. 2.4. Bottomland in rainy season
near Mamou, Guinea
Muddy waters flood the forest in a near
level reach of the stream. Very gentle
current, limited transport, and muddy
water.

places random obstacles cause a banking up of the water and a more concentrated and rapid flow that succeeds in tracing itself a channel. When it
is important enough or regularly repeated, in a constriction for example,
some erosion occurs, excavating a whirlpool or a short channel that merges
further down with a renewed broadening. Such pools remain filled with
water after the rainy season and are at the origin of the term *marigot*, which
etymologically only ought to be used to describe this type of watercourse,
different enough from that of a river. In such bottomlands the incision of
channels is practically impossible; the dominant morphogenic action is the
deposition of materials brought in by runoff from the interfluves, and in the
dry season the precipitation of dissolved substances in the ground water
zone, which is a place favourable to the formation of groundwater cuirasses.

The large rivers of savanna regions are characterised by important stage
differences. They flood vast areas of the valley floors and restrict themselves
at low water to the river channel where the current is slow. Such channels
have sharp banks, several metres high. They are always well protected by
vegetation, which takes advantage of the humidity even in the dry season.

FIG. 2.5. Gallery forest in the savannas of northern Ivory Coast near Korhogo
Marigot in flood, flowing below a canopy of foliage and between trees which effectively protect the banks. High load in suspension, but in spite of a rather swift current transport is practically limited to muds and a little sand on the bottom of the channel.

When the water rises 5 or 10 m (15 to 30 ft), as often happens, or even 14 to 17 m (45 to 60 ft) as on the middle Orinoco in Venezuela, the vegetation then submerged produces a comblike effect and causes deposition, especially of sands, on the top of the levees while the clays accumulate on the floodplain, which is often immense.

The following dynamic characteristics have been observed on all large intertropical rivers:

(*a*) A widespread deposition of muds on flood plains wherever topography allows. The fine alluvium may spread over many kilometres, thanks to the abundance of clay in the suspended load and the size of the floods.

(*b*) A strong contrast between periods of flood when deposition takes place on the flood plain, and periods of low water when sands and even pebbles are transported in the river channel, but massive transport is limited to sudden short sweeps, which allows the load in progress of migration to be attacked by weathering. The mobile channels generally reveal the contrasts in river regime by their sharp banks, whose height in part depends on stage differences.

(*c*) The ability of a river to continue its activities even with a very low gradient provided there is enough discharge. Clays in suspension can be evacuated as long as there is some current. Sands mainly move under the effects of violent spates, which noticeably increase the gradient of the water surface (it is approximately doubled on the middle Orinoco and on the Senegal at Kayes). Reaches with very gentle gradients are therefore perfectly functional.

ROCK CHANNELS

Rockbars are extremely frequent on tropical rivers, especially, it seems, in the forested zone. Contrary to what is the case in temperate latitudes, they do not necessarily reflect important breaks in the long-profile. For instance, the elevation of the Maroni (on the border between French Guiana and Surinam) at 125 km (78 miles) from the coast is only 20 m (66 ft) in spite of several dozen rapids. According to Bakker (1957b) the 100 m (330 ft) contour interval is reached only at a distance of 200 to 300 km (125 to 185 miles) from the river's mouth. Usually the rapids only modify the water flow at high stage. At low stage the rock outcrops appear in the channel and the water flows around them without forming important rapids. During flood, on the contrary, the rock outcrops cause a banking up of the waters and a violent overflow separating two well differentiated water levels. Such rapids may be frequent even on rivers whose gradients are as gentle as or slightly steeper than the gradients of midlatitude rivers with graded profiles. The small drops of elevation on the rapids are compensated by the near absence of gradient on the intermediate reaches. In spite of all the rapids the rivers of French Guiana and Surinam have an average gradient slightly gentler than or equal to those of the lower courses of the Loire and the Garonne below the 100 m (330 ft) contour interval.

Such small rapids are, of course, not the only ones, and many intertropical rivers clear more important breaks in their lower courses. Suffice to mention the large rapids or waterfalls of the São Francisco (at Paulo Afonso, Brazil), the Congo, the Bia (Ivory Coast), and the Zambezi. The only falls that may be compared to them in the mid and high latitudes are those of glaciated regions: the Niagara and the falls at the margins of the Canadian or Scandinavian shields. But they occur only where fluvial runoff has been reintroduced 5 000 to 10 000 years ago, whereas the huge falls of the large intertropical rivers have an origin that goes back to the Tertiary, to several million years.

The inequalities in the long-profile of intertropical and midlatitude rivers therefore correspond to two different orders of magnitude in their frequency and durability. Tropical rivers reduce the inequalities of their profiles with much greater difficulty than the rivers of the temperate zone. We may try to understand why through an examination of the nature of the rockbars.

Four kinds of inequalities or rockbars may be distinguished:

1. *Waterfalls* in which the water drops in a subvertical manner as it leaps across the rockbar. True falls are not nearly as common as rapids. Special topographic conditions are necessary for their existence: precipitous to sheer slopes that are almost perpendicular to the stream channel. Most falls

in the tropics are located on secondary streams flowing through tabular sandstones and quartzites. These rocks are the most apt to produce rather continuous escarpments. Thus falls are a common sight in the Fouta Djallon (Guinea) and in Venezuela south of the Orinoco. In the latter region is the highest waterfall in the world: the Angel Falls, which are 800 m high (2 600 ft). In most cases the streams have not cut a re-entrant into the rockbar, and there is no incision whatever at the site of the falls. As many such falls are very ancient, one is forced to admit that there is no sign of wear; the waters are generally clear, containing very little matter in suspension. Quartzitic sandstone which is almost immune to weathering precludes an important solid load. Another type of waterfall more common on large rivers occurs if a river drops into a canyon with sheer walls. This is the case of the Souma Falls in Guinea or the Paulo Afonso Falls in Brazil. In both cases a deep

Fig. 2.6. Souma Falls, Guinea
The river plunges over a quartzitic sandstone ledge without producing a notch. To the right, a tributary crosses the same outcrop in a channel cut into a zone of fractures.

canyon is entrenched in very resistant rocks, quartzites in the case of the Souma, migmatic gneisses in the case of the Rio São Francisco. While a secondary branch of the river descends the canyon at its apex in the form of a series of cascades, the main stream continues further along the margin of the plateau before catching up with the edge of the canyon over which it then drops in a single cataract. Such a pattern is typical of the tropics; for example, it is of common occurrence in the Sudanese zone, as on the Bafing and the Bakoy, tributaries of the Senegal. In other climatic zones the main branch of the river rushes into the canyon at its apex because the apex was cut by the stream itself. A classic example is the canyon cut by Niagara Falls. In the tropics the different pattern, which is the rule, results from other genetic conditions, which we shall study later.

Even on an important river gushing down a canyon, as the São Francisco, there is no trace whatever of a re-entrant cut by the falls. The waters simply leap without incising the rim over which they pass. The only processes which seem to function are those of sapping and of the pneumatic effect of water masses pounding on the rocky floor at the base of the cataract. They perhaps cause a widening of joints and in the long run a freeing of loose blocks, as is often observed at this place. In sedimentary rocks; for instance, in the Palaeozoic sandstones and quartzites of the Bafing and Bakoy basins, sapping in the long run occasionally causes a small retreat of the falls, which then assume an outline in the shape of an arc of a circle, the stream at low stage being concentrated in the central part of the arc where the retreat is at a maximum. Such circumstances are mainly encountered in areas where the hard capping stratum that causes the waterfall rests on less resistant beds that outcrop at its foot. This is the case of the Gouina Falls on the Senegal. These falls, it is true, are already in the Sahelian zone in an area where the sandy load is considerable and undoubtedly facilitates erosion at the foot of the falls.

2. *Rapids with potholes*, which are a very common type, are reported from West Africa (Tricart, 1955), Brazil (Tricart, 1959), and Formosa (Tschang Tsi Lin, 1957). The potholes often reach very large proportions of up to 2 to 4 m (7 to 13 ft) in diameter and an extraordinary density.

Let us take as an example the Felou Rapids on the Senegal near Kayes. The rocks are poorly cemented Ordovician sandstones that break easily under the hammer and form a large uniform surface without joints, dipping about 5° in an upstream direction and bevelled by a plane of erosion. It is presently incised by the river in the form of a narrow canyon, which canalises the stream at low or average water but not when in flood. It is then that the numerous potholes which pockmark the upstream part of the erosional plane function. Most of them are widest half way up, like amphoras, which enables neighbouring potholes to intersect and coalesce, with 'bridges' separating the openings. The bridges finally wear away and irregular oval cavities like bathtubs are the end result. Pothole depth is usually about 2 m (6 ft) but occasionally as much as 3 or 4 m (10 to 13 ft). The walls are always polished and occasionally moulded by drapelike folds. The material in which they are cut is always very fine grained. Of a total of 200 potholes examined, one single 6 cm (2 in) pebble was encountered. Usually the limited amount of material found in them is restricted to coarse sand and fine gravel. Silico-ferruginous concretions and cuirass fragments predominate. Harder than the bedrock they constitute excellent grinders. Not very heavy, they are easily lifted by the swirls, which cause the swelling of the potholes half way up; entrained by the water they describe complex spirals that produce the drapelike sculpturing. Because of the absence of coarse material potholes only form where the current is not too strong. Overflow in the form of a thin sheet of water on a rather gentle slope on the higher part of the erosional plane is an essential condition for their proliferation. Here irregularities in the stream bed allow the sand or pebbles to start swirling. On the other hand, at the lower end of the inclined plane of erosion where funnel-like conditions concentrate the flow of water this meagre bed load is forthwith swept away. No potholes form. To initiate them part of the load must saltate along the bottom. If the whole load is in swirling suspension it is impossible. Once a hole is formed it detains the material in the trap, which

FIG. 2.7. Coalescing potholes in sandstone; Faya Rapids on the Niger near Bamako, Mali

The intersecting potholes are separated by 'draperies'. Grinders, mainly composed of large ferruginous concretions, can be seen in one of the potholes.

allows the development of a pothole by chain reaction. As they coalesce potholes later produce channels, which concentrate the water and continue to be sculptured between spates.

Similar mechanisms are at work on the Niger at Sotuba near Bamako. But there is less development of potholes because the rock is more strongly cemented and not as easily sculptured. Potholes are smaller and more scattered. In other respects the rock is more jointed, facilitating the formation of potholes, which often occur at their intersection. But later they impede their development. Indeed when the bottom of the potholes reaches a plane of stratification a block is very often prised loose as the infiltration of water has caused the weathering and decementation of the block's lower surface. Channels, on the other hand, are more easily developed because of the presence of joints. The channel walls are subjected to an intense abrasion by sand moving in swirling suspension. They are eventually sculptured into 'draperies' because of intersection of niches that are excavated by eddies of lateral impulsion.

Fig. 2.8. Canyonlike river channel due to the coalescing of potholes at Sotuba Falls on the Niger, near Bamako, Mali
Ordovician sandstones. Potholes of lateral impulsion intersecting, forming draperies. Their development frees blocks that fall in the channel and gradually widen it.

Depending on the nature of the rocks, the relative part played by potholes and channels resulting from their coalescence or the exploitation of fractures is therefore rather variable. But lithology is not the only factor. Very fine potholes occur in the Sudanese zone, in savanna lands where the sand load, especially in sandstone regions, is considerable, and where heavier ferruginous pellets are more abundant. In the wet tropics where the debris derived from the interfluves is finer potholes are not as well developed and are of less common occurrence. On the other hand, in the subtropical climates where weathering is not quite so intense due to lower temperatures part of the year, they are again common, even in the wet forested regions as, for example, in the state of São Paulo, Brazil, or in Formosa.

3. *Polished rapids* predominate either on very hard rocks or in regions of very intense weathering. For example, the Comoé River near Abengourou, Ivory Coast, has only a few potholes as it crosses Precambrian quartzitic rockbars in a region of deciduous seasonal forest. The only potholes observed—perhaps only one for every 100 sq m (1 000 sq ft)—were not wider than 20 cm (8 in) and deeper than 30 cm (12 in). But potential grind stones are abundant due to the reworking of 'underbank gravels'. Loose pebbles lie scattered about the rapids. Characteristic features on the rapids are limited to polished surfaces, which are not unlike those of roches moutonnées in glaciated regions. Sharp edges on the quartzites have been dulled or rounded and their surfaces smoothed to the point that they are shiny after a shower. These aspects are due to a slow abrasion by fine sands carried in swirling suspension during floods. They are the result of an extremely slow abrasion. The rapids have, in fact, been exhumed from the cover of underbank gravels that occupies the valley floor.

In less resistant rocks abrasion produces a larger number of characteristic features. For example, in the massive fine grained granites at Salto, state of São Paulo, Brazil, the upper part of the large rapids is characterised by small canyon-like channels, 1 m wide (3 ft), located on a number of observable fractures. Their walls are sculptured and polished by abrasion with the help

FIG. 2.9. Granitic rocks polished by sand loaded currents at Salto, State of São Paulo, Brazil

Hydrodynamic sculpture during high water produces very disconnected but polished forms. Reflections are due to ferruginous varnish, which increases the resistance of the rock.

of currents of lateral impulsion. Between them irregular protrusions somewhat resembling lapiés but smooth, as if waxed, and shaped like curving
ridges with hydrodynamic and harmonious forms, intersect one another in
a loose network. Similar forms, only larger, are found on the upper tiers of
the Paulo Afonso Falls (Rio São Francisco) where the decrease in friction
produced by the falls causes the stream flow to accelerate. The microrelief
is sculptured by the abrasive action of sand brought by floods to turbulent
channels where the load is carried in suspension. The forms disappear where
the current is slow and the sand dragged along the bottom: they are then
replaced by polished surfaces as those found on the Comoé at Abengourou
(Ivory Coast). The very shape of the forms suggests a very slow sculpturing,
as all effects of polishing, but not quite as slow as in the Abengourou type of
simple polish. The greater abundance of sand in sandstone regions in the

FIG. 2.10. Rapids on a small stream at Mury in the
Serra do Mar, State of Rio de Janeiro, Brazil
Very steep mountain torrent. The stream is only
slightly incised, exploiting lines of weakness in the
gneiss and forming potholes. The gneiss outcrops in
the stream bed are polished. There are no pebbles.

wet–dry tropics, as at Paulo Afonso, on the one hand, or as at Salto in the cooler subtropics, on the other hand, also favour the development of a polish by hydrodynamic action.

4. *Rapids with scattered boulders* are characterised by a very irregular river-bed in which the stream flow divides or wanders over the surface of a bed-rock pavement strewn with boulders one metre (3 ft) in diameter or more, on the average. They may be slightly displaced during floods but are not carried away.

In the Brazilian Serra do Mar there are steep torrents (5 to 10°) that cascade down through accumulations of boulders. This is the case, for ex-ample, of the torrent that follows the road that leads to Itatiaia National Park. The boulders here are 2 to 6 m (6 to 20 ft) in diameter and are mostly composed of a syenitic rock different from the local gneisses. They were brought to their present resting place by Pleistocene earthflows or mud-flows and are now concentrated in the valley bottom as the matrix has been washed away. They form a stream pavement that the torrent cannot re-move but can only wear down very slowly by abrasion, occasionally with the help of small potholes. In another case, as at Mury, near Nova Friburgo, there are fewer, loosely spread boulders. The waters slide over solid flags of granite, exploiting the least weakness. Sudden elbows, often at angles of 90°, occur at the locus of a minor fracture. The transverse profile is in the shape

FIG. 2.11. Rapids with scattered boulders on the Bia River at Aboisso, Ivory Coast

Granite rockbar weathering into corestones which sometimes split up. There is some abrasion by sand. The loose bedrock boulders are scattered about the riverbed.

of a sharp **V**. The long-profile includes small cascades on the edge of fractures, flowing water on inclined flags, and occasional small potholes.

This type of rapids on a bedrock pavement strewn with boulders, in which the proportions of the two characteristic elements varies, is found on medium sized rivers such as the Bia at Aboisso, Ivory Coast, in the rainforest, or the Paraíba at Alem Paraíba, State of Rio de Janeiro, on the margin of a humid subtropical climate. The boulders are neither worn nor polished but instead superficially corroded. They seem to weather slowly under the effect of chemical processes induced by alternations of wetting and drying. Their numbers, in both cases, seem to be related to the lithology. The granites of Aboisso and the gneisses of Alem Paraíba are rather jointed and are easily dislodged along fractures. These rocks are rich in quartz and poor in biotite and therefore resistant to weathering. In such a case the prying loose of the blocks allows a much faster incision than in the preceding types of rapids. On the Paraíba there is a small entrenchment of the river in the bedrock below the base of the lower terrace.

Such rapids with scattered boulders on a bedrock pavement are common on rivers with gentle average gradients even under evergreen forest. Nearly

FIG. 2.12. The Bafing–Makana Falls on the Bafing River, Mali, seen from the southwest (drawing by D. Tricart)

The river flows on a bedrock pavement of Ordovician quartzites, dipping underneath sandstones upstream (upper lefthand corner). In the latter, the riverbed is wide and more or less graded. On the quartzites, small stripped surfaces are cleared in the flood zone, and the low water discharge is concentrated in a narrow channel with right angle turns, taking advantage of fracture zones. Some large, loose quartzite boulders are strewn over the stripped surfaces, but there is not a single bed of pebbles downstream.

all the *sauts* of French Guiana are of this type. They are also numerous on savanna rivers such as the Falémé, the Bakoy, and the Bafing, all tributaries of the Senegal. They seem to be mainly related to lithologic conditions, to rocks that detach themselves in blocks through the corrosion of fractures but are later difficult to weather by granular disintegration into products that are easily removed. The boulders thus form a constriction that impedes the regularisation of the long-profile.

The frequency of rapids and falls just described imparts special characteristics to the long-profiles of tropical rivers. Let us now see what happens to these profiles during the course of evolution.

Evolution of long-profiles

The channels of intertropical rivers are generally characterised by alternations of:

(*a*) Reaches with very low gradients frequently dropping to only a few centimetres per kilometre (or inches per mile) on platform regions. Gradients are even less in regions of subsidence where enormous quantities of Quaternary deposits are accumulating. The reaches are then lengthened concomitantly. The most typical of all is the Amazon. But even in regions of moderate erosion, such as shields or arches, the reaches have a very low gradient (e.g. the Maroni, on the border between Surinam and French Guiana). On all important rivers such quiet reaches have a well defined and generally graded channel. On small streams the channel is frequently uncertain, shifting on account of jams produced by fallen trees in forested regions, or widening on the floodplains of savannas and in certain areas of the deciduous seasonal forests.

(*b*) Rapids or falls forming steps in the long-profile. Differences in level between such breaks is frequently very small and noticeable only at high water, mainly on account of the water banking up behind obstacles that obstruct the current (as on rapids with scattered boulders). In such constrictions the channel is always poorly graded. Part of the flow is usually canalised in deep and narrow canyonlike channels that are too small to cope with flood waters. When spates occur the channels overflow and bank up the water over a vast surface before it is able to rejoin the main channel further down.

The main characteristic of intertropical rivers, therefore, is the contrast between these two types of channels, which alternate with one another over often very short distances. This contrast lasts during the course of evolution, as is shown by its endurance on rivers whose evolution can be dated, such as the Congo or the São Francisco. Irregularities in the stream profile caused by entrenchment during the Neogene persist to the present day, whether they are minor or as important as those of Paulo Afonso or Inga falls. For example, the polished quartzitic knobs of the Comoé, near Abengourou

79

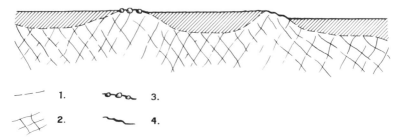

FIG. 2.13. Schematic profile of a tropical river
  1. Weathered mantle below the alluvium
  2. Fresh rock with density of joints
  3. Rapids with scattered boulders
  4. Rapids of fresh bedrock

(Ivory Coast) have only recently been exhumed from the valley fill, which is now a low terrace, probably dating back to the Würm. At Sotuba Falls on the Niger (below Bamako) incompletely developed potholes are mainly old exhumed forms. In several we have found pebbles and subangular rubble cemented by sands now transformed into ferruginous sandstone identical with the material of the ground water cuirass that frequently occurs in the low terrace. It appears, therefore, that the present above mentioned rapids are not the result of the last 10 000 years or so of erosion that has transformed the valley floor into a low terrace. The relief of the rapids was sculptured during the course of a much longer period, that which preceded it, and which seems to correspond approximately to the Riss-Würm interglacial (Eemian). We have gathered a series of similar observations on Brazilian rivers. It seems, therefore, that several tens of thousands of years are necessary to sculpture rapids with canyonlike channels in rocks that are mechanically not very resistant. Of course, even more time is needed for such channels to incise themselves sufficiently for regressive erosion to affect the higher quiet reaches.

A phenomenon that seems typical of the tropics appears to contribute to the retardation of the erosion of rapids. It is the formation of a coating that increases the superficial resistance of the rock. We have observed it on sandstones as well as on crystalline rocks. For instance, at Sotuba and at Tosaye, both on the Niger, sandstones are being transformed into very hard ferruginous quartzites affecting a thickness of up to 2 cm ($\frac{3}{4}$ in). This varnish is a brilliant black or violet due to the presence of manganese oxide. Among other places it forms on the walls of potholes, whose development is thereby impeded. It is much more resistant than the rock itself. In most cases, contrary to desert varnish, it does not veneer a partly weathered half decemented surface caused by the leaching of hydroxides, but a fresh rock; the products that form it are brought by the river's water and slowly penetrate into the rock. It is not uncommon that potholes are truncated by removal of the slab through which they are pierced; the bottoms that remain are then eventually put into a micro-inverted relief, their superficial coating having better resisted abrasion than the rock itself. Thinner but harder ferruginous coatings also occur on the fine grained granites of Salto Rapids (São Paulo, Brazil) and help explain the persistence of its disconnected forms.

FIG. 2.14. Potholes and ferruginous varnish on the Paulo Afonso Falls, Rio São Francisco, Brazil
Massive granites. Edge of a canyonlike channel with potholes of lateral impulsion intersecting one another in draperies.

Ferruginous coatings retard the incision of rapids by impeding abrasion, which is the principal agent of erosion. They play a stabilising role. It is therefore essential to note that they do not develop under all kinds of conditions. In West Africa from the semi-evergreen seasonal forest to the savanna, they especially characterise rivers with large stage differences. This distribution is easily understood: large stage differences are favourable to the formation of coatings in climates where insolation causes strong evaporation during low water stages. But as under the same climates a more abundant and coarser sand load causes a more rapid abrasion, there is at least a partial compensation, which helps to reduce the differences between rivers in the various intertropical climates. The nature of the rock also intervenes: on crystalline rocks we have observed coatings only on coherent, generally fine-grained rocks, which do not break up in blocks, as at Salto. Coatings are uncommon on rapids with scattered boulders. Iron hydroxide stains are frequently observed on the surface of the boulders, but they seem to help superficial disintegration and to impart a corroded and rough aspect to the rock. The formation of coatings seems to be a slow process. It can therefore take place only under conditions of sufficient stability and through a chain reaction later contribute to increase the stability.

There are then two extreme types of rapids from the point of view of the evolution of long-profiles on the scale of geomorphic time:

(*a*) Rapids on very coherent rocks that chemical actions are incapable of disintegrating. When the dry season is sufficiently long coatings form, which increases the rock resistance. The main erosional process is sand abrasion, which proceeds extremely slowly and sculptures grooves, contorted crests and potholes whose coalescence ends in the formation of canyonlike channels. Such rapids have an enormous endurance. At Félou Rapids, near Kayes (Mali), the entire Quaternary incision has been incapable of reaching the upstream reach whose valley floor has not been notched since the Pliocene.

(*b*) Rapids with scattered boulders, which correspond to rockbars weathering at a slower rate than the neighbouring rocks, which correspond to the quiet reaches. However, there is a certain amount of chemical weathering on this kind of rapids: joints are widened and slabs are detached, transforming themselves into boulders. The boulders themselves corrode and slowly disintegrate, not allowing enough time for ferruginous coatings to form. Generally, the effects of abrasion are limited on such rapids. In any case they are not predominant, as the condition of the rock surface demonstrates. Chemical weathering plays the decisive role. A progressive but slow reduction of the irregularities of the river's profile thus takes place.

The incision of rivers in the tropics is therefore very narrowly dependent on chemical actions. They limit the mechanical actions because of the difficulty of abrasion due to the too fine texture of the bed load. But they, too, prepare the formation of valleys. Moisture indeed concentrates in the lowest places where weathering will be most effective. It reaches deeper zones under the valley floor, down to several dozen metres, even a hundred (several hundred feet). It thus prepares a friable material without coarse components into which a river channel is easily cut. A river with shifting channel will ensue. As soon as it is important enough, it will grade its bed and lower its long-profile down to very low values that will still allow the transit of material owing to its reduced calibre. In this way quiet reaches are formed. They all correspond to rocks that are weatherable and to deep underground decomposition. They develop all the better as the long-profile is carved as a function of a stable base-level, which generally is the rockbar situated downstream. Difficulty of incision at rapids is an essential condition to the formation of quiet reaches. In the most favourable cases the hillsides have enough time to retreat sufficiently for floodplains to develop.

Differential weathering causes rocksteps and quiet reaches to appear as a stream course is progressively lowered. Quiet reaches develop where stream erosion is slower than the progression of weathering at depth, whereas rapids and falls form where stream erosion encounters an unweathered rockbar between two zones of deeper weathering. This general evolution enables the understanding of the following facts:

1. Rocksteps are most numerous in the lower course of rivers if they are incised close to the sea, for it is here that the tendency toward incision is greatest and, in non-consolidated formations, most rapid. The Pleistocene lowering of sea-level also played a part. Thus is explained the deep incision into the Neogene coastal sands of Ivory Coast and the northeast of Brazil. But the rate of incision sometimes overtakes the rate of weathering, especially in rocks that are rather resistant to weathering. It is then that the number of small rapids suddenly increases. The predominant type will be that of rapids with scattered boulders, such as the one on the Bia near Aboisso (Ivory Coast), or the ones of French Guiana, where a moderate uparching of the shield during the Quaternary seems to have facilitated this type of evolution. Such rapids, it seems, reflect a rather swift resumption of erosion.
2. Further upstream one frequently finds fewer small rapids with scattered boulders and more large rapids, even falls. Resumptions of erosion indeed propagate themselves with difficulty. This is why at a certain distance from the ocean streams are still at the level they reached during the Pliocene, not counting minor alternations of erosion and deposition due to climatic changes. A long evolution took place without change in the local base-level. Weathering has had the time to proceed deeply and intensely. Only the larger differences of weathering reveal themselves; for instance, a quiet reach in poorly cemented sandstones and waterfalls on quartzites, or a quiet reach in crystalline rocks and rapids at the crossing of a stratum of metamorphic quartzites or of a dike. The quiet reaches tend to be longer and rocksteps more prominent.
3. Lithologic differences are equally reflected in differential erosion and the development of stream profiles. Weatherable rocks produce quiet reaches. Hard to weather rocks that produce rapids and falls are mainly quartzites, whether metamorphic or not, and especially massive and ferruginous quartzites. Sandstones, when they are massive, even relatively weak, occasionally produce fine rapids, especially in the alternating wet–dry tropics where their disintegration is more difficult. This is the case of Félou Rapids, on the Senegal, at the margin of the Sahelian zone, and of Tosaye Rapids, in the bend of the Niger, at the dry margin of the Sahel. Even sandstones of mediocre mechanical resistance frequently produce rapids next to quiet reaches in crystalline rocks. In the latter rock composition is a factor in the case of rapids with scattered boulders, usually of recent age. When evolution is longer, it is mainly the massiveness of the rock that plays a role: rapids are then located on jointless coherent masses, akin to monolithic domes, whether granitic or gneissic, fine grained or porphyroidal.
4. Climate also has its effect, but in a manner as yet imperfectly understood. In the course of a long evolution, as on the Falémé, small differences in weathering seem to persist longer under savanna than under forest,

probably because the intensity of weathering is less. The middle Falémé, for example, commonly has a bed littered with boulders that match with the most resistant facies of a series of crystalline schists including intercalations of quartzite. The form of the river channel resembles to a certain degree that of the lower course of a rainforest river subjected to a resumption of erosion. There are then certain effects of convergence or of compensation, which remain to be worked out. Potholes and abrasion forms are as a whole better developed in climates where a slower rate of weathering produces a relatively large sandy load and, especially, pebbles derived from ferruginous concretions or the debris of cuirasses, both of which play an important role in the grinding of the giant potholes of Félou Rapids. The Sudanese climate because of the dry season, and the subtropical climates where weathering is slowed down are both favourable to them.

5. The location of large canyons is explained by differential weathering and not by headward erosion of a waterfall, contrary to what happens in other climatic zones. The canyon of several kilometres that is found below Paulo Afonso Falls, on the São Francisco, coincides with zones of fractures in an otherwise solid mass of migmatic gneiss. These factors explain the

FIG. 2.15. Canyon located on a fracture zone, Rio São Francisco below Paulo Afonso Falls, Brazil

In the foreground: the waters spill into the canyon, whose wall is a fault plane on the edge of the fracture zone.

angular pattern of the canyon. The main branch of the São Francisco only rejoins the canyon at a point approximately 1 km from its head; it reaches it almost at a right-angle and then literally throws itself down the abyss. In no way does the canyon result from a headward erosion of the falls. The latter are due to the pre-existent excavation of the canyon. Indeed the fracture zone with which it coincides has permitted the weathering of the rock and its removal as soon as a branch of the river placed itself on it. Certain weathered fracture zones intersecting the canyon, but not occupied by streams, are partially cleared at their mouth and form small branchlike re-entrants along the canyon walls. They are confluent with former distributaries that rejoined the canyon through cascading rapids rather than through a single cataract. Evidence of numerous distributaries indicate the successive approaches that finally enabled the clearing of the fractured area and the formation of the present falls. The basins of the Bakoy and the Bafing, in Mali, show evidence of a similar evolution. At high stage the rivers divaricate on a

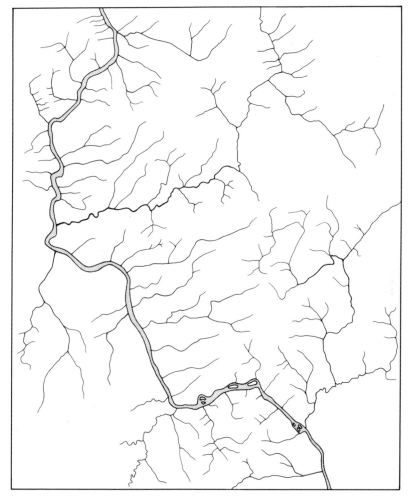

FIG. 2.16. Drainage pattern determined by fractures. South of Queluz, State of São Paulo, Brazil (after Teixeira, 1960). Scale approximately 1/220 000.

bedrock pavement of Ordovician quartzite, splitting in numerous channels before spilling over the obstacle. The chances are thus multiplied that one of the branches will encounter a fractured zone, which will then be cleaned out and transformed into a canyon that concentrates most of the water. The angular or even parallel pattern of stream courses indicates that intertropical rivers have frequently exploited fracture zones.

It appears, therefore, that the conditions of regressive erosion are very different in the humid tropics from what they are in other morphoclimatic zones. As a whole, stream erosion is very dependent on the chemical preparation of the material, which imputes a primordial role to differential chemical weathering. In producing rocksteps that block denudation every time stream erosion outpaces weathering, rockbars considerably slow down regressive erosion. In vast regions the evolution of interfluves has gone on for long periods of time as a function of the same local base-level represented by rapids or falls. It is such circumstances that produce the depressions and pediplains that are so characteristic of savanna regions.

The particularities of the zonal morphogenic system therefore cause a modification in the behaviour of rivers and the introduction of a peculiar evolution of the relief, which among other things is reflected by the development of special landforms.

The peculiar conditions of the tropics are also evident in the morphology of coasts, the special forms of which we will now study.

Particularities of the littoral processes and coastal landforms

The morphogenic processes responsible for the specific characteristics of the littorals of the tropics have divers origins. Some forms, like the familiar offshore bars, lagoons, and mudflats, are the direct result of the peculiar nature of tropical rivers. They result from the great abundance of fine debris carried down by the streams. Other forms result from the particularities of slope development, such as cliffs caused by landslides, or forms due to corrosion, especially through the crystallisation of salt. Still others are tied to specific biotic activities, such as the construction of coral reefs and the formation of algal crusts.

Particularities related to the abundance of fine sediments

The extent of coastal lowlands in the tropics is considerable, on the average larger than in other latitudes, periglacial regions excepted. Even the crystalline basement may be fringed by fluviomarine plains instead of ending in sea cliffs, as does most of the east coast of Brazil from Rio Grande do Sul to Paraíba, the coast of West Africa from the Casamance to Cameroon, and a great part of the coasts of India, Java, the Malay peninsula, and Borneo. Mountains or hills seldom reach the sea in the form of rocky coasts, as in the west of Ivory Coast. Rocky promontories are usually small and isolated between fluviomarine plains, as at Monrovia, Liberia, or at Cabo Frio and Ilheos in Brazil. They anchor the coast and frequently cause a change in direction, as at Cabo Frio or Cape Palmas (West Africa), but represent only an infinitesimal proportion of the coastline. There is an enormous disproportion between rocky coasts and lowland coasts. It poses the problem of the provenance of the alluviated material.

ORIGIN OF THE LITTORAL DEPOSITS

The enormous mass of sediments deposited in fluviomarine plains surely does not originate from a few sectors of sea cliffs through littoral drifting. As we have mentioned for the east coast of Brazil, the disproportion is enormous. Furthermore, long stretches of depositional coasts exist in regions where there is not a single cliff capable of furnishing the debris, except minor rocky promontories that produce a few reefs, as, for example, the whole coast of the State of Bahia to the south of the Bay of Todos os Santos (Brazil).

Sedimentological studies indicate the following sources of the sediments:

1. Streams, which bring mainly fine materials. In general fine sands and clays predominate as a result of the decrease in current in the lower stream course, affected by the Flandrian transgression.[1] Coarse sands are deposited in deltas where they form sand banks. Special circumstances are necessary to account for the presence of coarse sands and granules (2 to 4 mm), as, for instance, at the mouth of the Sassandra in Ivory Coast. This river clears important rapids 10 km (6 miles) from the ocean and then empties into a narrow estuary, where the current is strong during destructive spates that come down a long-profile rich in rapids for the last 100 km (60 miles). At that moment strong currents destroy the bar at the river's mouth and allow the arrival of granules onto the subaquatic delta constructed at the mouth of the passes. Most of the granules are ferruginous concretions of pedologic origin.

Deltas are more common than estuaries. Their very gentle gradients, even on large rivers, do not allow the transport of coarse sands out to sea. For example, the Senegal, although in the semiarid Sahelian zone, only transports medium and fine sands (medians of 200–300 microns) into the ocean in spite of a discharge exceeding 6 000 cu m (200 000 cu ft) at high stage.

2. Weathering of the continental shelf during marine regressions, especi-

[1] In general, the post-Glacial rise of sea-level. In particular, in Flanders, the last but one phase of this rise, coinciding approximately with the Atlantic period, about 5 000–1 000 BC. The last phase of the rise is known as the Dunkirkian, occurring in the Subatlantic period, starting approximately at the beginning of the historical period. It is characterised by three advances between the first and tenth centuries AD. The Dunkirkian marks the highest post-Glacial sea-level stage, 1 to 2 m (3·3 to 6·6 ft) higher than the present sea-level, occurring about 2 000 BC and particularly well recorded in the intertropical zone. Cf. Tricart: 'Les variations quaternaires du niveau marin', *Information Géographique*, **20**, no. 3, 1958, pp. 100–4 (KdeJ).

ally during the Würm. Even in regions of crystalline basement, as in Guinea, considerable areas of the continental shelf emerged down to depths of 60 to 80 m (200 to 265 ft) during the last glaciation. The duration of the regression, under climates assuredly somewhat different from the present ones, has been long enough to permit a considerable degree of weathering, which, judging from what may be observed on terraces and slopes dated to the same period, may have reached a thickness of 1 to 2 m (3 to 6 ft).

During the Flandrian transgression the ocean progressively invaded these plains, sometimes surmounted by residual reliefs, such as the Los Islands off Conakry. The rising sea removed the weathered products, even scoured part of the decomposed rock, churned it all up, pushing most of it, little by little, before it during its advance. Whereas the clays were dispersed, the sands piled up at the end of the transgression into enormous offshore bars (Dunkirkian), always very well developed on intertropical coasts, especially in regions where rocks rich in quartz liberate much sand (granites, gneisses, certain micaschists), as along the central part of the east coast of Brazil or in Ivory Coast (Tertiary sands).

We have two proofs of the importance of the last factor described, which is generally neglected by specialists of littoral problems, who tend to have their views too narrowly concentrated on the present littoral processes:

(*a*) The nature of the sediments, which is significant. Away from a river's mouth one can often find traces of two influences to which sand grains have been successively subjected. The terrestrial environment has left traces of ferrugination, either colouring the quartz in its mass or, more frequently, limiting the stains to fissures and cavities that look as though they have been made by pulling a corkscrew. One also finds the remains of pecked surfaces, which, it seems to us, are of chemical origin. Later, marine action has rounded off angles and smoothed surfaces, in short, given the grains a more or less subrounded and polished aspect (Le Bourdiec, 1958b). Also, in most cases marine action has broken the quartz grains weakened by oxidation, thereby increasing the fineness of the material, whose grade size is generally on the limit between medium and fine sands.

(*b*) The forces affecting the present littoral are also indicative. On nearly all shores where deltaic deposits are lacking or few, there is a tendency toward erosion. Dunkirkian beach ridges are undercut into microcliffs. Storms undermine coconut trees and cause huts to be abandoned. At times erosion causes a frontal retreat of the coast, at other times an intersection of old beaches by modern beaches aligned along a different angle, which implies certain changes in oceanographic conditions. The latter case is common in the states of Espiritu Santo and Bahia, Brazil. They must be explained, it seems, by post-Dunkirkian climatic changes (directions of wind and swell). In any case since the Dunkirkian the sedimentary equilibrium of the beaches has changed. A period of accretion has been followed by a period of generalised erosion, implying a decrease in the supply of sand. For us the explanation is simple and corroborated by a study of the material: as long as the transgression endured, the raking of the continental shelf provided a supply of new sands; once the sea-level became stabilised or slightly lowered, the supply practically ceased, resulting in a deficit made up by the erosion of the Dunkirkian beach ridges.

3. The biotic actions, which are very active in tropical seas. The common heavy swells that oxygenate the water, the high degree of luminosity and the transparency of the water away from river mouths permit the rapid development of numerous organisms, among which the corals are only a group. Many algae, some encrusting by secreting a calcareous coating, profit by such favourable circumstances. Both corals and encrusting algae are broken by swells upon death, adding debris to the sandy offshore on which they do not grow.

The proportion of biotic calcium carbonate is normally high in tropical littoral sands. At Salvador, Brazil, for example, it reaches 20 to 40 per cent on beaches with gneissic reefs in an area where the sea bottom is an abrasion platform. Ottmann *et al.* (1959) have demonstrated the importance of calcium carbonate in the beaches of Recife, Brazil. They arrive at similar percentages. For them the calcareous fraction is particularly high around a grade size of about 200 microns.

Of course important calcareous fractions are also found in littoral sands outside of the tropics, and in some parts of the latter the proportion of calcium carbonate may be low, as, for example, in Ivory Coast, where the luminosity is insufficient. But as a whole there is a great abundance of calcium carbonate in the tropics in spite of the importance of other materials, which, however, tend to diminish its proportion.

CONDITIONS OF DEPOSITION

The considerable masses of clastics deposited on tropical shores are mainly fines, for three reasons: the nature of the fluvial sediments, the materials weathered on the continental shelf during regressions and, in some cases, the biotic conditions. Clays and fine sands predominate. Coarse sands are uncommon: apart from the mouths of certain rivers they are mainly found on beaches where high swells progressively concentrate them, dispersing the finer clastics out at sea. For example, the grade size graphs made at Recife by Ottmann *et al.* give the following medians:

180 to 300 microns immediately below lowest spring-tide level
140 to 650 microns on the shore
170 to 180 microns on the storm beach

The storm beach sand is thrown up from the offshore during storms, and for this reason is finer.

The sands poor in carbonate of the Abidjan area, studied by Le Bourdiec (1958b) are medium to very coarse on the present shore, violently beaten by swells, but medium to mostly coarse on the Dunkirkian and Eemian[1] ridges, slightly inland. This difference in texture underscores the present tendency toward beach erosion following the period of Dunkirkian accretion. In Barbados, where the percentage of terrestrial debris is very small, beach sand is everywhere fine and highly calcareous.

An important characteristic of the sediments of intertropical shores is their *very high mobility*. It is due, first of all, to grade size, especially the usual absence of pebbles, which are less easily transported than sands or clays. But it is also promoted by other factors:

(*a*) high temperature, which seems to affect the mobility of clays as a result of an appreciable decrease in the viscosity of the water that increases its turbulence and thus helps colloids to remain in suspension. The suspended load of rivers in flood colours the sea over great distances. The Amazon is held responsible for the muddy shores that extend from its mouth all the way to French Guiana. But it is difficult to determine exactly how much more suspended matter is carried and just how much more mobile colloids are with increased temperature. Such measurements have never been made, and we remain in the realm of hypotheses.

(*b*) The tropical oceanic regime, which is generally characterised by a high and much more constant swell than in other latitudes. The ocean normally does not become unruly as in the temperate zone subject to cyclonic storms although on certain coasts hurricanes and typhoons accompanied by surging seas locally produce catastrophes that destroy beaches and modify tidal inlets. But the swell is powerful enough throughout the year thoroughly to work the sand of exposed beaches. This constant working of the sand contributes to its fragmentation and wear and explains the concentration by elutriation of the coarser fractions on beaches where there is a tendency to erosion. The coarse sand, in turn, causes the appearance of extremely convex beaches with upper slopes of 15 to 20°, not counting microcliffs. The swells are an extrazonal phenomenon, as Guilcher has pointed out for the coast of West Africa. They owe their origin in this region to the prevailing westerlies at 40° south latitude (the 'roaring forties'). In Senegal and in the West Indies swells are also high, receiving their impulse from the mid-latitudes of the northern hemisphere. They cause, almost without interruption, a beach and longshore drifting in the zone of agitation down to a depth of 20 to 30 m (65 to 100 ft). This migration is very constant as the swell is very regular. For example, in Senegal, it produces a permanent southward drifting north of Cape Verde. The movement of drift is more complicated in the west of Ivory Coast because of a wind shift from the northeast during the principal dry season.

As a whole, swells have an extraordinary effect on beach sands, which

[1] The glacio-eustatic transgression of the last interglacial as recorded in Germany and the Netherlands and corresponding to the Tyrrhenian II in the Mediterranean region and the Oujlian of M. Gigout (1949) in Morocco. Its highest stage was about 5 to 8 m above the present sea-level (KdeJ).

can hardly ever be said to be at rest. It explains the importance of littoral drifting, which at Vridi, Ivory Coast, is 800 000 cu m (2 800 000 cu ft) on the average since the opening of the canal in 1950. It favours the construction of spits and offshore bars along coastal plains. It is only on very indented rocky coasts, where random currents ceaselessly move the local sand, causing it to be highly rounded, as in certain coves near Rio de Janeiro or on certain jagged capes in western Ivory Coast, that no generalised drifting exists. Everywhere else the easy migration of sand allows the construction of enormous offshore bars anchored on widely spaced rocky promontories or even completely free.

Nevertheless, a certain number of antagonistic elements intervene that reduce the mobility and facilitate the deposition of the sediments:

Mangroves: on the shores of estuaries the brackish waters are occupied by mangrove, which tolerates divers degrees of salinity and, in the case of some species, even a temporary submersion by fresh water. These shrubs are remarkably well adapted to the environment: Avicennia seeds, for instance, are contained in fruits shaped like pointed bombs which when they fall stick in the mud, where they germinate.

Mangrove always grows on muddy ground subjected to frequent submersion usually by the tides but occasionally only by river floods, as certain

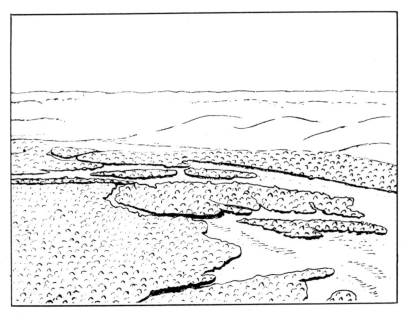

Fig. 2.17. Estuary gradually choked by mangrove near Taperoá, State of Bahia, Brazil
Mangrove has colonised the mud flats and restricted the open water to ordinary, irregular tidal creeks. The mangrove is indicated with a special symbol.

Rhizophora of the lagoonal shores of Ivory Coast. It plays a comparable role, which only differs in details, on tropical shores to that of Spartina in the temperate zone. Although the water circulates easily through it, a comb-like effect rapidly reduces the currents of stream floods or the pulsations of ocean swells, leaving only a light splashing, which even disappears in the heart of the mangrove. Important geomorphic consequences result from it:

On the one hand muddy shores are effectively protected and not subject to undercutting. Removal of the sediments already deposited, so important on the muddy tidal flats of the temperate zone, is very restricted here. Once occupied by mangrove the muds remain where they are as soon as they are slightly compacted. The reduced sub-mangrove currents return into suspension only the most recently deposited muds that have not yet agglutinated. Destruction of the mangrove, for example for the use of its wood, favours the degradation of the muds, which may have deleterious effects.

On the other hand mangroves form a continuous front that advances both by layering and insemination until it meets unfavourable conditions that block its progression, especially a too protracted submergence and the consequent asphyxiation of the young shoots. As soon as the bottom has a tendency to be built up it is occupied by new plants, which themselves in turn produce very effective sediment trapping mechanisms. Frequently rivers debouch in deep tidal creeks whose banks are supported by strong currents beyond the margins of the mangrove thickets where depths suddenly increase. These tidal creeks are swept by alternate currents that keep the muds in suspension. During floods or high seas some of the mud penetrates the mangroves and accumulates as the waters calm down. The mangrove thus rises little by little until it escapes the daily tidal submergence. It is then progressively replaced by marshy plant associations that tolerate seasonal flooding.

Mangroves thus contribute to the deposition of an important part of the fine sediments brought by rivers and tidal floods into estuaries. They trap fine sands as well as muds and are not characterised by a unique type of deposit.[1] They promote the rapid replacement of estuaries by deltas, which is a general phenomenon in the tropics.

Burrowing organisms play a part in mangroves and on beaches but in an opposed manner. In mangroves crabs play a role opposed to that of the plants by decreasing their role as fixing agents. Indeed they continuously 'plough' the recently accumulated deposits, keeping them friable. In this way, during submergence the muds are easily set back in motion, which keeps the waters turbid. But this effect, which is important from the sedimentological point of view, as it destroys all evidence of stratification, is not enough to modify the geomorphic evolution, which is that mangroves represent regions of rather rapid deposition.

---

[1] For this reason the term 'mangrovites' proposed by Bigarella in Brazil should be rejected.

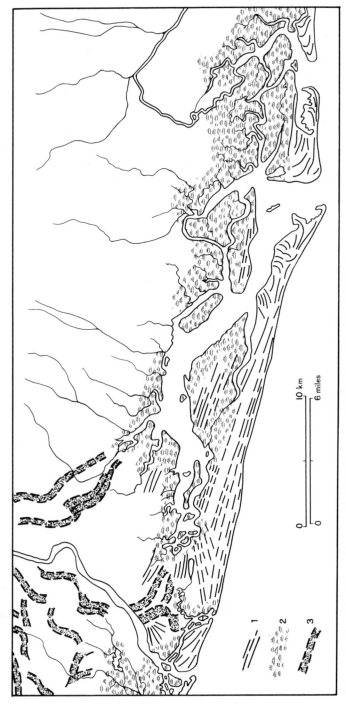

FIG. 2.18. Coastal plain to the east of the Río Lempa, El Salvador (after Gierloff-Emden, 1959, map 8)

1. Offshore bars alimented by the Río Lempa, which empties into the ocean in the west
2. Mangroves (fine sands and muds)
3. Former distributaries of the Río Lempa, corresponding to the delta of the river during the Flandrian transgression

On beaches all burrowing organisms cause the sand to be extraordinarily permeable, making it difficult to walk on, as in the area of Sassandra, Ivory Coast. The increase in pore volume may reach 10 per cent in a layer a few decimetres (a foot or so) thick. Wide and continuous cavities are thus created in which water infiltrates much more easily and rapidly than in ordinary intergranular interstices. The uprush, therefore, as each breaker comes in, forthwith penetrates into the sand, causing some bubbles and foam and a minimum of backwash, or even none at all where the water is spread thin. Under such conditions the sand washed up by the uprush is not returned but accumulates. True, when the ocean level is higher and this same part of the beach is in the zone of breakers, the sand is again removed. However, there is a net accumulation on the highest part of the beach, which between storms is not subject to wave action. On many very permeable beaches of western Ivory Coast, storm waves erode microcliffs into the Dunkirkian sands, whereas an upper beach ridge is formed between storms.

The formation of beachrock is even more important than the preceding factors. It is, as Guilcher (1961) has shown, essentially a tropical and subtropical phenomenon, reaching as far as southern California, Morocco and Israel. But it does not occur in every part of the tropics: there is none in Ivory Coast, for example, and it is rare in the Gulf of Guinea where we have observed only an isolated remnant, still being degraded, at Tema, Ghana. It is common on the east shores of Brazil north of the Tropic of Capricorn, in the West Indies and in Florida. It seems to characterise rather warm regions with a well marked dry season but without being semiarid.

Beachrock results from a cementation of beach sand by calcium carbonate. It produces beds that are unequally indurated and nearly always inclined in the same direction as the original deposits, and when it becomes eroded forms microcuestas facing inland. Whereas occasionally certain beds form real crusts, others may be cavernous. The cement is calcite and aragonite in variable proportions. It enmeshes shells, pebbles, quartz grains, and sometimes divers debris.

Near Salvador, Brazil, beachrock patches include glass fragments and metallic objects and therefore appear to be of recent origin. On the shores of central Brazil or at Tema, Ghana, beachrock is found on Dunkirkian beach ridges and results from the cementation of their material. It is now more or less eroded, forming reefs on the beaches, after which the city of Recife owes its name. Russell (1962) observed that beachrock outcrops always occur on retreating coasts. For us its presence is but a reflection of the general sedimentological disequilibrium that affects tropical shores since the end of the Flandrian transgression.

Beachrock occurs in the intertidal zone (which defines the shore). Occasionally (Russell, 1959) coral reefs form at its seaward margin. Frequently a narrow channel has developed between the inner face of the homoclinal cuesta which terminates the beachrock toward the land and the beach.

FIG. 2.19. Beachrock microcuesta at Ilhéus, State of Bahia, Brazil
In the background, cape formed by a basic intrusion in gneiss. There are no
sea cliffs. In the foreground, bed of Dunkirkian beachrock in process of
destruction. Its retreating front forms a microcuesta facing the beach. Lapiés
on the surface of the bed.

The thickness of the beds depends on the importance of the tidal range,
according to Russell (1962): it varies from 0·3 to 0·5 m (1 to 2 ft) in regions
where the tidal range is small, to 3 m (10 ft) where it is large, as in Puerto
Rico.

The origin of beachrock is not yet fully understood. It seems that two
series of factors intervene concurrently and vary in importance depending
on the situation. On the one hand, as Russell (1962) believes, the seepage of
fresh ground water into the calcareous materials of the beaches would cause
some dissolution and precipitation. On the other hand Guilcher (1961) proposes physicochemical explanations, especially the evaporation of water in
the beach sand at low tide, and biological explanations according to which
the decomposition of calcium carbonate associated with organic compounds
in shells would dissolve some calcium, which would precipitate under certain conditions. Bacteria and algae would also play a role. These two
explanations do not exclude one another, and we have seen certain beach-
rocks that would be hard to explain by seepage of fresh ground water. A
very definite climatic limit (35° north latitude in North Carolina, 33° in
California and the Mediterranean) seems to imply a biological phenomenon
requiring a certain minimum temperature. Evaporation of water in the
mass of the beach sand at low tide also seems to play a role, whatever the
origin of the water, and explains the absence of beachrock in climates that

95

FIG. 2.20. Beachrock at Ilhéus, State of Bahia, Brazil
View taken on the foreshore at low tide. Dunkirkian beach under the
houses in the distance, preceded by a subsequent depression at the foot of
the microcuesta formed by the beachrock, the inner face of which is
visible in the right foreground. The surface of the beachrock is bevelled
across the strata and pitted with small depressions.

are too permanently humid, such as along most of the Gulf of Guinea.
Finally, in Brazil, the formation of beachrock seems to be associated with a
dry climatic phase corresponding to the Dunkirkian.

Whatever the answer, the geomorphic role of beachrock is very important. It not only forms microreliefs on many a tropical beach but is also an
effective surf breaker and actively protects the upper beach from erosion.
The latter is only attacked at high tide during onshore storms. The water
column that then covers the beachrock is high enough to permit the passage
of the oceanic swell. This phenomenon is well displayed on the shores of the
state of Bahia, near Taperoa. The entirely sandy shore is fringed by discontinuous reefs of beachrock. Where they are interrupted, the coastline has
retreated 100 to 300m (300 to 1 000ft), forming concentric baylets hinged
on the extremities of the reefs. Where sands are cemented into beachrock,
which most readily happens to calcareous sands, the coast is protected
against the onslaught of the swell. The removal of the sand, which is easy
elsewhere on account of its grade size, is impeded. Coastal retreat is thereby
retarded.

LITTORAL DEPOSITIONAL LANDFORMS

The great abundance of material, its fine texture, which is the cause of its
great mobility, permitting easy transportation through beach and long-

FIG. 2.21. Beachrock at Ilhéus, State of Bahia, Brazil

View taken from a gneissic promontory. The beachrock forms a kind of
dike whose homoclinal attitude is clearly visible. The violence of the surf
is an indication of the protective role the beachrock plays in relation to
the Dunkirkian beach.

shore drifting, and the trapping role of certain phenomena (mangroves,
beachrock), all help to explain the great extent of fluviomarine deposits
along tropical coasts. Certain zonal influences also frequently impart special characteristics to the deposits, as, for instance, the complete absence of
pebbles. There are only sands and clays.

Sand bars are the key depositional forms in all regions where rivers actively bring large quantities of sediments to the ocean (humid tropics and
subtropics, mouths of large exotic rivers in the dry tropics, such as the Senegal delta). They are at the seaward margin of fluviomarine plains and form
the barrier behind which nearly all such plains are formed. The endless
stretches of their sandy beaches are a characteristic element of the littoral
landscape of the tropics.

They differ from the sand bars of higher latitudes in that they are seldom
associated with dunes. When dunes do exist they are generally poorly developed. They are practically absent along the entire coast of the Gulf of
Guinea. They are important only along certain coasts, as on Java or at the
mouth of the São Francisco, to the north of Bahía. But there they were
largely formed during the Pleistocene. The destruction of beaches by aeolian deflation, so important on arid, semiarid and temperate shores, is therefore unimportant in the humid tropics. The sand remains on shore within
the reach of marine forces, which remodel it continuously. The reasons for
the subordinate role of the wind are:

(*a*) A generally small tidal range does not uncover extensive areas of the beach. The frequently steep slope of convex beaches impedes aeolian deflation by decreasing the surface and by forcing the wind to push the sand grains uphill.

(*b*) The winds, very constant in direction, which strongly influence swells, are seldom violent and hardly ever attain the force of those of midlatitude storms. When it happens, during tropical cyclones, the rain neutralises their action.

(*c*) In the wet tropics the sand does not dry during low tide, especially as a small tidal range puts the upper foreshore out of reach of the waves for a very short period of time. Furthermore, dense vegetation, that even occupies the backshore, impedes deflation.

In Senegal well developed coastal dunes first appear in the Sahelian zone. In the state of Bahia magnificent dune fields, entirely stabilised by an impenetrable thicket of spiny palms, have formed during various dry intervals of the Quaternary, the last one of which occurred during the Dunkirkian. The abundance of sand at the end of a transgression may also have played a favourable role.

Composed of very mobile material moved almost without interruption, sandbars play an important role in causing the alluviation of fluviomarine plains:

(*a*) They shut off shallow bays by forming slightly curved spits tied to reefs, islets, or islands. They migrate on bottoms not deeper than 30 m (100 ft), which is the limit reached by heavy swells. As Abecassis (1958) has shown in Angola, their distal end is usually formed by a submarine screeslope. The spit at Lobito receives a yearly increment of 250 000 cu m (9 million cu ft) and advances 20 m (60 ft) per annum. Such spits are built in variable proportions by sands brought by streams flowing through savannas (especially if underlain by sandstone or granite), and by material provided by beach erosion and the action of shore drifting. At Vridi, Ivory Coast, all the sand in transit has the latter origin. The speed at which spits form explains that sandy beaches are common features along intertropical coastal plains. A series of exceptional conditions must coexist for bays not to be shut off by sandbars and to form tropical rias, such as Guanabara Bay, Brazil. Here fluvial alluvium is exceptional and is accumulated elsewhere, building up small fluviomarine plains to the west of Rio de Janeiro. Shore drifting is blocked by a series of rocky promontories where the water is too deep for drifting to take place. The cliffs formed by the flanks of monolithic domes hardly supply any detritus. The large drowned bay has therefore not been shut, but all small neighbouring bays have, forming the beaches of Copacabana, Ipanema, and the ones in the vicinity of Niterói.

(*b*) Spits are easily reworked and quickly modified in minor ways. They first appear with normal swells that produce a strong shore drifting. During storms long swells erode parts of the upper beach producing microcliffs. In this way narrow spits are sometimes breached, as, for example, has happened several times on the Langue de Barbarie at the mouth of the Senegal delta. Sandy washovers are thus formed on the lagoon side of the spit. Breaches, however, are rapidly filled by normal shore drift as soon as the length of the swell decreases. An identical mechanism may function at the mouth of a river obstructed by a bar of its own construction. When the river is low, shore drifting overcomes the river current, which then ceases to move the sand. Instead, a spit forms at the river's mouth, threatening to close it. To remain open a rather large lagoon is necessary to ensure a strong current at ebb tide, as at Grand Lahou, Ivory Coast, or at the mouth of the Senegal. Here, when the river is low, a spit bars the passes except for one or two metres of water (3 to 6ft) that continue to flow through a low pass thanks to tidal currents. Rivers that empty straight into the ocean, however, without first passing through lagoons maintain themselves with difficulty. They are frequently barred by spits even in rainforest regions such as the west of Ivory Coast. The water then seeps through the sandbar, but all

FIG. 2.22. Estuary shut off by an offshore bar 12 km (7 miles) west of San Pedro, Ivory Coast (drawing by D. Tricart from a photograph by J. Tricart)

A. Offshore bar
B. Aeolian sand blown from the offshore bar and forming an upper beach
   remodelled into beach cusps covered by low vegetation
C. Dunkirkian dunes covered by shrubs. This bar cuts off a tributary
D. Mangroves colonising the muddy shores of the estuary
The river has recently overtopped the bar and cut a passage through it. But when the river is low the outlet is closed by beach drift which is sufficiently permeable to let through the entire low water discharge.

the matter in suspension comes to rest behind it. When the flood arrives the river is banked up and re-establishes a channel.

The same phenomenon of course occurs in savanna regions, as Abecassis (1958) has shown in Angola. It has important consequences: the early stage of the flood, before the breaching of the sandbar, inundates the coastal plain and deposits alluvium, increasing aggradation. Furthermore, the mouths of such rivers are very unstable: in time they are often subject to abrupt migrations. The Senegal, for instance, has shifted its mouth some 150 km (90 miles) since the Dunkirkian. The Niouniourou River, in Ivory Coast, follows the sandbar for a distance of some 20 km (12 miles) a few hundred metres (1 000 ft or so) away from the ocean before being able to cross it, joining its waters to those of another river. This too influences the aggradation of coastal plains.

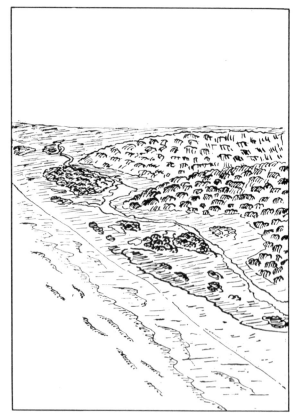

Fig. 2.23. River deflected along the shore by a sand bar east of Fresco, Ivory Coast (drawing by D. Tricart from a photograph by J. Tricart)
Sandy beach with a few minor dunes on the upper beach, followed by a marshy fill along the course of a wandering river confined by the base of low hills.

The construction of spits causes a rapid regularisation of coasts. Only abrupt coasts, where depths are too great for spits to form, preserve a morphology resulting from the invasion of the land masses by the Flandrian transgression (Guanabara Bay and Along Bay, North Vietnam). Everywhere else intertropical coasts are mainly regularised by deposition, the amount of which is out of all proportion to the reduction of promontories by marine erosion. We have explained why this is so.

In regions where fluvial alluviation is insufficient, offshore bars isolate large lagoons, which fill in slowly as waves rework the shore deposits and mangroves fix the muds if the water is sufficiently brackish. Such lagoonal coasts, which are well developed in Ivory Coast, El Salvador, and Rio Grande do Sul, and between Cabo Frio and Niterói in Brazil, have divers origins but always reflect a certain equilibrium between the construction of the offshore bars and fluvial aggradation.

(*c*) In some regions the construction of offshore bars is especially rapid because particular lithologies provide streams with a great abundance of sands, while lagoonal alluviation is slow because the same terrains provide a dearth of mud to the streams. This occurs, for example, in Rio Grande do Sul where erosion of thick sandstone series frees enormous quantities of sand which, under the impulse of a strong longshore current, migrate great distances to construct large offshore bars and isolate lagoons, which fill up slowly. A similar situation exists in El Salvador where streams carry a lot of sand but little clay because weathering has not yet proceeded deeply in a region of recent vulcanism. The Lempa River, for instance, has a solid discharge of 8 million tons a year, which contributes to the construction of well developed offshore bars. Lagoons persist where no important streams empty into them; they fill up very slowly as mangroves fix the fine sand mixed with some clay. At Cabo Frio lagoons exist in a coastal section little affected by alluviation because of the configuration of the coastline and the drainage system; the offshore bars were easily constructed, however, due to the sweeping of the continental shelf by the Flandrian transgression.

(*d*) In other regions lagoons have a tectonic origin, as in central Ivory Coast, studied in great detail by Le Bourdiec (1958b). Plio-Quaternary faults affect the margin of the crystalline basement, forming an uplifted block that drops toward the lagoons by way of an escarpment. The lagoons developed in sections down-faulted in Neogene sandy clays (the *Continental Terminal*) on the lower block during the pre-Eemian and pre-Flandrian regressions especially. During the Eemian and Flandrian transgressions the reworked *Continental Terminal* caused the construction of enormous offshore bars, which have isolated the lagoons formed in the highest part of the zone cleared by the regression. This explains the ria coastline on the northern margin of the lagoons. A continuation of the subsidence has helped the drowning. Furthermore, alluviation has been unequal: the Comoé, a mighty river originating in a savanna region, brings much alluvium and has constructed a delta that interrupts the lagoons. Elsewhere lack of alluvium assisted by continued subsidence has prevented the filling in of the lagoons: neither the Bandama nor the Bia have much alluviated the lagoons into which they empty.

But from the point of view of a hot and wet climate muds are generally the predominant sediments, making the persistence of lagoons difficult. The French Guiana coast in spite of subsidence hardly has any. In Surinam Plio-Quaternary deposits reach a thickness of 2 000 m (6 600 ft), according to Vann (1959). Although fine fluvial sands move swiftly along the coast and sandbars extend rapidly—1 900 m (6 000 ft) in thirty years for the spit to the west of the Courantijne River, or an average of 62 m (200 ft) per year—deposits are predominantly clayey. Mud flats, soon fixed by mangroves, form in front of the sandy beaches. When the shorter rainy season is poorly marked or absent, which happens about every five years, there is a shortage of mud, and some erosion, which produces clay pebbles, occurs. The same happens when the stabilising mangrove is destroyed. At a later date deposition proceeds as before. In French Guiana storms play the most important role in these morphogenic processes, according to Choubert and Boyé (1959), and there is, supposedly, since 1751, a recurrent cycle of eleven years related to sun spots. Storms from the north, coinciding with a reinforcement of the trade winds, bank up the ocean water against the shore and cause large chunks of mud to be torn loose. From February to May 1958 a mud flat close to Cayenne was stripped of an average of 90 cm (3 ft) of mud. In a year the removal of mud can cause the shoreline to retreat 200 to 300 m (660 to 1 000 ft) inland, whereas the mean seaward advance has been 10 to 15 km (6 to 9 miles) in a century. Tropical cyclones, of course, cause sudden changes, as Pagney (1958) has shown for another area: the island of Martinique. The sea-level can then be raised 1 to 5 m (3 to 16 ft), which causes a violent flooding of coastal plains. Trees are uprooted on sandbars, where they serve as breakwaters. In 1955 cyclone Janet moved back the high tide shoreline about 10 m on Diamond Beach. Old cliffs were swept away. Later shore drifting progressively rebuilt the former outlines in front of a microcliff. One can imagine what are the consequences of such phenomena on low coasts.

The result of these processes is the construction of vast fluviomarine plains, which are particularly wide in French Guiana where subsidence is compensated for by an abundance of sediments. The deposits form an alternation of low ridges of fine sand and amphibious mud flats. Water stagnates, forming shallow swamps flooded by 10 to 120 cm (4 in to 4 ft) of water during the rainy season. The accumulation of peat eventually buries minor relief features, a phenomenon very well described by van der Eyck (1957). Sometimes the peat is floating, as on a part of the coast of Nigeria. The role of vegetation is immense as Guilcher (1956a) has emphasised.

Important rivers have generally begun to construct their deltas in sheltered rias starting with the Dunkirkian, after which they advanced and protruded beyond the rias in the form of external deltas with intersecting sandbars and occasionally residual lagoons. The Paraíba delta, in Brazil, excellently described by Lamego (1955), may be taken as the type. It has a great deal of instability because of shifting tidal inlets and forking distributaries. The latter, as Pimienta (1956) has shown for deltas of southern Brazil, occur especially frequently on lower river courses with minor gradients because of the growth of vegetation which occupies the bottom as soon as it is sufficiently shallow. Sand banks thus become fixed, causing a forking of the stream flow and a progressively greater division of the waters. The lengthening of the channels produced by the shifting of the outlets produces floods that sweep vast surfaces, veneering them with clays, while swamps are filled with the help of the trapping actions of plants, as Porto-Domingues and Keller (1955) have shown for the state of Bahia. Such frequent channel shifts also increase the number of natural levees that accompany the distributaries. The levees create a tangled and complicated network of minor reliefs, generally very low because of the want of sufficiently long periods of stability. The topographic irregularities may later become almost completely obliterated by alluvium and organic filling in of the swamps, as Berthois and Guilcher (1956) have shown for the Ambilobé delta, Madagascar. The clays of coastal mud flats preserve the essential characteristics of the weathered products, according to Guilcher *et al.* (1958): in Ramanetaka Bay, Madagascar, there are kaolinites with some illite and geothite, all of which are derived from weathered basalt. The littoral environment only contributes calcite. Deposition is too rapid to permit the neoformation of other elements.

Low coasts, so widespread in the tropics, are, therefore, much influenced by the particular zonal conditions affecting the littoral processes, especially because of the very abundant supply of sand and clays produced by intense weathering. The subordinate role of physical weathering is reflected in the scarcity of pebbles. In spite of a stabilising role played by certain mechanisms (mangroves, beachrock), instability and rapid changes characterise the tropical fluviomarine plains. In nearly all cases they were constructed with sediments that are not supplied by marine erosion but are due to subaerial weathering (river alluvium, regolith of the continental shelf raked up by marine transgressions). We now consider the geomorphic significance of rocky coasts.

Characteristics of rocky coasts

A first characteristic of rocky coasts is that they are uncommon. High coasts are in many cases not even rocky. A beach and a narrow coastal plain, a few hundred metres wide (a thousand feet or so) frequently fringe them, preventing the formation of cliffs. This is a widespread phenomenon along the western half of the coast of Ivory Coast, which, nevertheless, we will take as the type of rocky coast. Rocky coasts are much more general on islands, especially small ones, than along continental coasts. The explanation is simple: there is less river alluvium that might form large deposits isolating outcrops from the sea.

Variety is also a characteristic of rocky coasts, opposing them to low coasts, which are less diversified. Three types of rocky coast may be distinguished:

Vegetated sea cliffs, which are characteristic of the wet tropics.
Bare sea cliffs, which are very common on limestones but also occur on other rocks in regions where evaporation is intense (e.g. Cabo Frio, Brazil); particular corrosion microforms such as visors and solution pools develop on them.
Coral shores, which are constructional.

VEGETATED SEA CLIFFS

Some sea cliffs even when precipitous (60–70°), as in western Ivory Coast, are entirely covered by vegetation. Only the lowest few metres, right above the coastline, are bare, as they are too exposed to sea spray and stormy seas. Above this bare strip there is a zone of shrubs in which spiny palmettos are dominant and make passage particularly difficult, as we have experienced at Monrovia. This zone is variable in width: it is wider at Monrovia (20 m: 66 ft), where some clearing has taken place, than in western Ivory Coast where it is usually not more than 5 to 10 m (16 to 33 ft). Above this zone are found dense shrubs with a few scattered trees and, lastly, the forest.

The coastal vegetation considerably influences the morphogenic processes and therefore has a typical zonal influence. Very few studies have been made of this narrow zone, indeed it has seldom been mentioned. Halophytic characteristics are obvious only in the lower spiny palmetto zone, where the sea spray of high breakers is abundant and provides a great deal of salt. But this zone is possible only in a much overcast and rainy climate, the only one that will wash away the salt and prevent it from playing an ecologically harmful role. Such a climate explains the extent of this type of coast: between Monrovia and Sassandra, where we have observed it in West Africa. (In Ghana, where the climate is marked by drought and a high evaporation, the cliffs are bare at this level.) We find it again in El Salvador on cliffs carved from volcanic breccias (the Mizata coast), on the Pacific coast of Colombia, which is very humid and overcast, and on certain rocky promontories between Rio de Janeiro and Santos, Brazil, especially near Angra dos Reis and Santos. It also seems to exist in certain parts of Indonesia.

The conditions governing the evolution of vegetated sea cliffs reveal similarities with that of ordinary hillsides. Weathering processes that progressively reduce the bedrock to a friable mass are similarly active here under the cover of vegetation. When the regolith has reached a certain thickness landslides occur, for example during a period of heavy rains, and the material is dumped on the shore. The forest is then left with a scar, which is slowly recolonised, first by formations of savanna grasses, followed by shrubs and lastly by trees. Western Ivory Coast displays all these stages, described in Tricart (1957a). Sea cliffs 50 to 100 m (160 to 330 ft) high, caused by a fault-scarp, which has broken the Monogaga Neogene surface such as to produce an en echelon disposition of the coast, display a series of semi-circular slip-plane scarps, which occasionally intersect in narrow crests. At

FIG. 2.24. Landslided sea cliffs 45 km (28 miles) west of Grand-Drewin, Ivory Coast
Residual relief of chemical erosion with a rather thick residual mantle in the coves between dikes of resistant rock. The action of the ocean produces landslides that prevent forest growth on the lower slopes and cause the coast to retreat in the bay heads.

the foot of each of them there are accumulations of boulders resulting from the spheroidally weathered material that was slided and then washed by the waves, which have not been able to remove the coarser fractions; the sand and clay, however, have been reworked and dispersed seaward or accumulated in the form of small beaches in coves. In closed coves the beach sand often is composed of a mixture of more or less rounded grains that have been subjected to the effects of the sea for a long period of time and angular grains, which still show traces of weathering, having only recently arrived in landslides. In the small bays on the southeast coast of Vietnam Milliès-Lacroix (1960) has observed analogous phenomena: the sand is not worn because it has only recently been produced by weathering.

Evolution is therefore similar to that of abrupt slopes under forest, where landslides are also important. It is particularly well realised in granites, gneisses, and lower micaschists that weather rather rapidly. In other rocks, such as the volcanic breccias of Mizata in El Salvador, weathering is retarded due to abrupt slopes and a high permeability that produces rapid and deep infiltration, limiting the effects of hydrolysis. This causes the cliffs to be very stable and to preserve their alignment along the original fault scarp. Ocean waves carve a nick at the base of the cliffs, which, because of weathering along joints, occasionally triggers debris slides that may extend upward the bare and abrupt basal part of the cliffs.

In any case, such cliffs never retreat rapidly and always remain aligned on the fracture that gave them birth. In this way the quantity of material they supply is small and can never explain the enormous sandy deposits that extend beyond them along the coast of El Salvador as well as along that of Ivory Coast.

BARE SEA CLIFFS

Bare sea cliffs are quite common on certain rocks or in certain climates. The lithologic factor is significant especially in the case of limestones, which nearly always produce bare cliffs because they do not provide much residual material as they are under chemical attack, and because this material, which is fine, is easily swept away when it does not fill up holes. All the limestone cliffs we have observed, mainly in the West Indies, are bare.

The climatic factor is important in relation to the harmful effect the salt of sea spray has on vegetation. In the alternating wet–dry tropics where evaporation is high, the dissemination of salt by surf spray, whose drops disappear by evaporation, is intense and prevents the growth of plants. Prevailing onshore winds during the dry season accentuate the effect, as at Cabo Frio, Brazil, where micaschists and gneisses in the coastal zone remain bare just as in Brittany. The difference with western Ivory Coast where rocks are similar is striking.

The crystallisation of salt plays an immediate and important geomorphic role on bare cliffs. We will first examine it as it affects crystalline rocks, which are not affected by solution processes, and, later, as it affects limestones, which because of solution processes have a more complex development.

Certain tropical coasts, not necessarily humid, are characterised by special microforms. They occur on rocks several metres above the coastline and are higher on promontories and exposed coasts. In bays the forms are not as well developed and restricted to a narrower fringe. The vertical zonation is as follows:

(*a*) On cliff walls about 5 or 6 m (16 to 20 ft) above high water mark there is a zone of honeycomb weathering and granular disintegration with a supply of debris generally composed of coarse sand, often rich in feldspar. The dark micas are frequently weathered and oxidise the adjacent rock. The pittings are small, a few centimetres wide (about an inch), smaller than tafoni. They are also less deep and only affect the surface irregularly. Occasionally there are flutings, but then the pittings are not as well developed or

FIG. 2.25. Sea spray corrosion pittings in micaschists at Cabo Frio, Brazil

Surface of a corrosion bench about 2 m (6·6 ft) above high water mark.

Forms are controlled by zones of weakness. The microstructure does not allow pittings to grow larger than 5 cm (2 in) in diameter.

FIG. 2.26. Cliff fashioned by corrosion due to sea spray at Cabo Frio, Brazil

Note lapiés and honeycombs in the higher zone and low cliff with overhangs (in the shade) due to the landward extension of the corrosion bench, whose level is slightly above high water mark.

disappear altogether. The flutings are identical to those of monolithic domes of the interior and often moist as a result of seepage from the plant covered weathered mantle above the cliff. They are well developed in the vicinity of Rio de Janeiro, where honeycombing is the exception. They are absent at Cabo Frio, in a climate that is much more sunny, drier, and very windy, and where pittings are well developed.

(*b*) At about 1 to 4 m (3 to 13 ft) above high water mark, pool covered rock benches are developed at the base of the cliffs. Landward they are bounded by a scarp that may be abrupt, sometimes joint controlled, as in the diabases of Cape Verde, and on which are found the pittings of the higher zone. In vertically jointed rocks the foot of the scarp forms a marked angle, some-times even a slight overhang. The surface of the bench is therefore extended by retreat of the cliff, especially at its base. Pools, a few decimetres in width (1 to 10 ft), appear on it; their bottoms, which are one or two decimetres (4 to 8 in) deep, become perfectly flat with time. The pools are separated by crests that are generally sharp and covered by lapiés. The growth of the pools causes their mutual intersection and the development of intercon-necting pools. In many a feature these coastal pools resemble the pools found on the monolithic domes of the interior. The main differences are a much greater density, causing intersections, and development on more level surfaces, which produces greater regularity.

(*c*) Seaward of the pools is the intertidal zone. The transition between the two is often a rather steep slope where the rocks receive the full onslaught of the waves. The surface of this zone is rough and indicates an origin by granu-lar disintegration. The biotites, with rust, constitute the weak elements. The intertidal zone itself, on the contrary, is characterised by less rough, more polished surfaces, with a beginning of fluting.

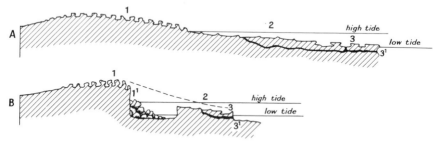

FIG. 2.27. Sketch showing the vertical zonation of littoral forms in aeolian sandstones of Morocco (from Guilcher and Joly, 1954)
A. Forms exclusively caused by dissolution
B. Forms produced by a combination of dissolution and mechanical effects (sapping of the cliff)
1. Zone of jagged lapiés and interstitial ponds with overhanging rims
1¹. High tide sea cliffs and talus
2. Corrosion bench with stepped pools
3. Zone of pool destruction
3¹. Low tide cliff with 'geyers'

This cliff profile thus indicates a zone of maximum erosion slightly above the highest seas, where the pool covered bench undercuts the cliff. At Cabo Frio small benches, a dozen metres or so wide (some 40 ft) with cliffs 3 to 5 m (10 to 15 ft) high, which represents a considerable amount of erosion, have developed in the micaschists of certain promontories since the Dunkirkian. At Cape Verde the bench, cut in diabases at the foot of cliffs a few tens of metres (about 100 ft) high, is 3 to 5 m (10 to 16 ft) wide, which is appreciable. The process is therefore very active. It stands in contrast with what one observes in the intertidal zone, which everywhere lacks clear-cut abrasion forms and only shows a chaotic accumulation of large boulders which, in the absence of storms equal to those of the temperate zone, remain immobile and are only slowly polished by grinding sand.

The corrosion forms that are found above the level of the highest seas are related to drenching by billows and spray. For this reason the height of the coastal zone which they affect varies with its exposure, increasing on capes beaten by waves, and decreasing in bays up to the point of disappearing in sheltered areas. In the zone of pittings moistening is limited and sporadic and therefore not very active. On the pool-covered bench the rocks are thoroughly drenched and subjected to alternations of washing and drying. The water gathers in the pools where it remains for several days and then evaporates, the salt crystallising in white coatings. At the foot of the bench

FIG. 2.28. Corrosion pool formed by sea spray during the Dunkirkian at Ilhéus, Brazil

Inherited form, not presently active but very well developed. Was made during a drier climatic episode as at present the climate is unfavourable to its formation.

drenching takes place at every high tide, and more or less thorough drying at low tide. The salt water penetrating the rock pores enters into contact with the minerals and then evaporates, precipitating its salt. Corrosion thus takes place, but the combination of processes is still poorly known: what is the role of the crystallisation tensions of marine salt, what is the role of hydrolysis? We cannot answer these questions. However, having reproduced the phenomenon in the laboratory, we have noted the very important role played by biotite. It swells and cleaves in tiny sheets, breaking up the rock at its contact and then weathers by setting free iron oxides, which colour it, just as we have observed in nature. Rocks rich in biotite, such as certain lower micaschists, are much more readily attacked than more basic rocks that have less or none of it, such as diabases. The latter are reduced to a powder, which swells, producing an aspect similar to that of a wall covered with saltpetre rot.

In the laboratory disintegration by spray corrosion has proven to be very rapid on crystalline rocks with a good microporosity. It has attained a rate comparable to that of frost action on marly limestones. The considerable development of certain benches since the end of the Flandrian transgression can therefore be understood.

The main factor is the alternation of wetting by sea water and drying. For this reason climate is important. We have never encountered such benches in the wet tropics, where evaporation is low: there are none in Ivory Coast. They begin to form at Monrovia, where there are only a few pools although corrosion is general, producing rough surfaces on rocks, as those that are normally encountered between the pools and the high water mark. This can be explained: a rough surface develops where the rock seldom dries and where salt does not have time to crystallise. The vertical zonation is excellently displayed near Dakar in rocks almost analogous to those at Monrovia. It is also found on the east coast of Brazil, to the northeast of Rio de Janeiro and near Salvador, in sunny climates. Some incipient forms, such as pittings, can even be found in mediterranean latitudes as far north as Catalonia. Dependent on high evaporation and rather high temperatures, this type of form is centred more on the Tropics than on the Equator.

The development of rock pool benches is a unique phenomenon, which results from the combination of two conditions, both of which are morphoclimatic: corrosion activated by periodic drenching by swell and high evaporation; and little erosion at the foot of the bench, in the intertidal zone, because of the lack of pebbles and the difficulty encountered by the ocean in clearing the large boulders that protect it—a consequence of the absence of violent storms.

In *limestones* the geomorphology is somewhat different even though the climatic and dynamic conditions remain the same.

Cliffs are often sheer, which is rare in the preceding case and always limited to the scarp overlooking the pool bench. Furthermore, the cliff-face plunges directly into the sea. On Barbados the vertical zonation is as follows:

(*a*) On top of the cliff, when it is not too high and sea spray reaches it regularly, there are very sharp lapiés formed by flowing water brought by high swells and spray.

(*b*) The cliff-face is a sheer wall formed by periodic rockfalls dependent on the joint pattern. In general this type of erosion has caused ancient notched levels to disappear; seldom are any preserved.

(*c*) The lower part of the cliff, which is approximately at high water mark, is eroded into an almost continuous notch, locally widened into sea-caves, especially on the most exposed coasts. Its depth is between 1 and 5 m (3 and 16 ft), its height between ½ to 3 m approximately (1 to 10 ft). The vault is concave downward. When the cliff is low, it is only slightly below the surface of the platform that overlies it, and because of its profile has been called *visor*.

(*d*) The foot of the cliff, submerged and vertical, is fringed by boulders derived from the cliff-face.

FIG. 2.29. Solution hollows on the boulders of an earthflow north of Hillcrest, Barbados
Large Miocene limestone boulders, remains of an earthflow that spread on the present littoral bench during the pre-Flandrian regression. Since the end of the Flandrian transgression two solution notch levels have formed: one Dunkirkian, the other current, both clearly visible, especially on the boulder at the left. Note the dissolution bench, which creates the illusion that the boulders were deposited on a level surface.

These forms are developed on Barbados in tabular, coral, Neogene and Quaternary limestones. They are the result of a rather rapid retreat of the cliffs because of the growth of the notch, which here plays the role of the

FIG. 2.30. Coastal lapiés at College Savannah, Barbados
Pleistocene coral limestones subjected to violent splashing (east coast).
Note vertical zonation: pools at left, followed by sharp, jagged lapiés in
the spray zone above high water mark.

FIG. 2.31. Stepped pools at College Savannah, Barbados
Lower part of the lapiés zone of the preceding view. Very flat pools due
to encrusting algae, forming steps that are unceasingly beaten by swells
(view taken at low tide).

wave-cut notch of classical geomorphology. The cause of the evolution is, therefore, the dissolution of limestone, which proceeds at a maximum rate at the level of the notch. It seems to us that the notch may be explained by the fact that the waves here unceasingly strike the basal cliff without ever submerging it permanently. They are rich in air bubbles that fill the water with a maximum content of CO_2 and, therefore, make it more aggressive. Moreover, the pneumatic effect is at a maximum.

Sometimes the littoral profile is more complicated: at the northeast extremity of Barbados where swells are highest, there frequently is in front of the notch a rock bench that is continuously kept wet by the waves. It is much better developed on promontories than in coves where, on the contrary, caves are common and large (due to the convergence of waves, increasing the pneumatic effect). This bench, a veritable *trottoir*, is produced

FIG. 2.32. Low limestone cliffs at Cummings, St Lucy, Barbados
Neogene coral limestone; the top forms an abrasion terrace of possibly Eemian formation. Note profile of cliffs and stacks: corrosion notch with trottoir and visor. Some stacks are completely abraded to the level of the trottoir.

by the combined effects of two processes: an abrasion related to the notched level to which it is associated, and a biotic construction of calcareous algae or sessile gastropods (*Vermetus*), which both prosper in the highly oxygenated water unceasingly agitated by standing waves. They coat the wavecut bench and project it seaward, like geyser pools and the travertines of calcareous springs extend their rims, but on a larger scale, as much as 3 m (10 ft). Similar trottoirs exist as far north as the Mediterranean. Progressive

FIG. 2.33. Limestone cliffs with caves east of Seaview, Barbados
At the top: dead cliff with Tyrrhenian (Mindel-Riss) littoral terrace,
clearly slanted. The coral constructions corresponding to this terrace are
intersected by the present live cliff in which two notched levels may be
distinguished: the present one and the Dunkirkian. Both grade into sea
caves. The first headland has been reduced by rockfalls.

extension of the trottoir seems to impede the development of the notch and,
finally, to stop it. Where the trottoir is wide, the notch seems to be stabilised,
and there are no loose boulders.

Gently sloping coasts have different forms, which, in particular, have
been studied by Guilcher and Joly (1954) in Morocco, and by Guilcher
(1957) in Portugal. We have also observed some on Barbados. These coastal
landforms have a climatic limit, as they are not found on the cool and cold
coasts of the midlatitudes; they make their first appearance somewhat north
of Lisbon, in what are already subtropical conditions, and are common,
among other places, in Morocco, Madagascar, the West Indies, and Bermuda. Just as spray corrosion benches they are centred more on the Tropics
than on the Equator. They comprise from top to bottom:

(a) In the spray zone: very sharp and jagged lapiés with small interstitial
pools similar to those found on top of the low cliffs of precipitous coasts.
(b) In a zone flooded at high tide: stepped pools etched out of the bedrock
(Portugal and Morocco). Bottoms are flat. Dividing rims, which may be 10
to 20cm (4 to 8in) high, resulting from the intersection of pools, are sharp
and sinuous. The lime is indurated because of a beginning of recrystallisa-
tion. On Barbados the pools are partly constructional and the crests are not
sharp, as they have been coated by encrusting algae. There is here, perhaps,
a difference between a subtropical and a tropical variety of warmer seas.

In Bermuda the Portuguese and Moroccan variety seems to exist at the foot of slightly notched cliffs, according to Taillefer (1957).

(*c*) In a zone continuously submerged except during spring tides: destruction of the higher zone of stepped pools, with cavities 25 to 80 cm deep (10 to 32 in).

(*d*) Close to low water mark: an abrupt cliff, with sea-caves and overhangs, under furious wave attack at low tide.

Guilcher and Pont (1957) have studied the mechanisms that produce littoral lapiés and pools. Alternations of wetting and drying cause simultaneously a dissolution of limestone and a freeing of sand in high quantities because of the presence of salt that is particularly effective on impure limestones, such as beachrock, or on aeolian sandstones. Porosity helps the processes. Swelling caused by the moistening of lithophytic blue algae, which act as wedges on the rock, also intervenes. Furthermore, after a certain period of time there is a liberation of silica due to the weathering of silicates. This silica recombines with calcium into a calcium silicate through the liberation of CO_2, which attacks the remaining limestone. Alternations of wetting and drying, therefore, play an essential role, which explains the formation of these specifically tropical forms. In cooler climates frost action intervenes and produces a different microrelief.

Revelle and Emery (1957) in research conducted at Bikini simultaneously arrived at other explanations, but these do not exclude the preceding ones: the water in the pools appears to be oversaturated with $CaCO_3$ during the day, but at night, as temperature drops, the pH decreases, and the photosynthesis of the algae having come to a stop because of respiration, the content of CO_2 increases, and there is no more saturation. There is some dissolution, principally in the film of water that penetrates into the micropores of the rock but also in the water enclosed between the rock and the algae which coat it, especially during low tide. Wide daily oscillations of temperature, alternations of wetting and drying and the overflowing of pools, which renews the water by replacing the water warmed in the sun by fresh water brought by the surf, would thus be favourable factors. Here again we find two climatic elements related to the distribution of these forms: a high evaporation and a large daily temperature range, both characteristic of warm sunny climates. To this must be added that pools are sometimes excavated by the action of sea-urchins, as we have observed in beachrock at Salvador, Brazil.

CORAL FORMATIONS

Coral formations are restricted to tropical seas; they thrive best in water whose temperatures are situated between 25 and 30°C (77 and 86°F). When the temperature drops below 18°C (64·4°F) they are excluded. Reef building polyps are associated with algae and therefore require light. They can only grow at shallow depths: less than 40 m (130 ft), usually less than 25 m

(80 ft). They die if exposed too long to the atmosphere and therefore always grow below low water mark. Finally, they need well oxygenated and agitated water. Waters carrying muds bury the polyps and kill them. Some species, however, fight back with their cilia and are able to live in turbid waters if the mud supply is not too great.

The construction of coral reefs results from two antagonistic actions: the biological phenomenon of polyp growth, which tends to extend the construction not only directly but also through the accumulation of shells and debris of various kinds in their intervals; and the normal littoral actions, which tend to destroy them. Indeed, coral reefs, built up to low water level, cause waves to break and thereby to attack and fragment the colonies. The most favourable environment for coral growth is the outer margin of the reef where the water is more oxygenated and nutritious. But the onslaught of the waves is also more effective here, breaking off numerous fragments, which drop to a detrital slope at the foot of the structure where they become progressively cemented together.

The configuration of coral formations implies many variants corresponding to this antagonism. In moderate seas construction progresses seaward, the top of the back-reef being composed of dead corals. Because shallow depths are the most hospitable, coral polyps construct overhangs to a depth of 3 or 4 m (10 or 13 ft) above the detrital slope. In certain very rough seas Lithothamnion algae establish themselves at approximately the mean sea-level, constructing a ridge at the outer edge of the reef above the coral formations. These encrusting algae do not require permanent submersion but only an uninterrupted wetting by spray, as we have already noted. Such ridges, which do not exist in closed seas (as the Red Sea), are the highest parts of the reefs, as Guilcher (1954a) reminds us.

Behind the Lithothamnion ridge the reef-flat is commonly a surface of erosion as, in many cases, the polyps have raised themselves during the Dunkirkian to a level slightly above that of present low tide, above which they cannot live. Here the corals are dead. There is here either a mechanical abrasion and formation of coral pebbles and sand, such as we have seen on certain Pleistocene marine levels on Barbados, or there is a corrosive action with dissolution of calcium carbonate, formation of pools, etc. Boulders detached by wave action are sometimes lined up in runs, and calcareous sand composed of the debris of marine organisms living on the reef (shells, foraminifera, algae) forms shallow layers in depressions. Where it is abundant it sometimes builds up into sandbars or hooked spits that prolong the extremity of the banks. When raised above sea-level they are occupied by vegetation. Such sand accumulations are known as *cays* or *keys* (U.S.). The very permeable materials are light in weight and the effects of sorting attain a remarkable degree of perfection, as Guilcher (1956b) has demonstrated in the case of the northwest coast of Madagascar.

Coral formations play a very important geomorphic role. They create islands or atolls that are nearly always situated on older islands, usually volcanic and subject to a moderate subsidence. In this way, coral growth could proceed at the same rate as subsidence and maintain an atoll in existence. At Takeroa, in the Tuamotu Islands, geophysical methods have found the basement to be at a depth of more than 1 000 m (3 300 ft), according to Ranson (1955). At Eniwetok, in the Marshall Islands, it has been reached at depths of 1 267 m (4 170 ft) and 1 401 m (4 600 ft), underlying a shallow water Eocene limestone. Some atolls are found on volcanic cones that are subsiding since the Eocene or even the Cretaceous. Besides these there are other, flat topped submarine cones whose surface is either an abrasion platform or a coral formation which could not keep pace with the rate of subsidence; they are called *guyots*. Guyots and atolls therefore have the same structural origin; only a difference in the rate of subsidence/rate of coral growth differentiates them. Guyots correspond to islands that at a certain moment subsided more rapidly than coral growth could keep up with, and thus have proceeded to sink to depths out of reach of the polyps. On atolls, on the contrary, coral polyps have been able to maintain themselves in spite of the vicissitudes of the glacio-eustatic oscillations of sea-level.

On the foreshore corals construct a kind of protective dike in the form of barrier reefs. In Madagascar this role is well illustrated by studies of Guilcher (1956b) on the northwest coast, and Battistini (1960) on the southwest coast. Beginning on the landward margin we find here the following geomorphic zonation:

(*a*) Sandy beaches alternating with mangroves, which develop due to the sheltering conditions produced by the barrier reef.

(*b*) An irregular channel, called the 'boat channel', which at Mikea is 2 km (1·2 mile) wide and 1 to 3 m (3 to 10 ft) deep and includes channels down to 5 m (16 ft) at right angles to the major passes that interrupt the barrier reef.

(*c*) The reef-flat of the barrier reef, which at Mikea is 2·5 to 3·5 km (1·5 to 2·2 miles) wide and is subject to abrasion and covered by pools excavated by sea-urchins. Passes, which in the case of atolls Ranson (1955) attributes to ancient valleys cut during the Pleistocene regressions, have been incised through them. In places such reef-flats display boulder runs that grade into sandbars. They are better developed in the northwest than in the southwest of Madagascar.

(*d*) The algal ridge.

(*e*) The external front of the barrier reef, which at Mikea is formed by an alternation of spurs 100 m (330 ft) long and 7 to 8 m (23 to 26 ft) wide, composed of living coral, and trenches narrowing down to tunnels, where the effect of the swell is furious and causes the formation of real canyons through regressive erosion. Detrital material lies at the bottom of the trenches. Lower down is the external face of the reef.

Barrier reefs place the littoral that they protect in a completely artificial position in relation to oceanic conditions. Waves are much smaller, reducing all forms of corrosion and erosion. The migration of debris is more difficult. However, the deposition of fine sands and muds is facilitated by immediate fixation by mangroves. The latter rim the shores behind the Madagascar barrier reefs.

In conclusion, let us try to point out the particularities of tropical littoral forms. Temperatures above freezing produce a specific environment in which the actions (swell, shore drifting) of certain processes operate in all climates where the sea is not frozen over. By this very fact the actions are, therefore, pluri- or polyzonal. The littoral forms themselves, however, are typically zonal, for they are the result of a combined effect in which zonal influences make themselves felt (mangrove mud-flats, coral formations, corrosion benches, etc.). By their very nature the forms are more eminently geographic than the polyzonal processes.

Nevertheless, the zonal distribution of specific littoral forms is not exactly the same as that of the landforms of the interior. Indeed, there the climate plays a role through the combined effect of temperature and humidity. Types of riverbeds reveal clear differences between regions of forest, savanna, and semiarid bush. The disappearance of gallery forests on the dry margins of the savannas, for example, is reflected in a much more intense erosion of river banks, which favours the formation of sand banks in the stream channel and the burial of minor rockbars. In the littoral environment conditions are different: high temperatures play a more important role, whereas the influence of humidity is reduced.

Atmospheric humidity, however, still plays a role in some cases:

(*a*) A high degree of evaporation is needed to dry rapidly the surfaces drenched by the seas, to permit the development of spray corrosion forms. Such forms, therefore, correspond to climates with long dry seasons or high insolation, which do not coincide with the forest/savanna division. Some forest climates have a high enough daily variation of humidity in the littoral zone for corrosion to function, as at Salvador, Brazil. Furthermore, corrosion proceeds without hindrance in semiarid and arid climates providing they are not foggy. It functions as far as the mediterranean regions and so extends beyond the intertropical zone.

The formation of beachrock requires partially identical conditions, but as far as groundwater contributes to its formation, it is excluded from arid regions. Tropical climates with a dry season are favourable, especially where wind increases evaporation. Beyond the tropics they occur in regions of humid subtropical climate (the Carolinas) or semiarid climate (California).

Large accumulations of fine sediments are, of course, favoured by perhumid tropical conditions, which produced the deep weathering of the continental shelf during the Pleistocene regressions as well as the great quantities of debris furnished by the streams. But exotic rivers, like the

Senegal, fulfil an identical role all the better as the large discharge originating in the wet–dry tropics serves to mobilise and transport seaward large quantities of fine materials from semiarid regions (dunes formed during arid intervals of the Pleistocene in the lower valley and delta). Coincidence with the climatic zones of the land masses is, therefore, also here, approximate.

(*b*) Temperature alone plays a role in the case of rock formations produced by organisms, such as coral reefs or algal ridges and trottoirs. Certain minimum temperatures must not be exceeded. Such temperatures permit corals to build reefs in all tropical seas and in the closed seas of the dry subtropics (Red Sea, Persian Gulf). Encrusting algae are less exacting and are found in the subtropical zone all the way to Provence. Winter cold is in both cases the limiting factor. Humidity, of course, does not play any role as these organisms take advantage of that of the sea.

The prolongation of high temperature effects into the subtropics, therefore, affects the littoral forms even more than it does the land forms. The barrier of the dry zone, so important on a large part of the continents, only has a limited effect here. There are some differences then in the zonality, which does not prevent it from playing a very important geographic role as far as geomorphology is concerned.

Bibliographic orientation

Fluvial morphology

Not a single general work, only dispersed articles or bits of incidental information in works dealing with other subjects.

AB'SABER, A. NACIB (1953) 'Geomorfologia de uma linha de quesdas apalachiana típica do Estado de São Paulo', *An. Fac. Fil. Catol.*, São Paulo, pp. 111–38.
Descriptive only. Some interesting photographs.

ANDRADE, G. OSÓRIO DE (1956) 'Furos, paranás e iguarapés. Anâlise genética de alguns elementos do sistema potamogrâfico amazônico', *Bol. Carioca de Geogr.* **9**, 15–50.
Good description of fluvial rias, to be replaced within the framework of a subsiding region. Long bibliography.

BAKKER, J. P. (1957) 'Quelques aspects du problème des sédiments corrélatifs en climat tropical humide', *Z. für Geom.* **1**, 1–43.
Basic work, but limited to Surinam.

BAKKER, J. P. and MULLER, H. (1957) 'Zweiphasige Flussanlagerungen und Zweiphasenverwitterung in der Tropen unter besonderer Berücksichtigung von Surinam', *Festschrift Lautensach*, pp. 365–97.
Repeats certain elements of the previous article.

CASTRO SOARES, L. DE (1949) 'Observacões sobre a morfologia das margens do Baixo-Amazonas e Baixo-Tapajós, Para, Brasil', *Proc. Int. Geogr. Congress*, Lisbon, vol. 2, 748–61.

GALVAO, R. (1955) 'Introdução ao conhecimento da área maranhense abrangida pelo plano de valorização econômica da Amazônia', *Rev. Brasil. de Geogr.* **17**, 241–99.
Exploratory study, as the preceding one.

MACAR, P. (1947) 'Les chutes de l'Inkisi (Congo occidental) et leurs divers modes d'érosion', *Ann. Soc. Géol. de Belgique*, **82**, 38–51.
Pioneer work.

MICHEL, P. (1959) 'Rapport de mission au Soudan occidental et dans le Sud-Est du Sénégal, II : dépôts alluviaux et dynamique fluviale', *SGPM*, Dakar, 9/59, 57 p., mimeographed.

MICHEL, P. (1962) 'Observations sur la géomorphologie et les dépôts alluviaux des cours moyens du Bafing et du Bakoy (Rép. du Mali)', *BRGM*, Dakar, 39 p., 16 phot., 11 pl., mimeographed.

These two studies by Michel are basic. Good descriptions, essential observations.

SETZER, J. (1956) 'A natureza e as possibilidades do solo no vale do Rio Pardo entre os municipios de Caconde S. Pe Poços de Caldas M.G.', *Rev. Brasil. de Geogr.* **18**, 287–322.

Description and analyses of subtropical alluvial formations from the pedologic point of view.

TRICART, J. (1955) 'Types de fleuves et systèmes morphogénétiques en Afrique occidentale', *Bull. Sect. Géogr. Comité Trav. Hist. et Sc.* **68**, 303–45, 21 pl.

Useful illustrations.

TRICART, J. (1956) 'Comparaison entre les conditions de façonnement des lits fluviaux en zone tempérée et zone intertropicale', *C.R. Acad. Sc.* **245**, 555–7.

TRICART, J. (1959) 'Observations sur le façonnement des rapides des rivières intertropicales', *Bull. Sect. Géogr. Comité Trav. Hist. et Sc., 1958*, pp. 289–313.

Numerous illustrations, detailed observations.

TSCHANG TSI LIN (1957) 'Potholes in the river beds of North Taiwan', *Erdkunde*, **11**, 296–303.

Subtropical environment. Good descriptions and photographs.

VOGT, J. (1959) 'Note sur la Lobo (Côte d'Ivoire)', Dakar, *SGPM*, 14 p., mimeographed.

WILHELMY, H. (1958) 'Umlaufseen und Dammuferseen tropischer Tieflandflüsse', *Z. für Geom.* **2**, 27–54.

Excellent study, to which we refer. Good photographs. Most examples from the Brazilian Pantanal.

Littoral forms

Here too dispersion is the rule. Guilcher's (1954) systematic treatment completely neglects the role of climate except as regards coral reefs which are excellently described. Concerning the latter, there is a voluminous literature from which only selected basic works (Guilcher, 1954, Ladd, 1961), to which we refer, and a few monographs which treat of the relationships between the construction of coral reefs and littoral evolution are listed here.

ABECASSIS, F. M. (1958) 'Les flèches de sable de la côte d'Angola', *Lab. Nac. Engenharia Civil, Mem.* 140, Lisboa, 9 p.

ANDRADE, G. OSÓRIO DE (1955) *Itamaracá, contribuição para o estudio geomorfológico da costa pernambucana*, Recife, 84 p.

Shows that here beach sands do not originate in streams. Spits, beachrock, mudflats.

ANDRADE, G. OSÓRIO DE (1956) *A 'ria' do Rio Formoso na costa sul de Pernambuco*, Univ. de Recife, Fac. de Fil., Secçao E, no 18, 13 p.

BATTISTINI, R. (1958) 'Structure et géomorphologie du littoral Karimbola (extreme south of Madagascar)', *Mém. Inst. Scient. Madagascar*, Sér. F, **2**, pp. 1–77.

Excellent description of a limestone coast, photographs.

BATTISTINI, R. (1960) 'Quelques aspects de la morphologie du littoral Mikea (SW coast of Madagascar)', *Cahiers Océanogr.* **12**, 548–71.

Limestone cliffs with notch, consolidated dunes, low tide wave cut bench hollowed out by sea urchins, barrier reefs.

BERTHOIS, L. and GUILCHER, A. (1956) 'La plaine d'Ambilobé (Madagascar), étude morphologique et sédimentologique', *Rev. Géom. Dynamique*, **7**, 35–52.

Good description of a delta, with sedimentological analyses.

BLUMENSTOCK, D., FOSBERG, F. R., and JOHNSON, C. G. (1961) 'The resurvey of typhoon effect on Jaluit atoll in the Marshall Islands', *Nature, Lond.* **199**, 618–20.

Sudden changes, most of which slowly disappear later.

CHOUBERT, B. and BOYE, M. (1959) 'Envasements et dévasements du littoral en Guyane française', *C.R. Acad. Sci.* **249**, 145–7.

GEIJSKES, D. C. (1952) 'On the structure and origin of the sandy ridges in the coastal zone of Suriname', *Tijdschr. Kon. Nederl. Aardrijkskundig Gen.* **69**, no. 2, 225–37.

GIERLOFF-EMDEN, H. (1958) 'Analyse de l'évolution et des conditions de développement actuel du littoral du Salvador', *Bull. Assoc. Géogr. Franç.* **278–9**, 2–22.
Dynamics of mangroves.

GIERLOFF-EMDEN, H. (1959) 'Die Küste von El Salvador. Eine morphologisch-ozeano-graphische Monographie', *Acta Humboldtiana*, Geogr.-Ethonol. Ser., no. 2, 183 p., 38 fig., 24 maps, 22 pl.

GIERLOFF-EMDEN, H. (1959) 'Die Küste von El Salvador', *Deutsch. Hydr. Zeitschr.* **12**, 14–24.
Useful illustrations.

GUILCHER, A. (1954a) *Morphologie littorale et sous-marine*, Paris, Orbis, PUF, 216 p., 40 fig., 8 phot. pl. Translated by B. W. Sparks and R. H. W. Kneese as *Coastal and Submarine Morphology*, London, Methuen, 1958, 274 p.

GUILCHER, A. (1954b) *Rapport sur les causes de l'envasement du Rio Kapatchez (Guinée Française)*, Gouv. Gèn. A.O.F., Serv. de l'Hydraulique, 29 p., mimeographed.

GUILCHER, A. (1956a) 'Aspects morpho-végétaux de côtes alluviales tropicales (Suriname et Nigeria)', *Norois* **9**, 95–8.

GUILCHER, A. (1956b) 'Etude géomorphologique des récifs coralliens du Nord-Ouest de Madagascar', *Ann. Inst. Océanogr.* **33**, 65–136.
Genetic conditions, detailed descriptions, numerous sketches.

GUILCHER, A. (1957) 'Formes de corrosion littorale du calcaire sur les côtes du Portugal', *Tijdschr. Kon. Nederl. Aardr. Gen.* **74**, 263–9.

GUILCHER, A. (1959) 'Coastal ridges and marshes and their continental environment near Grand Popo and Ouidah, Dahomey', *2nd Coastal Geogr. Conf.*, Baton-Rouge, pp. 189–212.
Lagoons due. to barred estuaries, grade size and morphoscopic curves.

GUILCHER, A. (1961) 'Le "beach-rock" ou grès de plage', *Ann. de Géogr.* **70**, 113–25.
Good summation; some bibliographic omissions.

GUILCHER, A. and JOLY, F. (1954) 'Recherches sur la morphologie de la côte atlantique du Maroc', *Trav. Inst. Sc. Chérifien*, Sér. Géol. et Géogr. Phys., no. 2, 140 p.
Good study on the vertical zonation of littoral limestone reliefs, leading to a general classification.

GUILCHER, A. and PONT, P. (1957) 'Etude expérimentale de la corrosion littorale des cal-caires', *Bull. Assoc. Géogr. Franç.* **265–6**, 48–62.
Basic.

GUILCHER, A., BERTHOIS, L., BATTISTINI, R., and FOURMANOIR, P. (1958) 'Les récifs coral-liens des îles Radama et de la Baie Ramanetaka', *Mém. Inst. Scient. Madagascar*, Sér. F, **2**, 117–99.

GUILCHER, A., BERTHOIS, L., and BATTISTINI, R. (1962) 'Formes de corrosion littorale dans les roches volcaniques particulièrement à Madagascar et au Cap Vert (Sénégal)', *Cahiers Océanogr.* **14**, 208–40.
Excellent study, basic.

KRAUS, R. W. and GALLOWAY, R. A. (1960) 'The role of algae in the formation of beach-rock in certain islands of the Caribbean', *Coastal St. Inst., Tech. Rep.* II E, 55 p.
In support of the phreatic hypothesis. Numerous tables and measurements.

LADD, H. S. (1961) 'Reef building', *Science*, **134**, 703–15.
Good summation.

LAFOND, L. R. (1957) 'Aperçu sur la sédimentologie de l'estuaire de la Betsiboka', *Rev. Inst. Fr. Pétrole-Ann. Combustibles liq.* **4**, 425–31.

LAMEGO, A. RIBEIRO (1955) 'Geologia das quadrículas de Campos: São Tomé, Feia e Xéxé', *Bol. Div. Geol. e Miner.* **154**, 60 p.
Good monograph about a delta, with aerial photographs.

LE BOURDIEC, P. (1958) 'Contribution à l'étude géomorphologique du bassin sédimentaire et des régions littorales de Côte d'Ivoire', *Etudes Eburn.* **7**, 7–96.
Solid monograph, beach studies, analysis of materials.

LE BOURDIEC, P. (1958) 'Aspects de la morphogénèse plio-quaternaire en Basse Côte d'Ivoire (A.O.F.)', *Rev. Géom. Dynamique*, **9**, 33–42.

MILLIES-LACROIX, A. (1960) 'La presqu'île tombolisée de Vung-Tau, étude des sables marins et de leur contexte physique', *Univ. Saigon, Fac. Sc. Annales*, 425–52.
Beach sand monograph of small rocky bays.

MORETTI, A. (1951) 'Fenomeni d'erosione marina nei pressi di Porto Torres (Sardegna)', *Riv. Geogr. Ital.* **58**, 181–97.
Description of corrosion forms in limestone.

OTTMANN, F. and J. M. (1959) 'Les sédiments de l'embouchure du Capibaribe, Recife, Brésil', *Trab. Inst. Biol. Marît. e Oceanogr.*, Recife, **1**, 51–69.
Tropical estuary with sediments brought in from a dry interior region.

OTTMANN, F., NOBREGA, R., COUTINHO, P., and OLIVEIRA, S. F. (1959) 'Estudo topográfico e sedimentológico de um perfil de praia de Piedade', *Trab. Inst. Biol. Marît. e Oceanogr.*, Recife, **1**, 19–38.
Interesting comparison of curves prior to and after decalcification of the sediments.

PAGNEY, P. (1958) 'Mouvements marins d'origine cyclonique. Leurs manifestations dans la mer Caraïbe, spécialement sur les côtes de la Martinique', *Bull. Assoc. Géogr. Fr.* **276–7**, 61–72.

PIMIENTA, J. (1956) *Evolution du delta lagunaire du Rio Tubarão à Laguna (Brésil méridional)*, XVIII Int. Geogr. Congr., Rio de Janeiro, mimeographed.

PORTO DOMINGUES, A. and KELLER, E. (1955) *Guidebook*, no. 6, *Bahia*, XVIII Int. Geogr. Congr., Rio de Janeiro, 254 p.

RANSON, G. (1955) 'Observations sur les îles coralliennes de l'archipel des Tuamotu (Océanie française)', *C.R. Somm. Soc. Géol. de Fr.*, 47–9.

REVELLE, R. and EMERY, K. O. (1957) 'Chemical erosion of beach-rock and exposed reef rock', *Geol. Surv.*, Prof. Paper 260 T.
Very detailed study, with measurements of processes.

RUSSELL, R. J. (1959) 'Caribbean beach-rock observations', *Z. für Geom.* **3**, 227–36.
Good summation.

RUSSELL, R. J. (1962) 'Origin of beach-rock', *Z. für Geom.* **6**, 1–16.
Completes preceding article. Considers only the phreatic explanation.

TAILLEFER, F. (1957) 'Les rivages des Bermudes et les formes de dissolution littorale du calcaire', *Cahiers Géogr.*, Québec, **2**, 115–38.

TEIXEIRA GUERRA, A. (1951) 'Algunos aspectos geomofológicos do litoral Amapaense', *Bol. Geogr.*, Rio de Janeiro, **9**, 167–78.
Very rapid construction of a fluviomarine plain.

TRICART, J. (1957) 'Aspects et problèmes géomorphologiques du littoral occidental de la Côte d'Ivoire', *Bull. IFAN*, Sér. A, **19**, 1–20.
Descriptions of sea cliffs overgrown by vegetation, useful photographs.

TRICART, J. (1959) 'Problèmes géomorphologiques du littoral oriental du Brésil', *Cahiers océanogr.* **11**, 278–308.
Numerous general problems are posed on the basis of a regional monograph. Photographs.

TRICART, J. (1962) 'Etude générale de la desserte portuaire de la "SASCA", I: conditions morphodynamiques génèrales du littoral occidental de Côte d'Ivoire; II: les sites portuaires', *Cahiers océanogr.* **14**, 88–97, 146–61.

TRICART, J. and SILVA, T. CARDOSO DA (1958) 'Observações de geomorfologia litoral do Rio Vermelho (Salvador)', *Est. de Geogr. da Bahia* (Salvador), 225–43.
Actions of sea-urchins on beach-rock. Corrosion.

VAN DER EYCK, J. J. (1957) *Reconnaissance Soil Survey in Northern Surinam*, 99 p., 9 pl.
Good observations on a fluviomarine plain.

VANN, J. H. (1959) 'The geomorphology of the Guiana Coast', *2nd Coastal Geogr. Conf.*, Baton Rouge, 153–87.
Good study of a low plain.

VERSTAPPEN, H. TH. (1953) *Djakarta Bay. A Geomorphological Study on Shoreline Development*, The Hague, Trio, 101 p., 10 fig., 26 phot.
Good monograph on a fluviomarine plain.

VERSTAPPEN, H. TH. (1954) 'Het kustgebied van noordelijk West-Java op de luchtfoto', *Tijdschr. Kon. Nederl. Aardrijks. Gen.* **71**, 146–52.
Good aerial photos with explanations.

VERSTAPPEN, H. TH. (1957) 'Short note on the dunes near Parangritis (Java)', *Tijdschr. Kon. Nederl. Aardrijks. Gen.* **74**, 1–6.
A coast much affected by a monsoonal climate.

3

Dynamic geomorphology of the forested wet tropics

The forested wet tropics, corresponding to the evergreen and semi-evergreen seasonal forests, are that part of the tropics where the most extreme tropical morphogenic conditions are realised. High temperatures and humidities combined occasion a particularly intensive weathering even in the case of the semi-evergreen seasonal forests where the dry season introduces a short interruption of the processes.

There are slight geomorphic differences between regions of evergreen and regions of semi-evergreen seasonal forest. Comparison of our personal observations in the rainforests of South America (Brazil and Venezuela) and those of Rougerie in the semi-evergreen seasonal forest of Ivory Coast suggest that overland flow[1] is more important in the latter. But there is no fundamental difference, only a matter of degree. Moreover the comparison is difficult: climate does not alone intervene. The floristic composition of the selvas is different and humus is more abundant in our South American examples; both have their repercussions on overland flow and, especially, on its morphogenic effectiveness. The intervention of man has been more important in Africa: former intervals of shifting cultivation could have been responsible for a minor modification of the land that now facilitates overland flow, as well as a certain amount of soil degradation that diminishes the permeability. Accurate comparable measurements, which are wanting, are therefore necessary to solve the problem. We must therefore be satisfied for the present to set forth the much more important common aspects that exist in the dense, tall forests, whether evergreen or semi-evergreen, and only incidentally introduce the minor details that differentiate them.

We examine first the characteristics of the bioclimatic environment peculiar to the tropical forest, they have a very great influence on the morphogenic processes, which we study later. Lastly, we analyse the landforms and their evolution.

---

[1] *Overland flow* as used in this work is one aspect of *runoff*, the most general term describing running water on the earth's surface; the other aspect is *stream flow*. Whereas overland flow is unconcentrated runoff, stream flow is concentrated or channelled runoff. *Rillwash* and *sheet floods* are forms of overland flow (KdeJ).

The bioclimatic environment peculiar to the tropical forest

The tropical forest is a product of climate, as we have shown above. But owing to the opulence of its vegetation it forms an exceptionally effective screen between the free atmosphere above and the ground surface below. The geomorphic climate is therefore really an original climate, or bioclimate, very different from the numerical averages provided by climatologists. We must therefore define it before making any morphogenic analysis. This is all the more important as the screen is destroyed or rendered ineffective by certain imprudent or inadequate methods of cultivation. Morphogenic conditions are then radically modified and the unbalance produced may be such as to prevent the regrowth of the forest after the land has been abandoned. The geomorphic problems caused by the intervention of man are very important in the tropical forest.

The tropical forest is indeed an effective screen, a real filter of climate. It owes this property to the density of the flora, which is three or four times superior to that of the forests of the temperate zone even though most of these are cultivated for a maximum production of wood. If the stratified aspect of the plants is not always as clear as formulated in textbooks, the tropical rainforest is always noted for the coexistence of plants of very different heights. Emerging from the mass, at 40 to 60 m (130 to 200 ft), are scattered tall trees which extend their umbrella shaped crowns to take a maximum advantage of the light. They are always some tens of metres distant from one another. Below, trees 20 to 30 m high (65 to 100 ft) form a continuous growth. Their foliage merges in an uninterrupted canopy. It constitutes the first effective screen. Below it there is shade and humidity reflected in the appearance of epiphytes and a particular insect fauna. Still lower are slender saplings which reach for light and whose lank outlines strike the observer in a cleared patch of land. At 6 to 7 m (20 to 23 ft) above the ground there is an almost total lack of vegetation. Here half light is unfavourable to photosynthesis, so that ecologic conditions are restrained. The underbush is thin, the ground usually bare. Lianas and seedlings grow upward as rapidly as they can to escape this milieu.

A layer filtering the climate is therefore found between 5 and 30 m (15 and 100 ft) above the ground. It completely modifies the conditions realised at the ground surface. Microclimatic measurements have been recently inaugurated at Adiopodioumé near Abidjan, Ivory Coast, in order to compare divers climatic elements at different heights above the ground. The period of observation is still too short to provide anything but working ideas, but the results are sufficiently clear so that they may be considered. In an as yet unpublished work Cachan (1960) shows that the rainfall at the ground is 50 to 95 per cent less than at the top of the tallest trees. Fine rains and rains of short duration are most affected. They sometimes do not even reach the ground. Heavy and continuous rains register the smallest loss.

Weather stations therefore exaggerate the irregularity of rainfall as to its occurrence on the forest floor. Heavy showers most apt to occasion overland flow are least affected. It seems that there may be even an accentuation of their destructiveness, for the drops dripping from trees are usually larger than those coming straight out of the atmosphere. They in fact form streamlets that concentrate their action on the same spot, which helps soil erosion. The lesser height from which the drops fall hardly constitutes an advantage: the water streaming from the trees generally originates high enough to cause the drops to reach the maximum velocity of free fall by the time they reach the ground. Their kinetic energy is only a function of their mass, which is increased by the concentration of the rain water in streamlets. Thus during heavy showers there is no marked diminution of the energy received on the ground by the impact of the raindrops. Indeed 95 per cent of the rain reaches the ground, but the energy is spread differently: it is concentrated in a restricted number of places that correspond to the aerial streamlets and is much less in the intervals. Splash erosion, therefore, strongly attacks the ground but only in spots; it is irregular.

Similar measurements have been made in Uganda by Hopkins (1960) in an environment that has not been clearly defined but which seems to be an open deciduous seasonal forest. It is interesting to compare the results with those of Cachan, although observations stretch only over a seven-week period. On top of a tower overlooking the forest the rainfall of 1 130 mm (45 in) per year is equal to that of Mpanga station (Ivory Coast) situated at an elevation of 300 m (1 000 ft). But only 66·4 per cent of the rainfall reaches the ground. Nearly the whole loss occurs below 9·20 m (30 ft) in the lower stories of saplings, shrubs and herbs. The low vegetation therefore seems to have a higher filtrating capacity than the high trees, but it is hazardous to be affirmative about observations that have been carried on over such a short period of time. Nevertheless it is apparent that the physiognomy of the plant formations plays an important role in the morphoclimatic conditions realised on the ground. Secondary forests characterised by a dense undergrowth are probably a more effective screen against rain than the virgin forest which is more open at the ground.

It is well known that the rainfall of the wet tropics is high, generally over 1 200 mm (48 in) per annum. What is less well known is that the nature of the rains seems to vary greatly from one region to another. But we still do not have precise measurements of rainfall intensity, even less of the diameter of the raindrops, which is so important in splash erosion. The remarkable measurements of Barat (1958) in Madagascar are exceptions. The only region well known in this respect is the Congo Basin, studied by Belgians and the object of an excellent publication by Bernard (1945). The intensity of the showers and the length of dry intervals are very uneven over sometimes rather small distances. In other regions, like upper Guinea, moderate rains predominate. Thus at Sérédou the 2 or 3 m (80 or 120 in) of annual rain fall in 177 days. Showers of less than 10 mm (0·4 in) constitute 53 per cent of the total rainfall, those less than 50 mm (2 in) 95 per cent. Only ten downpours dropping more than 50 mm, with a maximum of 90 mm (3·6 in), have been recorded per year. But all heavy showers with an average intensity of 5 mm (0·2 in) per minute (which is considerable) are very destructive: splash erosion is then at a maximum, and an intense overland flow strips the soils, as we will see later. This type seems to be most common in West Africa. Observations carried on by the University of Ibadan and published by Garnier (1953) confirm those of Sérédou, though they concern a region of semi-evergreen seasonal forest. The annual total is 1 200 mm (48 in), and the rainy season is slightly longer than seven months. The most intense rains come at the beginning and near the end of the wet season. Half of the showers last less than one hour but only produce 20 per cent of the rainfall. Showers lasting 2 to 6 hours produce 60 per cent of the total. Whereas showers dropping more than 12·5 mm (0·5 in) account for 70 per cent. Their intensity is generally high at the beginning, which seems to be a characteristic trait of the convective tropical regime. During one year of observation only one downpour had an intensity of over 100 mm (4 in) per hour for one quarter of an hour.

In Central America, in Java and probably also in Venezuela high mountains close to the sea cause orographic rain with different characteristics. Destructive showers are more numerous and their intensities higher. In El Salvador, for example, one to three showers of more than 100 mm (4 in) in a few hours are recorded yearly, and intensities of 5 mm (0·2 in) per minute are commonly reached between June and October. On the coast Gierloff-Emden (1958b) reports showers of 5 to 20 mm (0·2 to 0·8 in) in five minutes, of 50 to 80 mm (2 to 3·2 in) in one hour as being not exceptional; 103 mm (4 in) in three hours has been recorded. Yet the length of observations is about the same as at Sérédou: only a few years. It seems that a type of tropical climate with destructive rains of orographic convective nature and with more thunderstorms in mountainous regions close to the sea can be opposed to another type with less intensive showers found on the hills and plateaus of West Africa. The first is, of course, particularly favourable to splash erosion and overland flow.

The screen that constitutes the forest also modifies the humidity of the air. At Adiopodioumé (Ivory Coast), in a particularly cloudy climate, insolation causes the relative humidity to drop to 70 per cent at the top of the forest in fair weather, while at the ground it remains close to 90 per cent. The lack of wind, the transpiration of plants and the evaporation of water at the ground surface thus maintain an almost unchanging humidity in the atmosphere of the underwood. This balmy air prevents the soils from really drying up. The more hygrophilous plants of the undergrowth, such as coffee or cacao, can thus resist the desiccating wind that is the harmattan, whereas they are roasted if the forest has been thinned too much. The soil surface which is protected from high evaporation never dries up completely. Desiccation cracks do not form as in the temperate zone, doubly so as most soils are kaolinites, which are little susceptible to shrinkage. It does not even harden and so keeps a certain permeability which allows infiltration as long as the impact of the raindrops does not seal it and the intensity of the shower is not too high, causing a waterlogging. As long as the surface remains humid there is less overland flow than on surfaces directly exposed to the sun and on which thin, hard impermeable crusts develop. The subsoil may, however, become seasonally desiccated, which is important from the point of view of pedogenesis and runoff, but the drought proceeds not from the surface, but from some depth, due to the extraction of water by the roots of the plants. The drawing action of the forest is considerable and diminishes stream flow by an identical amount. The water consumption of the forest gives some idea of the quantity of water that infiltrates the soil. Lemée (1961) recalls that in Java this drawing effect reaches 2 000 to 2 300 mm (80 to 92 in) on a total availability of 4 200 mm (168 in) and in the Congo Basin between 1 230 and 1 510 mm (49 and 60 in) under forest and 950 to 1 100 mm (38 to 44 in) under savanna.

The forest screen also protects the soil from temperature variations, helping to prevent it from desiccation. Measurements made at Adiopodioumé show that the diurnal maximum passes from 50°C (122°F) on unprotected ground to 42°C (108°F) on ground covered by dried grass and only to 28°C (82°F) on ground covered by green plants. The temperature of the soil is therefore even more constant than that of the air registered in the shelter of a weather station. Measurements made in Guinea and reported by Aubert (1961) show a daily amplitude of 1·2°C (2·2°F) under forest at a depth of 20 cm (8 in), as against 13·6°C (24·5°F) under a neighbouring cuirass. At Kiendi near Bondoukou, Ivory Coast, on the margin between forest and savanna the same author gives the following data:

Table 3.1 *Temperatures under divers plant covers*

HOUR	UNDER BARE CUIRASS	UNDER GRASS COVER	UNDER FOREST
8.30 a.m.	29·8°C (85·6°F)	27·4°C (81·4°F)	25°C (77°F)
10.30 a.m.	44·3 (111·7)	37·4 (99·4)	26·8 (80·3)
12.30 p.m.	52·4 (126·4)	40·6 (105·0)	28·8 (83·8)
2.30 p.m.	43·2 (109·7)	36·8 (98·3)	28·2 (82·8)

While the daily amplitude reaches 22·6°C (40·7°F) under a bare cuirass and 13·2°C (23·7°F) under grass cover, it is only 3·8°C (6·8°F) under forest. It is too low to have any effect on the weathering processes, whether chemical or biological, which at all times operate under optimum conditions.

Such divers microclimatic characteristics vary from one type of forest to another depending on the importance of the screen cover. There are few

128

comparative data on this point, which is essential. Nevertheless our personal experience in West Africa indicates that the biological rhythm plays a primordial role. Things are quite different in the tropical rainforest, always green, and in the transitional forests, semi-evergreen and deciduous. Evergreen forest reigns where variations in humidity are small enough to allow each tree to have its own biological rhythm and to shed its leaves at variable intervals depending on the species. There never is a reduction in the plant screen and the conditions analysed above are fully realised. The Adiopodioumé station belongs in this category. The transitional forest, on the contrary, experiences seasons of biological drought during which the water supply is ill assured, which forces plants into a reduced metabolism. Trees then lose their leaves, which diminishes the effectiveness of the filter. The air of the underwood becomes drier, variations of temperature are increased and the upper layers of the soil dry up. Under a semi-evergreen forest the soil may thus harden, especially if it is clayey enough, waterproof itself and cause an important amount of overland flow with the first rains. Such conditions assure a transition between the deciduous seasonal forests and the savannas. In Africa the ancient practice of brush fires has forced a considerable southward retreat of the transitional forest, causing it to be replaced by a pyrophilous savanna with only a limited number of fire resistant species (shrubs with protective bark and plants with rhizomes or with seeds which need fire to germinate, etc.). In Latin America the lesser importance of Indian agriculture has not had as a consequence such an extension of the savannas, and the transitional forest (campo cerrado) continues to occupy vast areas. It is thus of specific morphoclimatic interest. It covers regions that in West Africa would be occupied by woodland savannas and park savannas. Savannas appear in West Africa where the annual rainfall drops below 1 200 to 1 500 mm (48 to 60 in) and when more than three consecutive months cause plants to suffer a water deficit. In Latin America numerous forests grow with annual totals of only 1 000 to 1 100 mm (40 to 44 in), and dense shrub formations, whose morphogenic role does not seem to be very different from that of the deciduous seasonal forests but, on the contrary, far removed from that of the African savannas, grow with only 600 to 700 mm (24 to 28 in) as, for example, in the west of the state of Bahia. There is here an important problem which ecologists, phytosociologists and geographers should study as a team.

Even if their characteristics still remain to be specified by varied and continued systematic observations, the underwood of the intertropical forest does constitute a peculiar morphogenic environment, highly original in spite of minor differences due to divers types of forests and rainfall regimes. Let us now analyse the chemical weathering that takes place under the tropical forest.

Chemical weathering under tropical forest: the weathered material

The microclimate at the base of a tropical forest produces a predominantly biochemical weathering of the lithosphere. Physical weathering is inhibited by the small range of temperature and humidity. Moreover it can only act very locally on the bedrock where it is exposed, i.e. on stream rapids and rock hills such as monolithic domes. There, of course, the forest is inter-rupted and there is no screening effect. Temperature variations become important. In the sun rocks heat up to 50 and 60°C (122 and 140°F) and sometimes even more when they are dark and dry. When it starts to rain they are suddenly cooled, for the water that wets them is only at 22 to 25°C (72 to 77°F). A sudden drop of 30 to 40°C (50 to 70°F) can thus occur. In heterogeneous rocks, like granite or sandstone, it may cause fissures to widen or crystals to separate, both of which increase the permeability and allow the infiltration of a little water. These temperature variations also cause alter-nations of wetting and drying on the rock surface. They are specially im-portant where thermic actions have increased the permeability. Rain water infiltrates the pores, dissolves certain products, notably iron oxides, and hydrolyses the micas, especially the dark micas. Sun and heat reappear-ing, the solutions rise to the surface by capillarity and evaporate. Varnishes may thus form, particularly in sunny climates and on rapids where the rocks are submerged and well saturated during floods but emerge later. Varnishes are particularly apt to form on rapids in regions of transitional forests where there is a marked dry season. Ferruginous and resistant, they protect the subjacent rock. Varnishes are seldom encountered on mono-lithic domes. Each time hydric and thermic alternations make themselves felt on a rock into which water can penetrate, the actions result in a granular disintegration, no matter how slow.

Apart from these exceptional cases the bedrock is everywhere covered by a weathering mantle sufficiently thick to prevent the variations of tem-perature and humidity of the air from reaching it. They only affect, and very moderately under the forest screen, the upper soil horizons, which are produced by biochemical processes. It is the nature of the latter that deter-mine the physical or mechanical processes of runoff, creep, and landslides, all of which make possible the migration of matter downslope. They must therefore be studied first.

The soil

The intensity of biochemical processes characteristic of the wet tropics results from a convergence of factors:

1. The supply of a great quantity of organic matter on the ground, which plays a major role. It results, on the one hand, from the very particularities of the tropical forest, i.e. the density of the vegetation and its rapid growth

due to favourable climatic conditions, and, on the other hand, from economic conditions: the tropical forest is natural even if it is secondary; it is
not cultivated as that of the midlatitudes; plants grow spontaneously.
Numerous trees undermined by insects are blown down by the wind and
rot where they drop, increasing the supply of organic matter.

2. The rapid decomposition of organic matter, which is assured by ecologic
conditions that are close to the physiological optimum of micro-organisms.
Very constant temperatures and sufficient moisture create a real culture
medium. Low temperatures, which slow down metabolism, and high temperatures above 50°C (122°F), which are sterilising, do not exist. The
gradually increasing supply of litter produced by species which shed their
leaves each according to its own rhythm allows the persistence of a thin
layer of litter on the ground. It is a condition for the unrelented decomposition of organic matter at a high rate (Birch and Friend, 1956). At the
tropical agricultural experiment station of Yangambi in the ex-Belgian
Congo, under an annual rainfall of 1 850mm (74in), d'Hoore (1961)
reports that the yearly decomposition of organic matter in the litter produced by a secondary forest reaches 50 per cent. The coefficient of decomposition of humus reaches 68 to 76 per cent as against 6 to 12 per cent under
beech forest in California. The quantity, however, amounts to 150 to 250
t/ha/yr (60 to 100t/acre/yr).

3. The presence of a certain amount of humus in the topsoil assures a soil
structure favourable to infiltration. The soil forms aggregates and remains
relatively permeable. Its filtration capacity is higher than that of identical
soils devoid of humus, which rapidly lose permeability, become puddled
and cause more overland flow. This essential property of course vanishes
with cultivation, which causes the humus, if not replaced by manure, to
disappear. For this reason the forest once cleared becomes the locus of
intense erosion.

4. The conservation of enough moisture in the soil, which prevents its
hardening and allows it to remain permeable. Water infiltrates as long as
rainfall is not too intense; under heavy showers overland flow takes place,
but its coefficient is modest. At Adiopodioumé, on gentle slopes it is true
(less than 10°), it remains 1 to 3 per cent. A rainfall of 193mm (7·7in) was
necessary to increase it to a maximum of 7·5 per cent. Such figures must be
considered minimal, for the intensity of the rainfall is relatively moderate
and the substratum is sandy and very permeable.

The soil of the tropical forest is thus imbued with important quantities
of water full of organic substances and carbon dioxide provided by the decomposition of the litter. This factor and the persistence of moisture in the
deepest horizons are favourable to weathering. They produce a rapid dissolution of the soluble products and explain the high leaching rate of
tropical soils. Chemical fertilisers are of little use, for their ingredients are
rapidly washed down beyond the reach of plant roots. Leaching is of great
biologic and geomorphic importance:

In the topsoil mineral salts indispensable to plants, particularly the alkaline earths, are abundant, for they are provided by the decomposition of litter originating in fallen leaves, branches, and tree trunks. Deeper down a whole interval is impoverished by leaching, and it is necessary to go down very far, close to the unweathered bedrock, again to encounter higher contents of soluble salts. This is why most plants of the intertropical forest have root systems that are superficial, tracing close to the ground surface. In this way they can particularly easily recuperate the precious alkaline earths derived from the decomposition of the organic matter. Lemée (1961) indicates that the greater part of the root system is concentrated at a depth of 20 to 30 cm (8 to 12 in) in argillaceous soils, and 2 to 3 m (6 to 10 ft) in the sands of Peruvian Amazonia. Enormous trees do not have taproots and do not penetrate more than 2 m (6 ft) into the ground. They can only stand upright thanks to plank buttresses but are nevertheless prone to uprooting by high winds. Line squalls (*tornades*) cause serious tree falls. Of course, where soils are thin, conditions are more favourable: roots reach richer horizons close to bedrock. For this reason mountains, in which intense denudation keeps soils thin, are more favourable to vegetation than plains or low hills where the residual mantle is tens of metres thick and plants depend mainly on the recuperated salts of the debris of their predecessors. One can therefore well understand why mountain slopes, even very steep slopes, are overgrown with magnificent forests. The thin soils that typify them, contrary to the case in the temperate zone, are not an obstacle to their development. On the contrary, they are favourable to it. In certain regions where forest encroaches on savannas as a result of a recent climatic oscillation, it begins by colonising the slopes, from the more moist base upward. The crests, especially cuirassed plateaus, remain in savannas, which are a common landscape in the Fouta Djallon (Guinea) where the slope forest occupies the screes formed during the dry climatic phases of the Pleistocene, and where the grassland plateaus have received the vernacular name of *bowal* (plural: *bowé*). The same disposition has been described in the Congo and in East Africa. There is a narrow relationship between geomorphic evolution and the dynamics of plant formations. This relationship is particularly obvious in the zones of contact between different biotopes and in pioneer fronts, but nevertheless it is present everywhere, even in the heart of the dense forest, where the distribution of plant species is often determined by geomorphic subdivisions (Lemée, 1961).

The presence at a shallow depth of a horizon (1 to 2 m: 3 to 6 ft thick) particularly rich in roots plays an important role in the circulation of soil solutions. It is thinner in an argillaceous regolith (a few decimetres: a foot or so) and forms a barrier with a limited water capacity but quickly saturated during showers and therefore productive of overland flow. But because it is clayey the adsorbed water is abundant and can supply plant needs for a relatively long period of time. A very slow infiltration of the subjacent clays maintains seepages even after a relatively long dry interval. A semi-evergreen seasonal forest can thus thrive in regions where the total rainfall is moderate thanks to a high field capacity and the slowness of infiltration, which maintains a high water table. In a sandy regolith (acidic rocks, sands, sandstones and quartzites) the much smaller field capacity forces roots to draw upon a thicker soil layer: 2 to 3 m (6 to 10 ft) thick. But longer roots are not enough to resist periods of drought, for the water, infiltrating much more rapidly, is quickly out of reach of the roots and does not come to rest until it reaches a depth of several tens of metres (50 to 200 ft). The water supply of this horizon of radicular pumping therefore depends essentially on percolation of the rain water falling on the ground above. If dry intervals become too long there is a deficit. Roots then extract the water from the soil itself, away from the fine, especially argillaceous, particles that compose it and around which the water forms tiny films. The volume of this reserve, which corresponds to the field capacity, therefore plays a vital role. It allows the evergreen forest to bridge a more or less long dry period, the length of which depends on the clay content of the soil. The more clay there is, the higher is the field capacity. For this reason, as Rougerie has indicated in Ivory Coast, a forest can grow with only 1 200 mm (48 in) of rain on argillaceous soils, whereas it requires 1 500 mm (60 in) on sandy soils.

There is, then, between 0·5 and 2 m (2 and 7 ft) depth a horizon which is noted for its high variations in moisture. Roots dry it up as soon as rainfall decreases. Products in solution are precipitated every season and concretions form, preferably close to roots where the variation in water content is highest. Such mechanisms are, of course, most active under a semi-evergreen seasonal forest. At Yangambi, Congo, d'Hoore and Fripiat (1948) have shown that the clays are dispersed by iron hydroxides at the contact of the roots. The deflocculated horizons are thicker in depressions and below termitaries. The process is more active under secondary than under 'primary' (climactic) forest and least under cultivated land. The concretions thus formed are mainly ferruginous. Some are composed of quartz grains, the size of fine sands, cemented together into more or less hard aggregates, the major part of which resist disintegration during humid phases. They really are *ferruginous aggregates*, less resistant and less regular than ferruginous concretions. As they increase the sandy fraction of the soil, pedologists usually call them *pseudo-sands*. They are characteristic of forest soils, whereas the more regular, lustrous, hard, and nodular concretions are common in savanna soils, where desiccation is much more intense. The illuviated horizon, the B horizon of pedologists, is thus characterised by small ferruginous concretions, which mainly form in an argillaceous environment, and also by a mottled aspect. They form *mottled clays* with splotches and streaks ranging from grey (poorly aerated compact zones constituting reducing micro-environments), to brown (precipitation of hydrated iron oxides, 'limonite'), to red (precipitation of less hydrated sesquioxides, 'hematite'). This horizon is generally found between a depth of 2 and 5 m (7 and 16 ft). In the topsoil, under the litter which provides abundant humic acids, humates form and iron is leached. Here the soil has a typical yellow-greyish colour, common in all wet tropical regions of little eroded crystalline or metamorphic rocks. The thickness of this zone is generally less than 2 m (7 ft). But when the forest is destroyed and erosion becomes more active, this 'pallid' zone is stripped off, exposing the mottled zone. We have often observed this evolution in hilly regions of Brazil where the forest has been cleared to make way for cultivated fields, and where the latter have later been replaced by pasture.

Such soils, characterised by a leaching of iron oxides and clays, which increases the proportion of sand, belong to the category of podzolic soils. They are known as *tropical podzols* and when the characteristics are less obvious as *tropical podzolic soils*. They never develop directly from the parent rock but from the subsoil. They occur, as all phenomena of podzolisation, only under clearly acidic conditions. Their development is particularly typical on acidic rocks with a sandy, therefore permeable, subsoil. But with time they can form on all non-basic rocks, even those that produce a rather high proportion of clays, such as micaschists. On basic rocks too little percolation impedes leaching and the environment does not become acidic enough to permit podzolisation. Only red ferruginous soils rich in clay and with ferruginous aggregates develop. Segalen (1948) has described a typical profile with podzolic tendency under primary forest in Madagascar:

A_1 horizon, 10 to 20 cm (4 to 8 in): grey crumbly horizon with numerous more or less decomposed organic remains.

A_2 horizon, 20 to 150 cm (8 to 60 in): pallid zone impoverished in iron and, generally, in clay; beginning of podzolisation.

B horizon, beyond 150 cm (60 in): very compact red clay.

These horizons can be considered as absolutely typical.

Litter, pallid zone and mottled clays constitute the soil proper. Below begins the subsoil or lithomargin, that part of the regolith which results from the weathering of the bedrock but by processes in which vegetation only intervenes indirectly. The great thickness of this subsoil is one of the originalities of the intertropical forested zone. But one must be careful not to confuse subsoils and soils, as unfortunately happens all too often. The subsoils are indeed a veritable parent material and do not have good agronomic qualities. One should not be deluded: tropical soils are thin and fragile. Once eroded, the subsoil can still be cultivated, but with pitifully poor returns. It is by empirical adjustment to these conditions that the traditional agrarian civilisations practising non-irrigated agriculture in the tropics have adopted extended periods of fallow and have always spared some trees, ensuring a rapid recolonisation of the fields after they have become abandoned. Indeed, due to lack of manure cultivation cannot last more than two or three years. The soil which loses its humus becomes all the more easily eroded as it is no longer protected by vegetation after the destruction of the forest. Segalen (1948) emphasises that under 'primary' forest (in fact, climactic) the A horizon has a good structure. The humus retains water and protects the aggregates and risks of erosion are negligible. But under secondary forest, following only two or three years of cultivation, the organic matter has been partially destroyed and the humus has not yet been reconstituted. The soil suffers a beginning of seasonal desiccation, and contraction cracks apt to gully start to appear. If cultivation continues the danger increases. Erosion spreads rapidly and threatens to eliminate the soil. It is in order to avoid such a trend that the field is abandoned and given the opportunity to reconstitute the humus. The shortening of the fallow period as an effect of the introduction of colonial agriculture or the increase in the density of population upsets the balance and threatens the system. Unrestricted clearing and annihilation of the forest to replace it with cultivated fields without fallow, as happens so often in Brazil, produce a catastrophe and a squandering of the soil capital. The utter ruin of regions, such as the Val Paraíba, cleared only a century ago and from which the pallid zone has disappeared nearly everywhere, bears witness to it. In chapter 5 we study the anthropic landforms that result from it.

The subsoil

The subsoil, as Ruxton and Berry (1957) have shown, also has a characteristic profile. It is particularly clear on acidic igneous rocks, which have a great extent in the humid tropics. Together with these authors we should distinguish four different zones:

Zone I, at the surface, is in fact the soil in which the texture of the parent-rock is no more to be recognised. The passage from the soil proper to Zone II, which is the zone of decomposed or actually rotten rock but with preserved texture, poses an important problem. It takes place amidst mottled clays in which roots do not penetrate and which is generally not reached by termites. In Zone I quartz veins are dislocated and the original texture is completely blurred. Sometimes certain hard minerals, especially quartzes, tend to orient themselves according to the slope. When they are abundant enough and roughly aligned, pedologists call them a *stone-line*. It reflects an important disturbance. In some cases stone-lines are caused by palaeo-climatic oscillations (cf. chapter 5). The mottled clays, for their part, are characterised by a mixing which, exceptions excluded, is not of biological origin. For Leneuf (1959) it is caused by swelling produced in the last phases of weathering when kaolinite is crystallised into well formed crystals that replace the earlier formed gibbsite. Quartzes reach a maximum of fragmentation and corrosion in the zone of mottled clays.

Zone II is the zone of decomposed or rotten rock. It is the result of isovolumetric weathering (Millot and Bonifas, 1955). The original texture of the parent rock is perfectly preserved, as is demonstrated by distinct crystal bands of gneisses, various kinds of lineaments, veins, and joints. The rock has lost its consistency and crumbles when touched. It is transformed chemically. Feldspars are kaolinised, micas transformed into clay, and quartzes, which persist, are friable and break. Sometimes the surface of quartzes has a farinaceous aspect and shows traces of corrosion and fluffs of crystals of neoformation. Such *farinaceous quartzes* have become corroded when silica was dissolved during the initial stage of weathering, whether alkaline or neutral. The liberated silica was partly removed in solution, partly recombined to alumina to form kaolinite, and, lastly, has in part been reprecipitated *in situ*, forming the siliceous fluff covering the surface of quartz grains. These mechanisms render quartzes fragile and cause them to break. For this reason the dimensions of quartz grains decrease upward in the profile. At the base of Zone II their size is hardly smaller than that of the crystals of the parent-rock. In Zone I, on the contrary, the fragments are smaller than 300 to 400 microns. A cementation into ferruginous aggregates is necessary to reconstitute the sandy fraction in the form of 'pseudo-sands'. Morphoscopic study, according to the methods that we have developed from those of Cailleux, makes it possible to follow in detail all these transformations and to find their more or less obliterated traces in the detrital deposits derived from the parent material. To produce a completely decomposed rock in which the friable products have remained undisturbed a thorough *in situ* weathering must have taken place without even producing a change in volume. Elements have migrated only in the form of ions, mainly through hydrolysis. As the density of the rotten rock is much inferior to that of the fresh bedrock (one third or half), there is an important removal of substances in solution, while the remaining elements augment in volume to compensate for the loss. It is on granites and gneisses that Zone II is best developed and the loss of matter greatest. These rocks also produce the least argillaceous, therefore the most permeable, products of weathering through which water percolation is highest. The circulation of water is less on shales or schists as their soils are more argillaceous and less permeable. Loss of dissolved matter is less and difference in density smaller. The zone of rotten rock is also thinner, 2 to 10 m (7 to 30 ft), as against 10 to 50 m (30 to 160 ft) for granites and gneisses.

Zone III, in which weathering is also isovolumetric but has not yet affected the totality of the rock mass, nuclei of fresh rock surrounded by successively more weathered aureola are buried in the rotten rock. The latter develops along avenues favourable to water circulation, mainly joints in crystalline rocks. This zone of *spheroidal weathering* is all the more perfectly developed as joints are clearcut and sufficiently spaced (for example 2 m: 7 ft). This ensures good water circulation and, at the same time, the persistence of sufficient masses between them, so that residual corestones develop. Between too closely spaced joints, the entire zone decomposes almost at once. The joint pattern determines the thickness of the zone. It may be 10 to 20 m (30 to 60 ft) in granites jointed in cubelike fashion. It is non-existent in massive rocks without joints, as in certain crystalline domes. There is then a transition, only 20 or 30 cm (8 to 12 in) wide, from the perfectly friable rotten rock to the completely unweathered bedrock, which may be used, for example, for constructional purposes. Rather than a matter of chemical composition, it is the jointing that is decisive, for it controls the penetration of the solutions causing the weathering. Diffuse weathering takes place in a well drained environment when joints are well developed. The characteristic mechanism, according to Leneuf (1959), is *bleaching*, in which the feldspars lose their transparency, turning chalky and drab, and becoming fragile. It accompanies the fissuring of the crystals, not only of plagioclase and microcline but even of quartz. Bases are dissolved and in well drained environments alumina may be liberated in the form of gibbsite. The gibbsite later recombines with silica to form kaolinite, whose crystals grow and are reorganised, becoming more and more characteristic towards the top of the profile all the way to the base of Zone I. In well drained oxidising environments iron hydroxides, which occur at the sites of old joints or on the surface of more or less reduced corestones, are precipitated. A mixture of goethite, gibbsite and kaolinite is characteristic of this zone of weathering under a wet tropical climate with good drainage. When due to lack of jointing drainage is poor, Zone III is not only reduced and bleaching restricted to the surface of the fresh rock, but there is no goethite or gibbsite; only a little montmorillonite mixes with the kaolinite as a result of the incomplete leaching of the bases. When serious erosion is unleashed through

FIG. 3.1. Rotten gneiss at Nova Friburgo, State of Rio de Janeiro, Brazil
Near the surface, slided material with weathered corestones in an earthy matrix. Below, *in situ* rotten rock in which the banding of the gneiss can still be recognised. The soil that is loaded into the truck indicates the consistency of the material.

137

destruction of the natural vegetation occasioned by a climatic oscillation or, artificially, through the hand of man, it happens that Zones I and II are completely removed and that the zone of spheroidal weathering (Zone III) becomes exposed. Castle koppies and chaotic accumulations of corestones (or woolsacks) then make their appearance, as in certain places around Hong Kong. Such an evolution is facilitated by a thin Zone II, which is found on granitoid rocks with very widely spaced joints. In Surinam the Votzberg region displays a Zone III, 10 m thick (30 ft), with partially ex-humed corestones 6 to 10 m (20 to 30 ft) long (Bakker *et al.*, 1957).

FIG. 3.2. Rotten granite in the Sierra de Paranapiacaba, state of São Paulo, Brazil

In spite of the steepness of the hillside, the unweathered rock is hidden by the great thickness of the regolith. Near the surface, the darker layer, 2 or 3 m (6 to 10 ft) thick, is the soil. It thickens through colluviation down slope in the distance. Leaning trees indicate soil creep. White traces in the rotten rock are veins, which prove that the rock is indeed *in situ* underneath the soil that slides and creeps.

Zone IV (at the base) is characterised by a simple widening of the joints due to weathering. This zone may reach depths of 100 or 130 m (300 to 400 ft) below the surface, as in the neighbourhood of São Paulo, Brazil. In widened joints water circulates and frequently deposits ferruginous coatings. Blocked at the base by the fresh bedrock it forms an irregular phreatic zone. Through wells this water supplies the factories of the industrial capital of Brazil.

The above described succession, with differences due to lithologic factors, can be found in the regolith of very divers rocks. The isovolumetric decomposition of Zone II is generally less well developed than on granites. On basic rocks there is nearly always a loss of volume, so that above the zones of widened joints and spheroidal weathering there are clays that have not kept the texture of the unaltered rock but have become compacted. They are, moreover, much thinner than in the type profile on granitoid rocks with sandy subsoil, for water percolation is much more difficult. These clays merge at the base of Zone I with the zone of mottled clays, which, here too, is not very characteristic. Only Zone III is well developed, with corestones that peel like onion rings when joints are rather densely spaced. The result is a I/III regolithic profile. But if joints are too few the base of the subsoil rests directly on the fresh rock marked by cryptolapiés and poor diffuse drainage. The result is an even more incomplete profile: I/IV. The same is true of massive granitoid rocks: they are in fact the cause of monolithic domes. I/IV profiles with cryptolapiés are also the rule on limestones, which are covered by a thin layer of more or less sandy clays that are the insoluble residues of the limestone. On certain sandstones and even on quartzites one may find I/III profiles with more or less angular spheroidal weathering at the base of an amorphous regolith. In gneisses complete profiles are frequent, but the corestones are more irregular than in granite and the zones usually thinner. In micaschists, which are very

FIG. 3.3. Exhumation of landforms of differential chemical weathering, between Teresópolis and Petrópolis, Brazil

Elevation: 1 100 m (3 600 ft). Gneiss with widely spaced joints. Joints guide weathering, which produces some enormous corestones.
Mechanical erosion, principally landslides that occurred during periods with a more contrasted seasonal rainfall than at present, have partially removed the regolith and exposed Zone IV, of widened joints.
Vegetation: anthropic savanna.

friable, there are no corestones; they are replaced by veins or friable nuclei incompletely decomposed but having undergone a beginning of bleaching. The profiles, too, are generally thinner.

We must now examine the rate of weathering. In the past there has been a tendency to believe that rock weathering is very rapid in the wet tropics. Some teachers, short of metaphors, have even said that tropical corrosion attacks rocks as water dissolves sugar. Freise's inaccurate measurements are at the origin of what today appear to be completely false conceptions.

At the foot of the Venezuelan Andes and in the area of Salvador, Brazil, in climates presently producing evergreen or semi-evergreen seasonal forest, we have never observed typically tropical, highly weathered regolith in late or middle Pleistocene deposits. The Pleistocene crystalline rubble only indicates migrations of iron hydroxides and certain weathered pebbles, which have become friable and show a beginning of bleaching or are even completely decomposed. On the contrary, in the lower pre-Eemian Pleisto-cene near Salvador or in the oldest terraces of the Andean piedmont, before the penultimate cold period in the mountains, we have encountered com-plete profiles of kaolinitic weathering with mottled clays below the soil (Zone I) and underlain by completely decomposed pebbles, forming a typical Zone II, which gradually merges with the zone above. It does not seem that the embryonic appearance of the profiles of the more recent surface deposits is due to an absence of kaolinitic weathering, which every-body believes is presently proceeding, but only due to an insufficient dura-tion of the process. According to our observations it seems that several hundred thousand years are necessary for this, or a minimum of 100 000 years if one admits that processes were interrupted or slowed down during drier periods.

Leneuf, without any reference to similar field observations or geomorphic arguments, arrives at analogous figures through strictly theoretical calcu-lations. By applying Henin's (1952) formulas of underground drainage, which requires numerous hypotheses, he arrives at a duration of 20 000 to 77 000 years to reduce one metre of granite to mottled clay in the wettest part of Ivory Coast. Still, the calculation does not take into account the concept of critical mass. Furthermore, for granite to be transformed into mottled clay, a good drainage in the underlying subsoil is necessary, which requires a minimum thickness much greater than one metre. If the initial weathering phases are slower than admitted, it is the whole evolution that takes longer, the slowest processes impeding the others. There is, so to speak, a retrograde chain reaction that effectively prolongs the duration of the mechanism, bringing the figures more closely in line with those sug-gested by the geomorphic evolution.

The slowness of tropical weathering is also attested in another case, on the island of St Vincent in the West Indies, where Hay (1960) has studied very young soils formed on volcanic ash 4 000 years old. Here the andesitic glass has been transformed into a brown clayey soil already 1·80m (6ft) thick, and representing the initial stage of a yellow soil with halloysite. Yet though the glass is highly decomposed, chemical weathering is in its initial stage: all the unstable minerals, such as anorthite, labradorite, augite, hypersthene, and olivine, are still intact. There is still a long way to the kaolinitic or gibbsitic stage of weathering, which is normal in the climate of the area. Much more time is needed for it to begin, in spite of optimum physical conditions presented by this permeable and finely divided material, in spite, too, of the freshness of the material and the abundance of minerals that can be weathered.

The constitution of a complete and typical wet tropical weathering profile is therefore a slow process which has required a considerable part of the Quaternary. As we have observed in the field, such profiles characterise topographies whose present morphology dates back, in its major outlines, to the early Pleistocene or the Pliocene, epochs which would mark the starting point of the development of the regolith, either subaerially or under a permeable cover of detrital sands, such as the Barreiras Series near Salvador, Brazil. This formation has indeed fossilised the bare and coherent rock surfaces, including the spheroidally weathered gneiss under a cover of sands and crystalline conglomerates. At a depth of 20 to 30 m (65 to 100 ft) below this permeable material an intense kaolinisation has taken place, perhaps facilitated by the spongelike effect of the Barreiras Series. Not only are the gneiss corestones completely decomposed at its base but even the subjacent bedrock for several metres (5 to 15 m: 15 to 50 ft). This evolution began in the early Pleistocene when valleys were incised in the piedmont slopes of the upper Barreiras. Comparable profiles may be observed under the almost contemporaneous formations of the São Paulo and Taubate basins, in east-central Brazil. Such cryptoweathering under a permeable cover with spongelike qualities seems to be frequent and important. It should not be forgotten in the case of certain weathered profiles affecting the European ancient massifs under the sedimentary cover fossilising the post-Hercynian surface.

The soil and the subsoil, whose combined thickness is nearly always more than 10 m (30 ft), except on hard-to-weather quartzites, are the lithologic environment in which the superficial weathering processes take place. There is in this regolith a real dissociation between the divers elements of the morphogenic system. The chemical processes, which are the more intense, are in the vanguard and mould a contact between the subsoil and the fresh rock at a depth of several tens of metres, sometimes 100 m (60 to 300 ft). Büdel (1957a, b) has correctly drawn attention to this fact but by giving it an unsatisfactory designation, as he speaks of a *doppelte Verebnungsfläche*, or a 'double surface of planation', which does not really correspond to the facts. It is rather a *double surface*, as planations are rare under dense forest. As a result of the high intensity of weathering, the lithosphere is attacked on two fronts: a subterranean front (except on monolithic domes), which is that of chemical weathering, and a subaerial front, where mechanical processes operate at the expense of the regolith. The outline of the two fronts is different. Merged on rock outcrops, they may be separated by tens of metres, even 100 m (330 ft) in some depressions or under gently rolling topographies. The regolith which separates them forms a stock protected by the forest. If the forest disappears as a result of a natural climatic oscillation or due to the action of man, this stock may be rapidly mobilised by uncontrollable mechanical processes. The concept of a *double morphogenic front* is therefore more realistic than that of a double surface. A considerable dissociation of the two fronts is characteristic of the tropical forest, where it reaches its greatest amplitude and where it is most influential on the genesis of landforms.

Now that we have studied the genesis of the first, or subterranean, front of chemical weathering, we turn to that of the second, or subaerial, front of mechanical erosion.

Mechanical denudation under tropical forest: minor landforms

The mechanical attack of the regolith is dependent on its nature, which is itself conditioned by the parent rock. Lithologic influences therefore only indirectly affect the relief. They are reflected in the relative importance of three outstanding morphogenic mechanisms of the forest environment: landslides, creep, and overland flow.

Landslides

Because of deep weathering mass movements are sometimes of considerable importance in regions of consolidated rock that would be stable if the rocks were unweathered. However when weathering is at its peak the predominance of kaolinites limits the role of mass movements. Kaolinites indeed have two properties unfavourable to sliding:

1. They have a small shrinkage coefficient, so that they hardly ever crack under the effect of drought. Contraction cracks are most favourable to the rapid infiltration of water to depths of several metres, therefore to a good water supply necessary to acquire the liquid state indispensable to the triggering of sudden landslides. But such conditions are seldom realised under forest where desiccation is infrequent. Moreover, one should distinguish between evergreen forest where, in principle, the soil never dries up, and semi-evergreen seasonal forest where, on the contrary, it dries up seasonally.
2. They have a high liquid limit and a high degree of plasticity. They therefore need a high water content to pass into the liquid state, which implies massive infiltration.

Complete, well drained, almost exclusively kaolinitic weathering profiles

(I-II-III-IV) are as a whole rather unfavourable to sliding. Landslides are infrequent on old platform regions that are slowly being dissected and where weathering produces profiles rich in kaolinite. In Ivory Coast, for example, slides are exceptional in spite of steep slopes on the flanks of residual landforms. Conditions are more favourable to sliding in more actively dissected regions where weathering has not had the time to reach the kaolinitic climax and where illites are common even on acidic rocks. Illites indeed have a much lower liquid limit than kaolinites. Slides may be extremely important in them, even chronic, as in the granitic batholith east of the Peruvian Andes, near Huachón. At an elevation of 600 to 1 000 m (2 000 to 3 000 ft) narrow ridges display concave hillsides sculptured into enormous scoopshaped scars by repeated landslides. Fresh scars not yet reoccupied by vegetation cover about 10 per cent of the total land surface.[1] Such, too, is the *selva nublada* of the Venezuelan Andes where, precisely, the regolith is rich in illite. A whole series of favourable factors account for these conditions, which set in motion chain reactions:

Steep slopes and a strong tendency toward dissection, the result of present tectonic uplift, maintain a rather thin regolith (5 to 10 m: 15 to 30 ft) that is easily saturated with water.
A rapid geomorphic evolution does not allow enough time for kaolinites to form but instead illites, which are mechanically less stable. Landslides periodically clearing the slopes of course contribute to it.
The topographic position on a mountain front is favourable to the production of intense and prolonged rain, which occasions the saturation of the regolith.

In a regolith on granitoid rocks landslides generally take place at the base of the subsoil. The regolith is more sandy here and consequently more permeable. Although its liquid limit is higher than in the overlying clays, its water supply is much better. It benefits from the relative impermeability of the bedrock, which holds the ground water, and the oblique influx of rain water from the upper slope. A sharp contact between the fresh rock and the regolith, with Zones III and IV reduced or absent, is another favourable, almost indispensable, factor. Such landslides occur after protracted rains, accumulating several hundred millimetres (10 to 30 in) in two or three days, thus providing an enormous amount of infiltration. Hillsides on rocks with few joints where Zone II lies directly on the unweathered rock are the ideal site for their development. The basal contact of the regolith is sharp, relatively smooth and inclined. Such landslides also play an important role in the formation of monolithic domes by clearing the fresh rock and exposing its surface. The mechanism is of great practical importance, for such slides, which may affect several thousands of cubic metres, may wash out roads and railroads on hillsides or destroy settlements at their foot. Santos,

[1] According to a verbal communication and aerial photographs kindly provided by A. Dollfus.

Brazil, was thus devastated in March 1956, after torrential rains which, according to Pichler (1958), reached a maximum intensity of 250 mm (10 in) in ten hours. A residual mantle more than 20 m thick (66 ft) was stripped away. At Santa Terezinha, where the transition from the bedrock to the subsoil is more gradual, rock strata separated by widened joints (Zone IV) parallel to the slope were also affected. In complete profiles (I–II–III–IV) localised landslides often occur in places where there are counterslopes of fresh bedrock or of Zone IV, enclosing a thick pocket of regolith above. Subterranean waters then accumulate behind the ledge and the liquid limit is more easily attained. In such a case the slide normally clears the ledge, creating a niche above it.

As in the temperate zone, one may distinguish two varieties of landslide. Those that occur at a sharp contact between Zone II and the unweathered bedrock and remove the overburden. They form massive slides along a pre-existing slip plane parallel to the slope, in the manner a dumptruck empties its load. If the residual mantle has a complete profile sliding occurs at the base of Zone II, in the rotten rock, or at the top of Zone III, affecting some products of spheroidal weathering. Special conditions are necessary for the slip plane to cut through Zone IV; such a movement also usually assumes a different form, that of a slide by rotation or slump, in which the upper layers of the mass in movement drop almost vertically without being disturbed, forming a kind of terrace against the arcuate slip-off scarp, while the lower part of the mass tends to rise more or less.

Landslides determined by pre-existing slip planes generally affect thinner masses. The rock surface they expose on steeply inclined slopes is surrounded in a roughly rectangular manner by small irregular slopes. Actively dissected mountains on which sheet erosion impedes the formation of a thick regolith are favourable to this type of slide. Garner (1959) has reported and described several examples from the Sierra de la Costa of Venezuela.

Slumps, on the contrary, predominate in the thick regolith of hilly regions, such as the 'half-orange' convex hills of east-central Brazil (Val Paraíba, southern Minas Gerais). They form typical terracettes with walls 70 to 80° steep (when fresh), 5 to 20 m (20 to 60 ft) high and arranged in arcs of a circle, with at their foot a jumble of earth. In this mass the rather impermeable superficial clays are eroded, producing infiltration and often secondary slides. Good descriptions of such slumps in the Val Paraíba have been made by O'Reilly Sternberg (1949). In Ceylon, in November and December 1957, according to Cooray (1958), particularly abundant and persistent rains triggered the materials of an old slump in the form of a mudflow that lasted seven minutes and covered a distance of 3 km (2 miles). Several other mudflows occurred, caused by the bursting of genuine mud pockets. Some have cut steep flat-bottomed ravines in the regolith of the hillsides.

Landslides may occur under a cover of natural vegetation, but they are more common on cleared land. In Brazil they are particularly numerous in cultivated areas and especially on the poor pastures of abandoned land. We have not yet been able to obtain documented explanations of this fact. It seems, but this remains hypothetical, that the destruction of the vegetation, by considerably increasing soil temperatures and by eliminating the confined humid atmosphere of the forest, favours an intense desiccation. The drying up of the soil causes small mud cracks (kaolinitic clays have a small shrinkage coefficient), which increases permeability and allows the concentration of a high enough water content at the base of the regolith to permit the liquid limit, which is high in this case, to be reached. The degradation of the soil, however, by affecting its structure, gradually decreases its permeability, causing an increase in overland flow. Nevertheless, shortly after the destruction of the forest, there frequently is an accelerated precipitation of iron oxide in the soil, which increases the proportion of aggregates and thus creates a texture more favourable to percolation. But these three phenomena (destruction of the forest, texture, and permeability) are each opposed to one another, and precise observations and measurements under various conditions are necessary to understand their respective importance.

The danger inherent in landslides makes this kind of mass wasting an important topic of applied tropical geomorphology. The considerable energy released calls for the adoption of precautionary measures. Precise geomorphic cartography helped by exploratory bore holes or geophysical cross-sections makes possible the location of the most dangerous zones. One should above all avoid the cutting of trenches, which frequently causes unexpected undermining. Many railroad washouts in Brazil have been caused by trenches which have occasioned slumps; they were imprudently cut in the thick regolith at the base of hills. Once started, a slump is extremely difficult to control. It presupposes a good drainage, which is almost impossible because drainage channels are interrupted by earth movements and collect water, which is dangerous. Landslides on pre-existing slip planes can be more easily stabilised by underpinning their lower limb with a retaining wall anchored in the bedrock. To block the base of a slump, the construction of a wall should be avoided, for even if it is provided with weep-holes it blocks drainage and facilitates the acquisition of the liquid state and, consequently, of further movement. A stepped, permeable and flexible gabionade, rebuilt periodically if necessary, would be preferable.

Creep and tropical solifluction

Creep affects a superficial layer several decimetres thick (8 to 40 in) or even as little as 10 to 15 cm (4 to 6 in). A mass movement, it consists of a slow downward displacement caused by several processes each of a different nature:

(a) Swelling at the base of Zone I, in the mottled clays, admitted by Leneuf, seems to be one of the factors. As with all changes of volume, it necessarily results in a downward migration, in the same way as expansion due to frost. It explains the similarities between periglacial solifluction and a type of *tropical solifluction* in which various loose rocks, such as cuirass fragments, 'float' from veins, and residual pebbles derived from rock ledges and plateau remnants form stone-lines, as quartz veins are beheaded or the rubble of plateau remnants or terraces is dispersed down slope. This kind of tropical solifluction is common in perhumid forested regions even if there are no traces of important palaeoclimatic variations. Geomorphologists have not yet attached any importance to it, and it even has escaped the notice of such a conscientious observer as Rougerie. The study of its distribution and its relationship with neighbouring non-consolidated deposits

FIG. 3.4. Stone-line at Poços de Caldas, Minas Gerais, Brazil
Syenite weathered into yellow-ochre clay, truncated by a surface over which is spread a rubble of more or less weathered syenite, which originated upslope. A silty colluvium covers the rubble pavement.

makes it possible to differentiate products of superficial solifluction from rubble deposited during episodes of overland flow due to deforestation, which are another cause of the stone-line.

(*b*) The momentary attainment of the liquid state in the superficial clays because of heavy rains, as, for example, on the appropriately named Morne Savon of the island of Guadeloupe, described by Lasserre (1961). In an area of basalts and a climate of high orographic rainfall the argillaceous soils become liquid muds, into which one sinks to a depth of 20 cm (8 in). Superficial solifluction seems to play a part in the formation of stepped microplanations on the muddy soils in spite of the interference of overland flow, which, for its part, produces miniature earth pillars capped with stones and concretions.

Rougerie (1960a) has also described momentarily liquefied soils in Ivory Coast. When it rains the superficial soil layer may become saturated and liquid locally. Miniature slides, similar to the more important ones we have studied above, then occur. They form 5 to 10 cm (2 to 4 in) scarplets and small niches. On steep slopes (30° and more) they sometimes produce a series of miniature terracettes cutting up the soil in steplike fashion, each step being a few centimetres above or below the next. But these microforms are usually effaced by overland flow, so that their existence is shortlived although quantitatively they play an important role in mass wasting.

(*c*) Burrowing animals, lastly, are an important cause of creep. The debris they extract from the soil is always ejected downslope. These creatures are particularly numerous in sandy regions, where they can dig more easily. We have observed their action in the *Continental Terminal* in the neighbourhood of Abidjan. Steep slopes (40°) are mainly their work, as there is very little overland flow on them. The soils have not had time to become sufficiently clayey to waterproof themselves enough to cause more overland flow. Overland flow, however, is predominant on gentle slopes (less than 10°). Debris of all kinds, extracted from the holes of insects, larvae, and small mammals, is toppled downward on slopes that are steeper than talus. Soon it is stopped by an obstacle—a dead branch, twigs, a tree trunk, or a creeping root—to form an obstruction that is all the more important as it is durable. Behind twigs the accumulation hardly builds up to a centimetre, as twigs rot too fast. Behind creeping roots accumulations of 2 or 3 cm (1 in) are the rule; as the downslope face of the root is bared, a step 5 to 8 cm (2 to 3 in) high appears. Lastly, behind stumps obstructions may build up to a height of 10 to 20 cm (4 to 8 in). Such high values indicate an important transit of material, but one which can only be maintained on steep slopes formed by resumed erosion (here caused by a combination of tectonic movements and glacio-eustatic oscillations of sea-level), and where not enough clay is formed for overland flow to become important, as on sands. It must result in a parallel retreat of slopes, with development of flat-bottomed valleys, kept marshy by springlets at the foot of the slopes. The gradient of the hillsides is about 40°. Such a topography is typical of highly dissected sandy regions.

North of Brazzaville the walls of the cirques of the Batéké Plateau, described by Sautter (1951), have formed particularly rapidly because of the undermining effects of springs. These amphitheatres are located on the margin between forest and savanna, in poorly consolidated sandstones several hundred metres thick. Their streams are fed by important springs; their slopes are steep, 32–34°, approximately equal to the maximum slope of talus. Overgrown by forest these cirques have flat floors, joining the walls at a clear break in slope, implying a parallel retreat of slopes. The drawing off of fine and dissolved elements, causing compaction above the springlets, must accelerate creep and facilitate the retreat.

Overland flow

Overland flow sculptures the microrelief as soon as the soil contains enough clay, i.e. on practically all rocks except certain quartzites, and even on sands when slopes are gentle.

Indeed weathering is active enough on sands under forest to produce some clay as well as minute amounts of siliceous debris, of a few microns only, which cement the interstices between the remaining quartz grains. The result is an almost impermeable concrete, easily waterlogged during rains, as happens on gently sloping surfaces in the neighbourhood of Abidjan, where weathering is at a maximum. Paradoxically, overland flow on sands is more intense on gentle than on steep slopes. It is least on tropical podzolic soils: under forest the small amount of litter is enough to prevent the soil from puddling, and its leached horizon is relatively permeable, which favours infiltration and accelerates the leaching process. It is especially on granitoid rocks that the most characteristic podzolic soils are to be found.

Overland flow is in most cases the predominant agent in the sculpturing of slopes. It has the double advantage over other processes that it is more universally active and acts in a more continuous manner. Landslides, although quantitatively important by the masses affected, are generally catastrophic phenomena, occurring every now and then and concentrating their effects on small surfaces. For example, in the area of Kandy, Ceylon, the mudflow studied by Cooray (1958) occurred in a material that had itself slid some fifty years earlier. Very often the same location is affected at much longer intervals of time. Excluding steep sandy slopes, overland flow, on the contrary, occurs every time there is a beating rain, i.e. dozens of times each year, in many cases.

Overland flow under tropical forest has been carefully described and analysed by Rougerie (1960a), who has emphasised its great importance and has convinced us on this point. One must, indeed, well understand that conditions are very different from those under temperate forest, where the litter is thicker, the rain usually not as violent, and grass common. Nevertheless Rougerie's observations must be replaced in their morphoclimatic context. Most of them were made in a semi-evergreen seasonal forest and in a region where man-induced changes in the vegetation have been shown to be more important than originally believed. In the evergreen tropical forest which has never been cultivated, as in French Guiana (Cailleux, 1959), and in certain parts of Atlantic Brazil, overland flow is insignificant. Measurements on experimental plots conducted at Adiopodioumé, near Abidjan, by ORSTOM[1] (Dabin, 1957) also reveal very little overland flow: from 1 to 3 per cent under forest, with a maximum of 7·8 per cent during a downpour of 193 mm (7·7 in). The terrain is composed, it is true, of Tertiary sands, but the slope is of a few degrees only, therefore optimum for overland flow on this kind of soil.

It is possible, however, that the net overland flow as it leaves the plot is less than the effective overland flow at a given point on the plot, as overland flow is essentially discontinuous and infiltrates the soil every few metres. This makes the measurements more complex and their interpretation more difficult. The problem is far from being resolved. Considerable differences appear, depending on the regolith, the climate and the type of plant cover. They must be worked out in detail through precise and patient observations, of the kind made by Rougerie.

Overland flow under tropical forest often originates in streamlets dropping from trees. The soil is then violently beaten by big drops, which succeed in isolating argillaceous particles and putting them in suspension in spite of the high cohesion of the material. The sheet of muddy water that leaves such an emplacement meets with numerous obstacles in its path: creeping roots, trunks and obstructions caused by debris or litter. It often proceeds below masses of fallen twigs and plant remains. It is important to note that their presence should not be taken to mean that there is an absence of overland flow. Overland flow frequently moves the litter, accumulating it against obstacles further down.

The multitude of obstacles and the microheterogeneity of the soil surface cause overland flow to be essentially discontinuous in space. The rivulet produced at the place where a streamlet drops from tree branches often divides against an obstacle a little further down, and its waters then spread in two different directions. Whether it is that the soil aggregates have a better structure or that the voids are due to the rotting of roots or the work of burrowing animals, the water infiltrates and the rivulet expires as if exhausted. At the next shower the growth of leaves and branches may have moved the locus of the aerial streamlet; dead twigs will have rotted and new products will have been extracted from the soil by burrowing animals. The path followed by the elementary rivulet will be a new one. Such discontinuous overland flow is therefore changing all the time and ends by sweeping, about equally, the whole surface. More durable obstacles, such as stumps and big creeping roots, however, block the debris in the same place for years, creating obstructions of vegetal and mineral debris, as in the case of soil creep.

It is clear that overland flow is the more discontinuous as showers are less

[1] Office de la Recherche Scientifique et Technique pour les Pays d'Outre-Mer (KdeJ).

intense; this is an important factor of differentiation within the wet tropics, and, for example, opposes West Africa and the Andes Mountains. But whatever the climate the role of the slope is primordial. The steeper it is the faster the overland flow (or rillwash), enabling it to escape infiltration and to overcome obstacles with greater ease. Overland flow thus becomes less and less discontinuous. The same shower that produces rills of 1 or 2 m (3 or 6 ft) long on a 5° slope produces rills some 10 m (33 ft) long on a slope of 10°. There is then a critical limit, variable according to the intensity of rainfall and the nature of the more or less permeable soil, above which stream flow (or channelled runoff) makes its appearance. This runoff is mostly continuous, capable of overcoming obstacles and of joining the drainage net. It therefore entrains all the way down to the regional river the muds provided by the impact of raindrops. It cuts a bed and gradually makes a gully, which canalises the runoff of the next shower. Such continuous and concentrated runoff, on the condition that it recurs often enough, inscribes itself on the topography of the hillside in the form of a stream course. For this to occur it must be periodically washed clean, and function often enough to prevent its obstruction by creep.

The elementary stream course, therefore, has a high physiographic significance. It is the sign of a certain coefficient of runoff often enough realised, which depends on the intensity of the rainfall, the degree of slope and the nature of the soil. In each area it appears above a certain critical limit, or threshold, whose aspects are multiple. There is first of all a surface threshold. A certain discharge of water is necessary for overland flow to overcome obstacles and to fight infiltration. This rate of flow is realised at the end of a maximal surface, which marks the critical limit at which concentrated runoff begins. Actually, it varies from one shower to another. Although the existence of a gully facilitating the concentration of runoff helps to lower this critical limit through a mechanism of chain reaction (autocatalysis), we rediscover here the concept of static frequency of a certain intensity of rainfall. There is, therefore, also a minimal frequency limit of shower intensity, a strictly climatic factor. Finally, in each particular case, soil permeability and degree of slope may be regarded as constants, as parameters that do not vary within the scale of human history. The plant screen belongs to the same category as long as it is left in its natural state. The destruction of the forest, eliminating the protective screen, permits a more generalised pelting of the soil by raindrops together with soil degradation, particularly a decrease in permeability by the destruction of soil aggregates. Deforestation results in a rapid increase in runoff, which takes an exponential character, for the degradation of the soil proceeds progressively after the forest has been removed. Such an exponential increase of the runoff coefficient goes hand in hand with the multiplication of elementary stream courses which, in the clearest case, ends in badlands.

In a given region whose forest cover has not been modified, the essential factor on uniform soils is therefore the slope. There is a critical limit beyond which channelled runoff develops, which is reflected by a sudden increase in the density of elementary watercourses. It seems to occur when slopes reach gradients of 10°, at least on the clayey soils derived from the basalt of the mountains of Central America. But this is only one specific value. It would be necessary and interesting systematically to determine the value of this threshold under varied conditions.

Below the threshold of channelled runoff, the major part of the slopes is sculptured by discontinuous overland flow, which, moving its locus of activity from one shower to the next, in the long run brings about a generalised but minor sheet erosion. It reworks in a kind of perpetual slopewash the superficial soil layer, giving the slope a continuously changing and different microrelief, but not affecting the progression of the weathering front. When the latter passes the rate of denudation, the thickness of the regolith increases. Profiles are then complete and typical.

Beyond the threshold of channelled runoff, a dense system of elementary stream courses cuts up the slope and becomes quickly organised in an efficient drainage network. The intermediate spaces, subject to overland flow, are small enough so that the muds removed by rills can join the nearest elementary stream course. Sheet erosion then becomes intense and balances the rate of weathering. Sometimes it is faster and the thickness of the regolith decreases progressively. In general, modifications of the plant cover are the cause of such an acceleration of erosion (through a climatic oscillation or the action of man). Under natural conditions landslides more than any other process remove the regolith, and when they are chronic prevent it from reaching a maximum development. Even under deciduous seasonal forest, where it is more important, overland flow does not predominate over pedogenesis, as is proved by the general existence of the soil cover. But kaolinitic weathering is slow. If such a situation is to persist the volume of debris removed by overland flow must remain modest. Under secondary forest studies made by ORSTOM in the experimental basin of Nion in the Man area of Ivory Coast show an important flood lag after rainfall despite a high proportion of cultivated land. The vegetation allows infiltration to continue for a long time after the rainfall and reduces the amount of overland flow. The flood lag after showers is twenty-eight hours, as against ten in a Sudanese basin of the same dimensions. The soils, although argillaceous, permit a considerable amount of subterranean drainage, which is perfectly compatible with the discontinuous overland flow described by Rougerie.

The dynamic geomorphology of the forested humid tropics is thus characterised by an enormous predominance of the biochemical processes, which produce a particular regolith of exceptional thickness (the latosols). It is on this residual mantle that the mechanical agents of the external forces exert themselves. The presence of a high proportion of clay, associated with the climatic particularities and the peculiar physiognomy of the vegetation, gives the primacy to discontinuous overland flow (which benefits from the absence of grass in the underwood), to landslides, to creep, and, in certain cases, to chemical undermining. The relative importance of these divers processes varies as a function of the local slope, which is determined by the tectonic evolution and regional geomorphology (resumptions of erosion), and as a function of the facies of the regolith, especially of its clay content, for the nature of the clays remains remarkably constant. The result is a number of landform types, some of which are quite original.

Types of wet tropical landforms and their genesis

In a survey which must necessarily be brief, we must exclude certain rocks, such as the limestones, which evolve predominantly through dissolution and produce original landforms in the wet tropics where the climate limits the dissolving action to the limestone surface only. At the other extreme, quartzites produce an atypical relief, as these rocks are practically immune to chemical weathering. They bear only thin soils, not thicker than those of the temperate zone, so that the bedrock commonly outcrops. The micro-relief is directly influenced by structure. They are resistant rocks *par excellence*, and most of them produce clear structural forms, such as are commonly found in the temperate zone. They generally end up in culminating positions through inversions of relief. We are therefore mainly concerned with landforms produced by poorly consolidated sandstones, shales and schists, and granites and gneisses, which are, as a matter of fact, the most widely spread rocks in the humid tropics.

Landforms produced from poorly consolidated sandstones

The case of poorly consolidated sandstones is the simplest. Observations that we have made in Ivory Coast and Brazil (state of Bahia) indicate that a certain law of 'all or nothing' operates, and that, paradoxically, overland flow plays an important morphogenic role on gentle slopes.

Sand grains are resistant to chemical weathering. However, if sufficiently intense and prolonged, chemical weathering will fragment the grains. Humic ferruginous solutions penetrate into fissures and are precipitated, the iron combining with the silica in the form of ferruginous quartzite. It may be remobilised by humic solutions. Part of the silica is thus removed, and the fissures are widened. In the long run this process of ferrugination—leaching of sesquioxides of iron ends in corroding the quartz grains and even coarser clastics such as pebbles. The latter take on a saccharoidal aspect. Quartz grains are thus fragmented into very small, siltsize particles. They are removed by percolating waters and fill the voids of the formation whose permeability they reduce. If poorly consolidated sandstones are sufficiently weathered, especially if they are ferruginous to begin with, they produce a relatively compact regolith, hardening when dry and impermeable enough to permit overland flow.

But the chemical attack of sand grains is slow. It can therefore take place on gentle slopes not affected by sheet erosion or resumed erosion. It does not occur on steep slopes with gradients close to the angle of repose of talus, where creep is active. Overland flow must remain discontinuous if denudation is not to overtake the slow rate of weathering. The process produces wide valleys with sides not steeper than 5° at the most; they may debouch into narrow valleys caused by the recent resumption of erosion. Upon deforestation these soils are fragile; their impermeability causes intense runoff that rapidly gullies the soil. This is what happened when large rubber plantations were first installed in Ivory Coast. Now, as soon as the forest is cut, terraces are built in order to avoid a burst of accelerated erosion before the growth of new trees.

In regions subjected to intense resumed erosion narrow incised valleys are opened up. Indeed the regolith is no match for the incision of stream courses, which grade themselves rapidly in accordance with their base level. They form steep sideslopes. There is no slow weathering of the sand grains, which immunises them against runoff. The slopes develop by the double action of infiltration, which causes chemical undermining, and creep, which produces slopes slightly steeper than talus. In so far as springlets permit the evacuation of materials brought down to the base of the valley sides by creep, the slopes retreat parallel to themselves, which results in steep-sided valleys with flat bottoms terminated by an amphitheatre if the interfluve is large enough to produce important springs at the head of the valley. Other incised valley heads pass after a sudden break in slope onto the rolling level of the plateau, fashioned by discontinuous overland flow. This type is most common around Abidjan; the other seems to predominate in Congo (Brazzaville).

Such morphologic differences seem to be the result of the geomorphic evolution. In poorly consolidated sandstones valleys are easily cut and their streams rapidly graded. Resumptions of erosion propagate themselves swiftly, producing the characteristic contrast between steep slopes and gentle undulating surfaces, which Delhaye and Borgniez (1948) have also observed in the ex-Belgian Congo, where, for example, the Pleistocene terraces are very clearly developed. As to the formation of amphitheatres, it is necessarily a slow process because of the intervention of chemical undermining. It can take place only with a stable base level, which prevents the incision of the valley. Areas, such as the Batéké Plateau (north of Brazzaville), protected from resumed erosion by rockbars that refuse to be worn down, offer such conditions, which are not realised near Abidjan.

Landforms in schists and shales

On schists or shales, on the contrary, runoff plays a very important role, whereas mass wasting is comparatively unimportant. These rocks, of course, weather into clay. When there is mica iron oxides are abundant, and the clays become thoroughly deflocculated. Their plasticity is thereby reduced and their liquid limit rises to such high values that it is seldom reached. Ferruginous clays are therefore unfavourable to flowage. Only small slippages take place in the topsoil because, more heterogeneous, it is sufficiently permeable to absorb a large quantity of rain water and thus assure transition to the liquid state. They produce scarplets a few centimetres high. Larger slides are uncommon and usually take the form of earthflows, or mudflows when a pocket of liquid mud is formed in the subsoil behind a rockbar that blocks the subterranean drainage. But slowness of infiltration allows this to happen only when rains are exceptionally protracted, or if an area has been deforested and contraction cracks have formed in the soil.

Overland flow is highly favoured by the general impermeability of the soils. It is on schists that Rougerie has noted its clearest effects on gentle slopes (5 to 10°) in Ivory Coast.

As de Leenheer *et al.* (1952) have shown, deflocculated clays are more responsive to overland flow than poorly consolidated sandstones. Rillwash occurs on slopes of only 2 to 4 per cent under semi-evergreen seasonal forest at Yangambi, Congo (Kinshasa).

The texture of the drainage net is denser than in granito-gneissic regions because the surface threshold of the appearance of stream flow is much smaller. The denser drainage net is one of the facts that enables the identification of shale and schist regions on vertical aerial photographs. Sheet erosion on hillsides is intense, causing a progressive lowering of the land. The rate of denudation is relatively rapid, resulting in slopes with low average gradients, usually between 5° and 20°. The narrow spacing of stream courses is reflected in relatively short hillsides, which increase the efficiency of discontinuous overland flow. The result is a monotonous relief of low hills or ridges, without marked convexity. Overland flow brings much material to the base of the hillsides, causing a colluviation reflected in the concavity at the base of the slope. Marshy bottomlands, except in the case of local alluviation, are not characteristic and therefore quite different from those found in granito-gneissic regions.

Evolution is close to Davisian theory, due to the preponderant role of runoff: reduction of slope gradients and progressive lowering of interfluves characterise regions where the incision of stream courses is not too rapid.

As a high proportion of the water runs down the hillsides, infiltration is moderate, and percolation is slow in an essentially clayey regolith. A considerable part of the soil water is absorbed and dissipated by plants. Because of this a local drying of the soil close to the roots is important and favours the formation of ferruginous concretions, which are particularly abundant in schistose regions. All told, little water is available for the weathering of the bedrock, so, though the latter is easily decomposed, weathering is slow. The morphogenic balance[1] is detrimental to a thick regolith because of the combination of rather active sheet erosion and slow progress at the weathering front. Zone IV, of widened fissures, is encountered at a depth of a few

[1] For this important concept see Jahn (1954) or Tricart (1957b) (KdeJ).

metres only (Zone III is absent because of a too high density of joints). The same has been observed on andesites and basalts in El Salvador. Soil water reserves are small, which poses serious problems of water supply and causes a regular seasonal drying up of secondary streams.

A rather thin and impermeable regolith (composed of clay and inclusions of broken, half decomposed rock fragments) lacking in conditions favouring the concentration of soil water or the formation of a slip plane explains the minor role of deepseated mass movements in schistose regions.

Geomorphic conditions are somewhat different in shales, especially in the geosynclinal flysches of the large fold mountains, such as the Andes, the Sunda Islands and Vietnam. Such shales and flysches are less compact than schists and have little iron, for there is no mica. Weathering into clay is therefore easier. Above all, there are intercalations of siltstones (or fine sandstones) that permit the circulation of water and saturation. Because of them landslides are important if slopes are steep. They nearly always originate at the contact of a siltstone sufficiently thick to serve as an aqueduct and shales that are altered by the soaking that results from this association. Landslides generally take the form of slumps, with scarplets and terracettes, as the limit of plasticity is overstepped. Of course, if the bedding is parallel to the slope, the mechanism is much facilitated.

In flysch regions where valleys are incising themselves unimpeded by thick beds of coherent sandstones, slumps are the predominant erosional process of steep slopes (Lee Chow, 1956). They allow the valley sides to preserve average gradients of 30° to 50° in spite of rapid vertical corrasion by streams. There is both retreat of slopes parallel to themselves and an increase in their height, an evolution evoking a series of larger and larger interlocking Vs but keeping the same angle.

Landforms in granite and gneiss

The morphogenic system of the wet tropics produces typical landforms on granites and gneisses.

The weathering of granite and gneiss results in a thick permeable mantle in which water percolates freely. When the topsoil has a good structure, there is little overland flow. For this reason, in a region like that of Rio de Janeiro, which has known a perhumid tropical climate for a long time, and where evolution has proceeded under relatively constant conditions, there are hills with convex sides, shaped like hemispheres, called *meias laranjas* or half oranges by the Brazilians. Junction of their flanks with neighbouring depressions is always sudden, through a very prominent break-in-slope, whose clearness demonstrates the insignificance of colluviation and, therefore, of discontinuous overland flow. Nearly all the water infiltrates, offering optimum conditions to deep weathering. This explains the considerable thickness of the regolith. Values of 40 to 50m (130 to 160ft) for Zones I to III are common around Rio, Salvador, and Recife.

The more clayey superficial layer that forms the soil is subject to some creep, which helps to disturb the horizons, as is well shown by remnants of quartz veins that are frequently beheaded in the direction of slope. But soil creep does not cause any accumulation of debris at the foot of the hills, as is shown by the presence and persistence of a basal break in slope. On the contrary, another mechanism appears to function at the base. Examination of terrace remnants in the area of Rio clearly shows it (Tricart, 1958). The terrace slopes form convex surfaces covered with scattered stone-lines. From the evidence it seems that a chemical undermining of the regolith by the draw of springlets at the base of the slopes produces a sag that causes the convex profile. The pull of gravity sets off a mass movement of the soil, whose velocity progressively increases downslope. As the materials affected are argillaceous, and weathering has reduced quartz grains to fine powder, they can be wholly evacuated by the discharge, however small, of the springlets. They contribute to the filling in of the marshy bottomlands in the rainiest season, during which the mechanism functions best. This fragile equilibrium is quickly destroyed if the vegetation is modified or removed. Runoff then develops and an important colluviation takes place at the foot of the hillsides, which become concave.

Fɪɢ. 3.5. Bottomland in gneiss on the Rio Altéia, near Cantagalo, state of Rio de Janeiro, Brazil

Convex ridges with very thick regolith. The base level of the bottomland, which is covered with fine alluvium and is at an elevation of 700 m (2 300 ft), is controlled by an unweathered gneissic rockbar crossed by rapids. There is a sharp break of slope between the depression and the convex slopes; it is accentuated by chemical undermining and the tapping action of springlets, which produce minor re-entrants in the hillsides. A graded meandering stream flows through the bottomland. The area was deforested.

The great importance of infiltration and the existence of a more permeable subsoil makes possible a considerable accumulation of underground water. The regolith acts like a veritable sponge and partly withholds variations in stream flow, as Rochefort (1958) has shown in the case of the Rio Paraíba in Brazil. Springlets function during the rainy season and continue to restitute water entrapped in the regolith for several weeks after the last rains, sustaining the low water levels. It is owing to this mechanism that an important entrainment of dissolved materials can take place. But isovolumetric weathering, which is the rule in this material, limits their influence on the relief to chemical undermining at the foot of the hills, producing convex slopes.

Sometimes slumps occur in regions of 'half-oranges'. They are uncommon, however, for this type of relief is composed of sufficiently jointed rocks for Zone IV to be well developed. Most slumps are due to deforestation and to accelerated infiltration of rain water in the superficial clays cracked by desiccation as a result of the elimination of the forest microclimate. Sometimes the presence of a fresh rockbar blocking the drainage of subterranean water at the base of the regolith impounds enough water to cause the liquid limit to be passed. Such phenomena are frequently caused by aplite dikes, which are very resistant to weathering.

In granito-gneissic regions of intense weathering, the relief of valley bottoms is most irregular. Valleys are poorly graded, with marshy widenings alternating with narrowings frequently caused by rockbars with or without rapids or cascades. Such widenings take the shape of depressions, almost without gradient, often 1 or 2 km (0·6 to 1·2 miles) in diameter on small streams. Their bottom, flooded or swampy in the rainiest season, is filled with grey kaolinitic clay mixed with siliceous precolloids and a little organic matter. They are the products of chemical weathering of the surrounding hills. Such deposits are normally several metres thick. Below there is the rotten rock, which may reach a great depth. Many depressions, often the most typical, are located at valley heads, as the sandstone amphitheatres of Sautter. Frequently depressions follow one another at different levels (like beads on a string) separated by minor rockbars even without a marked narrowing of the valley. Narrow passes occur only when sills produce important obstructions, with cascades or foaming rapids. In such cases the valley constrictions are directly dominated by monolithic domes. Such depressions, like the sandstone amphitheatres, are formed by a slope retreat that proceeded without a concomitant decrease in the slope or deepening of the valley. They require a stable climate and a blocking of regressive erosion by very resistant rockbars. Aplite dikes, hornfels, or simply micaschists or more massive rocks generally cause the barring of such depressions.

The evolution of regions of half-oranges is characterised by a very high infiltration, a minor role of runoff, and a great predominance of deep weathering and underground drainage. Such characteristics depend on a good development of Zone IV and a great thickness of the regolith. A sufficient density of joints is required, as we have already seen. When the rock is massive or affected by rare concentric fractures, like onion rings, half-oranges give way to monolithic domes or sugar loaves (when high and narrow). In Brazil we have seen all the conditions of the transition as a function of jointing. The petrographic particularities of different granites, it seems to us, do not play an important role. An incomplete weathering profile, without Zone IV, with a sharp contact between the unweathered and the rotten rock, and constituting a good water reservoir is ideal for the triggering of landslides, which bare the bedrock surface. They partially clean the hillsides and succeeding one another transform the hill into a monolithic dome.

Once the bedrock is exposed, evolution proceeds in a different manner. Weathering practically ceases. Water flows down the steep rocky surface without infiltrating. Minerals can decompose only if joints widened by weathering are capable of retaining loose material. But this occurs only at an advanced stage of evolution and then only helps to enliven the rock dome with a few isolated trees. Usually such joints are so inhospitable that only a few aerophytes or xerophytes can thrive on them (cacti or euphorbs). Water reserves are indeed practically nil, which moreover impedes the course of weathering. Having arrived even at this stage of the reconquest of vegetation, the supply of organic matter is still practically negligible, so that weathering marks time not only because of local environmental drought but also because of the absence of humus or litter. Biochemical actions are at a practical standstill, and the only type of weathering is physical, as described in Chapter 1. Thus evolution is very slow, and the longevity of monolithic domes is assured.

Depending on the degree of jointing, itself largely determined by the final positioning of the plutons, granito-gneissic regions can therefore produce two very different types of landforms: meias laranjas and monolithic domes. They may be found together in a series of transitional varieties of which the environs of Rio de Janeiro and Teresópolis offer remarkable examples. They can also succeed one another, the half-orange hill being the initial form and the rock dome resulting from the stripping by landslides of the rock nucleus. Monolithic domes generally occupy culminating positions, as the near absence of jointing hampers weathering, which progresses at a slower rate than on half-oranges. As Ruellan (1945) has shown, they are, all told, landforms of differential chemical weathering, having escaped, and this all the better as weathering is slowed down once they are stripped, the gradual lowering of neighbouring hills through intersection of their convex slopes.

One should therefore think of monolithic domes as nuclei having resisted weathering because of a near absence of jointing. They have progressively acquired relief and been denuded, as a result of exhumation, through erosion of the regolith, which is supplied in much greater abundance by the neighbouring rocks. Birot (1957) has correctly emphasised the importance of jointing in the genesis of rock domes and in explaining away their detailed forms. In the area of Rio de Janeiro rock domes are composed of augen-gneiss of late feldspathisation with curved joints spaced at intervals of 1 to 5 m (3 to 16 ft) and sometimes considerably more, according to our own observations. Intermediate depressions are sunk into easily weathered bio-tite gneiss or highly jointed augengneiss and granulites. The flanks of the rock domes are determined by the shape of large joints. For Brajnikov (1953) differential chemical weathering causing rock domes would have an ancient origin: it would simply have exhumed a granitisation front intruded into schists and gneisses. The rock domes would correspond to protuberances of the front later dislocated by faults. The joints responsible for their present detailed forms would be fractures corresponding to tension surfaces de-veloped during the positioning of the pluton accompanied by an augmenta-tion in volume and recently revealed by an unloading effect consecutive to the erosion that has exposed it. Similar ideas have been upheld by Choubert about the monolithic domes of French Guiana.

In regions of resistant crystalline rocks the influence of joints and fracture zones are frequently revealed in the drainage pattern, which becomes angulate, associating a complex dendritic pattern with rectilinear stretches adjusted to fractures.

Other types of landforms

Other lithologic environments are less common in the wet tropics and sel-dom produce large uniform landscapes. It is therefore difficult to state what, precisely, is their influence on landforms. In any case, the effects of differ-ential erosion are more apparent than in the case of the preceding rocks.

Such, for example, is the case of basic igneous rocks, which often form plutons in the midst of acidic igneous rocks. In Ivory Coast Rougerie has shown that gabbros often produce inselbergs and stand in relief in relation to gneisses and micaschists. The same is usually true of basalts and diabases, and frequently also of syenites. On all these rocks the regolith is generally thin and always argillaceous. Cryptolapiés are common. And Rougerie (1956) has reported intense overland flow on the slopes of Orumbo-Boka, Ivory Coast, under semi-evergreen seasonal forest. Trees are said to be uprooted on slopes steeper than 15° and small slides are said to occur when gradients reach 47° to 50°. Sheet erosion would have attained 1 to 2 cm in eighteen months. Overland flow, however, remains discontinuous although gullies appear on slopes of 10° only, but they disappear downslope. Such intense erosion is surprising on residual reliefs that exist only because of their resistance to erosion. Perhaps the period of observation was marked by exceptionally heavy rainfall? Or are they relict palaeoclimatic land-forms?

Volcanic rocks are less resistant than limestones, at least on Haiti where Butterlin (1953) reports that volcanics (rock type not specified) are overlain by nummulitic limestones folded with them. The limestones produce scarps above depressions, notably over anticlinal valleys, that are eroded into the volcanic rocks. 'The volcanic rocks, not very permeable and deeply weathered through the action of the tropical climate, play the role of marls in the Jura' (p. 290). All this lacks precision both as concerns the thickness and nature of the regolith and the climatic conditions. Nevertheless the gross fact of such selective erosion is interesting by itself.

In the state of São Paulo, Brazil, under a subtropical climate, França (1956) reports that argillaceous sandstones are less resistant than basalts. The latter cap the sandstones, forming a cuesta (that of Botucatú) and stream rapids that isolate reaches easily eroded from the argillaceous sandstones and shales. Calcareous sandstones, however, form canyons with some small scarps and pinnacles.

The landforms we have just described and explained by way of examples develop in typical fashion only under stable climatic conditions. A change in climate causes a change in vegetation; both, in turn, cause a change in the morphogenic processes, which threatens the morphoclimatic equilibrium. The landforms may therefore be qualified as *climactic*, i.e. they are characteristic of a sufficiently advanced adjustment to a particular morphoclimatic environment. They presuppose a certain palaeoclimatic stability. The Brazilian littoral of Rio de Janeiro, Salvador, and Recife seems to have provided these conditions (Tricart, 1958). For this reason soils are well developed and the regolith is very thick. This is not everywhere the case. In West Africa lower Ivory Coast has known several times, and even quite recently, drier phases that have introduced savanna type plant formations in the presently forested zone. The morphoclimatic evolution has been interrupted. The regolith has been eroded and its reworked products built up into river terraces. The landforms are therefore less typical. They preserve important traces of former morphoclimatic conditions in the form of inherited landforms richer in stream courses that favour runoff. The regolith that was severely eroded now reconstitutes itself but is much thinner than that of Brazil in spite of a flatter relief.

One cannot rely, therefore, whether here or in other bioclimatic zones, on the present mechanisms to explain all the land forms. Many of them have been influenced by different palaeoclimates, whose importance has varied according to location within the wet tropics. In general savanna type plant formations have alternated with the forest. But before examining the consequences of climatic oscillations, we must first study the morphogenic processes and landforms of the alternating wet–dry tropics.

Bibliographic orientation

It is hard to avoid a certain amount of arbitrariness in the choice of references cited below and at the end of Chapter 1. The more general works have, in principal, already been listed in the latter, so that only those works which treat more specifically of the perhumid forested

160

regions will be found here. It will, however, be useful to refer back to the bibliography of Chapter 1.

The bioclimatic environment

AUBERT, G. (1961) 'Influence des divers types de végétation sur les caractères et l'évolution des sols en régions équatoriales et subéquatoriales ainsi que leurs bordures tropicales semi-humides', UNESCO, *Humid Tropical Zone*, Abidjan Symposium, 1959, pp. 41–7. Interesting data on climate and soils as a function of vegetation.

BARAT, C. (1958) 'La Montagne d'Ambre (Nord de Madagascar)', *Rev. Géogr. Alpine*, **46**, 629–81. Important climatic data.

BERNARD, E. (1945) *Le Climat écologique de la cuvette congolaise*, Brussels, Publ. INEAC, Congo, 240 p. Basic. One of the best climatic analyses, in an ecological sense.

BONNET, P., VIDAL, P., and VEROT, P. (1958) *Premiers résultats des parcelles expérimentales d'étude de l'érosion de Sérédou en Guinée forestière*, Sérédou, station expérimentale du quinquina. Mimeographed, 38 p. Analysis of pluviometric conditions.

CACHAN, P. (1960) *L'étude des microclimats et de l'écologie de la forêt sempervirente en Côte d'Ivoire*, IDERT-UNESCO, Adiopodioumé, mimeographed, 10 p. Basic. First comparative climatic observations in the various storeys of the forest.

FLORENCANO and AB'SABER (1950) 'A Serra do Mar e a mata atlântica em São Paulo', *Bol. Paulista de Geogr.* **4**, 61–70.

GARNIER, B. J. (1953) 'The incidence and intensity of rainfall at Ibadan, Nigeria', *C.R. IVe Réunion Conf. Int. Africanistes de l'Ouest*, Abidjan, p. 87.

GIERLOFF-EMDEN, H. G. (1958) 'Erhebungen und Beiträge zu den physikalisch-geographischen Grundlagen von El Salvador', *Mitt. Geogr. Ges. Hamburg* **53**, 7–140.

HOPKINS, B. (1960) 'Rainfall interception by a tropical forest in Uganda', *East Afr. Agric. J.* **25**, 255–8.

LAUER, W. (1951) 'Hygrische Klimate und Vegetationszonen der Tropen mit besonderer Berücksichtigung Ostafrikas', *Erdkunde* **5**, 284–93. Basic. Excellent map showing the length of the dry season in Africa and South America, classification of types of intertropical vegetation, vegetation and climatic maps of East Africa. Good bibliography to which we refer.

LEMEE, G. (1961) 'Effets des caractères du sol sur la localisation de la végétation en zones équatoriale et tropicale humide', UNESCO, *Humid Tropical Zone*, Abidjan Symposium, 1959, pp. 25–39. Excellent summation, showing the relationships between vegetation and geomorphology.

Reconnaissance Soil Survey of Liberia, US Dept. Agr., Washington, 1951, 108 p. Important climatic data, especially on relative humidity.

TRICART, J. (1961) 'Les caratéristiques fondamentales du système morphogénétique des pays tropicaux humides', *Inform. Géogr.* **25**, 155–69.

Weathering mechanisms and pedogenesis

In conformity with the position adopted in the text, we will not include here problems which can only be discussed in their relation to climatic oscillations: bauxites, the stone-line.

AUBERT DE LA RUE, E. (1954) *Reconnaissance géologique de la Guyane française méridionale*, Paris, Larose, 128 p., 22 pl., 5 maps.

BAKKER, J., MULLER, H., JUNGERIUS, P., and PORRENGA, H. (1957) 'Zur Granitverwitterung und Methodik der Inselbergforschung in Surinam', *Dtscher Geogr. Tag*, Würzburg, pp. 122–31.
Descriptions, study of weathered products.

BAYENS, J. (1938) *Les Sols d'Afrique centrale, spécialement du Congo Belge*, Brussels, Public. INEAC, hors-série.

BERLIER, Y., DABIN, B. and LENEUF, N. (1956) 'Comparaison physique, chimique et micro-biologique entre les sols de forêt et de savane sur les sables tertiaires de la Basse Côte d'Ivoire', *Rapports VIᵉ Congr. Sc. Sol*, Paris, vol. v, 499–502.

BOYÉ, M. (1960) 'Morphométrie des galets de quartz en Guyane française', *Rev. Géom. Dyn.* **11**, 13–27.
Mode of fragmentation of quartz, corrosion. Precise study, very useful.

BRANNER, J. C. (1896) 'Decomposition of rocks in Brazil', *Bull. Geol. Soc. Amer.* **7**, 255–314.
Old but useful observations.

BRANNER, J. C. (1913) 'The flutings and pitting of granites in the tropics', *Proc. Amer. Phil. Soc.* **52**, 163–74.
Is mostly concerned with monolithic domes.

CAILLERE, S. and HENIN, S. (1951) 'Etude de l'altération de quelques roches en Guyane', *Ann. Inst. Agron.*, Série A, **2**, No. 4, July–August.

CAILLEUX, A. (1959) 'Etudes sur l'érosion et la sédimentation en Guyane', *Mém. Serv. Carte Géol. Fr.*, pp. 49–73.
Study of the proportion of divers minerals in the weathered profile and in alluvium. Analysis of morphogenic processes, description of the relief.

COLLIGNON, J. and ROUX, C. (1955) 'Indications concernant les caractères physico-chimiques de quelques eaux douces du Moyen-Congo', *Bull. Et. Centrafricaines*, **9**, 5–14.

CORNET, J. (1896) 'Les dépots superficiels et l'érosion continentale dans le bassin du Congo', *Bull. Soc. Belge Géol.* **10**, pp. M44–116.
Early, sometimes penetrating view; interesting.

CRAENE, A. DE and LARUELLE, J. (1955) 'Genèse et altération des latosols équatoriaux et tropicaux humides', *Bull. Agron. Congo Belge*, **46**, no. 5, 1113–1243.

DENISOFF, I. (1957) 'Un type particulier de concrétionnement en cuvette centrale congo-laise', *Pédologie* (Ghent), **7**, 119–23.

DOUGLAS, I. (1967) 'Erosion of granite terrains under tropical rain forest in Australia, Malaysia and Singapore', *Int. Ass. Sci. Hydrol.*, Bern, *Symp. River Morph.*, pp. 31–9.
Measurements are interesting but should be used with care. Northeast Queensland has a long dry season.

ERHART, H. (1947) 'Les caractéristiques des sols tropicaux et leur vocation pour la culture des plantes oléagineuses', *Oléagineux*, **2**, no. 6–7.
Recent illitic soils and old kaolinitic soils whose age is unfortunately not determined for want of geomorphic study.

ERHART, H. (1954) 'Sur les phénomènes d'altération pédogénétiques des roches silicatées et alumineuses en Malaisie britannique et à Sumatra', *C.R. Acad. Sc.* **238**, 2012–14.

GLANGEAUD, L. (1941) 'Evolution des minéraux résiduels et notamment du quartz dans les sols autochtones en Afrique Occidentale Française', *C.R. Acad. Sc.* **212**, 862–4.

HAY, R. L. (1960) 'Rate of clay formation and mineral alteration in a 4 000 year old volcanic ash soil on St Vincent, B.W.I.', *Amer. J. Sc.* **258**, 354–78.

HEINZELIN, J. DE (1952) *Sols, paléosols et désertifications anciennes dans le secteur nord-oriental du Bassin du Congo*, Brussels, Public. INEAC, 168 p., 8 pl.
Basic. Has opened a new path in pedology.

HOORE, J. D' (1961) 'Influence de la mise en culture sur l'évolution des sols dans la zone de forêt dense de basse et moyenne altitude', UNESCO, *Humid Tropical Zone*, Abidjan Symposium, 1959, pp. 49–58.
Important comparative data on the mineralisation of humus.

HOORE, J. D' and FRIPIAT, J. (1948) *Recherches sur les variations des structures du sol à Yangambi (Congo Belge)*, Public. INEAC, Sér. Sc., no. 38, 60 p.
Influence of vegetation on soil structure.

LAPLANTE, A. and ROUGERIE, G. (1950) 'Etude pédologique du bassin français de la Bia', *Bull. IFAN*, **12**, Sér. A, pp. 883–904.

LEINZ, V. and VIEIRA DE CARVALHO, A. M. (1957) 'Contribução à geologia da Bacia de S. Paulo', *Bol. Fac. Fil. Univ. S. Paulo*, no. 205, 61 p., 3 maps.
Good monograph about weathering. We do not agree to consider the weathered mantle overlain by detrital Neogene as fossilised by this formation, rather its weathering seems to be subsequent to the deposition of the Neogene.

LENEUF, N. (1959) *L'altération des granites calco-alcalins et des grano-diorites en Côte d'Ivoire forestière et les sols qui en sont dérivés*, ORSTOM, Paris, 210 p., 15 phot. pl.
Essential work.

LOVERING, T. S. (1959) 'Significance of accumulator plants in rock weathering', *Bull. Geol. Soc. Amer.* **70**, 781–800. Reviewed by A. Cailleux (1960) in *Rev. Géom. Dyn.* **11**, 173–4.
Basic. Goes far beyond the bounds of our study.

MILLOT, G. and BONIFAS, M. (1955) 'Transformations isovolumétriques dans les phénomènes de latéritisation et de bauxitisation', *Bull. Serv. Carte Géol. Alsace-Lorraine*, **8**, 3–20.
Explains disintegrated rocks.

NIZERY, A. and CRESPY, S. (1949) 'Un laboratoire pour l'étude de la résistance des matériaux en climats tropical', *Rev. Gén. Electr.* **58**, 455–99.

NYE, P. H. (1954) 'Some soil forming processes in the humid tropics: I. A field study of a catena in West African forest', *J. Soil Sc.* **5**, 7–21.
Important.

ROCH, E. (1950) 'La genèse de certains sables rouges en Afrique Equatoriale Française', *C.R. Acad. Sc.* **230**, 670–1.

ROUGERIE, G. (1959) 'Latéritisation et pédogénèse intertropicale', *Inf. Géogr.* **23**, 199–206.
Brief but useful summation.

RUXTON, B. and BERRY, L. (1957) 'Weathering of granite and associated erosional features in Hong-Kong', *Bull. Geol. Soc. Amer.* **68**, 1263–92.
Basic.

RUXTON, B. and BERRY, L. (1961) 'Weathering profiles and geomorphic position on granite in two tropical regions', *Rev. Géom. Dyn.* **12**, 16–31.
Interesting comparisons with arid climates.

SCHOKALSKAJA, S. (1953) *Die Böden Afrikas*, Berlin, Akad. Verlag, 408 p. (pp. 181–218).
German translation of a Russian work. A secondhand work which is far from being up to date on a number of important questions.

SEGALEN, P. (1948) 'L'érosion des sols à Madagascar', *Conf. Afr. Sols*, Goma, pp. 1127–37.
Descriptions of type profiles.

SETZER, J. (1955–56) 'Os solos do municipio de São Paulo', *Bol. Paulista de Geogr.* no. 20, 2–30; no. 22, 26–54; no. 24, 35–56.
Numerous analyses, but difficult to use for want of information on the bedrock.

TEIXEIRA GUERRA, A. (1952) 'Importáncia da alteração superficial das rochas', *Bol. Geogr.* **10**, 42–7.

WAEGEMANS, G. (1951) 'Introduction à l'étude de la latéritisation et les latérites du Centre Africain', *Bull. Agron. Congo Belge*, **42**, 13–56.

WILHELMY, H. (1958) *Klimamorphologie der Massengesteine*, Westermann, Braunschweig, 238 p., 137 fig.
Honest summation, useful bibliography, good photographs.

Morphogenic processes

Among the above mentioned works one should also use Aubert de la Rue (1954) and Cailleux (1959), both listed above, and, of course, Rougerie (1960), a basic work listed in Chapter 1.

BAKKER, J. P. (1957) 'Quelques aspects du problème des sédiments corrélatifs en climat tropical humide', *Z. für Geom.*, pp. 1–43.

BRAJNIKOV, B. (1953) 'Les "pains de sucre" du Brésil sont-ils enracinés?' *CR Somm. Soc. Géol. Fr.*, pp. 267–9.

COORAY, P. G. (1958) 'Earthslips and related phenomena in the Kandy district, Ceylon', *The Ceylon Geogr.* **12**, 75–90.

DABIN, B. (1957) *Note sur le fonctionnement des parcelles expérimentales pour l'étude de l'érosion à la station d'Adiopodioumé (Côte d'Ivoire)*, Dakar, Secrét. Permanent Bureau Sols AOF, mimeographed, 16 p.
Results of measurements on runoff and erosion.

FREISE, F. W. (1933) 'Beobachtungen über Erosion an Urwaldgebirgsflüssen des brasilianischen Staates Rio de Janeiro', *Z. für Geom.* **7**, pp. 1–9.

FREISE, F. W. (1933–35) 'Brasilianische Zuckerhutberge', *Z. für Geom.* **8**, 49–66.

FREISE, F. W. (1935) 'Erscheinungen des Erdfliessens in Tropenwälder; Beobachtungen aus brasilianischen Küstenwälder', *Z. für Geom.* **9**, pp. 88–98.

GOUROU, P. (1956) 'Milieu local et colonisation réunionaise sur les plateaux de la Sakay (Centre-Ouest de Madagascar)', *Cahiers d'Outre-Mer*, **9**, 36–57.
Excellent maps of plateau indentions.

KAISIN, F. (1949) 'L'érosion et la stabilité des tranchées en climat tropical', *Bull. Soc. Belge Géol.*, pp. 292–7.

LAMOTTE, M. and ROUGERIE, G. (1952) 'Coexistence de trois types de modelé dans les chaînes quartzitiques du Nimba et du Simandou (Haute Guinée Française)', *Ann. de Géogr.* **61**, 432–42.

LASSERRE, G. (1961) *La Guadaloupe, étude géographique*, Bordeaux, Union Française d'Impression, vol. 1, 448 p. (p. 116).

PALMER, H. S. (1927) 'Karrenbildung in den Basaltgesteinen der Hawaiischen Inseln', *Mitt. Geogr. Ges.* Wien, **70**, pp. 89–94.

PICHLER, E. (1958) 'Aspectos geológicos dos escorregamentos de Santos', *Noticia Geom.* 2, 40–4.

RICH, J. L. (1953) 'Problems in Brazilian geology and geomorphology suggested by reconnaissance in summer of 1951', *Univ. S. Paulo, Fac. Fil., Ciencias e Letras, Bol.* 146, 80 p.
Data and interesting ideas on monolithic domes.

ROUGERIE, G. (1950) 'Le pays du Sanwi. Esquisse morphologique dans le Sud-Est de la Côte d'Ivoire', *Bull. Assoc. Géogr. Fr.*, pp. 138–45.

ROUGERIE, G. (1958) 'Existence et modalités du ruissellement sous forêt dense de Côte d'Ivoire', *CR Acad. Sc.* **246**, 290–2.

ROUGERIE, G. (1960) 'Sur les versants en milieux tropicaux humides', *Z. für Geom.*, Suppl. 1, pp. 12–18.
Useful series of bibliographic references.

RUELLAN, F. (1945) 'Evolução geomorfológica da baía de Guanabara e das regiões vizinhas', *Rev. Brasil. Geogr.* **6**, 445–508.

RUXTON, B. (1967) 'Slopewash under mature primary rain forest in Northern Papua', in J. N. Jennings and J. A. Mabbutt, eds, *Landform Studies from Australia and New Guinea*, Cambridge University Press, pp. 85–94.

SAUTTER, G. (1951) 'Note sur l'érosion en cirque des sables au Nord de Brazzaville', *Bull. Inst. Et. Centraf.*, n.s., no. 2, 49–61.

SCHNELL, R. (1948) 'Observations sur l'instabilité de certaines forêts de la Haute Guinée Française en rapport avec le modelé et la nature du sol', *Bull. Agron. Congo Belge*, **40**, no. 1, 671–6.

TEIXEIRA GUERRA, A. (1952) 'Contribuição ao estudo da geologia do território federal Amapá', *Rev. Brasil. Geogr.* **14**, 3–26.

TEIXEIRA GUERRA, A. (1954) *Estudio geográfico do território do Amapá*, Rio de Janeiro, C. Nac. Geogr., 366 p.

TRICART, J. (1958) 'Division morphoclimatique du Brésil atlantique central', *Rev. Géom. Dyn.* **9**, 1–22.

Landforms and land-forming processes

Many works cited above, which treat forms as well as processes, especially Cailleux (1959), Choubert (1957), Rougerie (1960), and Sautter (1951), are relevant in this section also. Further, reference should be made to:

ALIA MEDINA (1951) 'Datos geomorfológicos de la Guinea continental española', *Consejo sup. Invest. Cient.*, 64 p., 9 pl.
Rather mediocre.

BAKKER, J. P. (1957) 'Die Flächenbildung in den Feuchten Tropen', *Dtscher. Geogr.-Tag.* Würzburg, **31**, 86–8.
Rapid presentation with insufficient distinction between semihumid and perhumid tropics.

BERRY, L. and RUXTON, B. (1960) 'The evolution of Hong-Kong harbour basin', *Z. für Geom.* **4**, 97–115.
Differential weathering resulting in hills and domes.

BIROT, P. (1957) 'Esquisse morphologique de la région littorale de l'état de Rio de Janeiro', *Ann. de Géogr.* **66**, 80–91.
Good summation on the origin of monolithic domes.

BÜDEL, J. (1957) 'Die Flächenbildung in den feucten Tropen und die fossiler Rolle solcher Flächen in anderen Klimazonen', *Dtscher. Geogr.-Tag*, Würzburg, **31**, 89–121.
Interesting theory of a double planation surface. But many theoretical views do not coincide with the facts and apply much better to savanna regions than to the forest.

BUTTERLIN, J. (1953) 'Données nouvelles sur la géologie de la République d'Haïti', *Bul. Soc. Géol.*, 6ᵉ sér. **3**, 283–91.

CAHEN, L. and LEPERSONNE, J. (1948) 'Notes sur la géomorphologie du Congo occidental', *Ann. Musée Congo Belge*, Tervuren, sér. in-8°, Sc. Géol., i, 95 p.

DELHAYE, F. and BORGNIEZ, G. (1948) 'Contribution à la connaissance de la géographie et de la géologie de la région de la Lukénie et de la Tshuapa supérieures', *Ann. Musée Congo Belge*, sér. in-8°, Sc. Géol., iii, 155 p.
Good descriptions and analyses, comparison between forest and savanna.

FRANÇA, A. (1956) 'La route du café et les fronts pionniers', *XVIII Int. Geogr. Congr.*, Rio de Janeiro, *Excursion Guidebook* no. 3.

FREISE, F. W. (1936–38) 'Inselberge und Inselberg-Landschaften im Granit- und Gneiss-gebiete Brasiliens', *Z. für Geom.* **10**, 137–68.

FREISE, F. W. (1939–43) 'Der Ursprung der Brasilianischen Zuckerhutberge', *Z. für Geom.* **11**, 93–112.

KING, L. C. and FAIR, T. (1944) 'Hillslopes and dongas', *Trans. Geol. Soc. S. Afr.* **47**, 4 p.

LEE CHOW (1956) 'The silt problem in Taiwan', in UNO, *Proc. Reg. Conf. on Water Res. Dev. in Asia and the Far East*, Bangkok, pp. 265–70.
Very high suspended loads due to chronic landsliding.

LEENHEER, L. DE, HOORE, J. D', SYS, K. (1952) *Cartographie et caractérisation pédologique de la catena de Yangambi*, Public. INEAC, sér. Sc., no. 55, 62 p.
Description of present runoff and slope wash.

LEHMANN, H. (1936) 'Morphologische Studien auf Java', *Geogr. Abh.* **3**.

MARTONNE, E. DE (1940) 'Problèmes morphologiques du Brésil tropical atlantique', *Ann. de Géogr.* **49**, 1–27, 106–29.
Especially of historic interest. Occasionally perspicacious views backed up by poorly established facts.

ROCHEFORT, M. (1958) 'Rapports entre la pluviosité et l'écoulement dans le Brésil subtropical et le Brésil tropical atlantique (Etude comparée des bassins du Guaíba do Sul)', doct. diss., University of Strasbourg, Fac. of Lettres, *Trav. et. Mém. Inst. Hautes Etudes Amérique Latine*, vol. 2, 279 p.

ROUGERIE, G. (1951) 'Etude morphologique du bassin français de la Bia et des régions littorales de la lagune Aby', *Et. Eburnéennes*, no. 2, 108 p.

ROUGERIE, G. (1956) 'Etude des modes d'érosion et du façonnement des versants en Côte-d'Ivoire équatoriale', *1st Rep. Comm. Evolution of Slopes, IGU*, Amsterdam, pp. 136–41.

SAPPER, K. (1935) *Geomorphologie der feuchten Tropen*, Leipzig, Teubner, 150 p.

STRAUCH, N. (1955) 'A bacia do Rio Doce', *Cons. Nac. Geogr.*, Rio de Janeiro, 195 p.
A few observations on monolithic domes.

TEXEIRA, D. (1960) 'Relêvo e padrões de drenagem na soleira cristalina de Queluz (São Paulo), notas prévias', *Bol. Paulista Geogr.*, no. 36, 3–10.

VERSTAPPEN, H. (1955) 'Geomorphologischen Notizen aus Indonesien', *Erdkunde*, **9**, pp. 134–44.

WENTWORTH, C. K. (1943) 'Soil avalanches in Oahu, Hawaii', *Bull. Geol. Soc. Amer.* **54**, 53–64.

WHITE, L. S. (1949) 'Process of erosion on steep slopes on Oahu, Hawaii', *Amer. J. Sc.* **247**, pp. 168–86.

4

Dynamic geomorphology of the alternating wet–dry tropics

In tropical climates with a long dry season the forest gives way to shrub (campo cerrado) or grass (savanna) formations. The effect of these plant formations on relief is not the same as that of the forest because they poorly protect the ground against variations of temperature and humidity as well as against overland flow. The landforms reflect these differences clearly. In regions like the northeast of Brazil where Quaternary oscillations of climate have been minor, tropical evergreen forest (mata) grows on steep convex slopes separated by closely spaced bottomlands. The whole landscape is formed by high hills. In the agreste,[1] on the contrary, valleys are widely spaced, slopes flaring, interfluves have lost height and are transformed into wide flattened hills sometimes grading into gently inclined slopes, true pediplains above which inselbergs rise. The difference between agreste and caatinga is in fact more subtle than that between agreste and mata. There is a certain morphodynamic kinship between the realms of savannas and campos cerrados and that of the dry regions of the world.

We will first study the morphoclimatic environment of campos cerrados and savannas subjected to an alternating wet–dry tropical climate. We will then discuss the problem of cuirasses, which find in this environment ideal conditions for their development. This will be followed by a discussion of the mechanical processes, such as the work of termites and overland flow that contribute to the land-forming processes.

Characteristics of the morphodynamic environment

The rainforest materially modifies the soil climate by creating a humid hothouse atmosphere. Hydric and thermic variations are very small. It corresponds to an extremely specialised environment which if it ceased to exist would exclude the forest, because if the forest creates its own ecologic conditions it also demands them. This is why isolated patches of savanna in the midst of rainforest, as in lower Ivory Coast, are surrounded by a sheer forest wall without any transition. The forest terminates abruptly along a

---

[1] The *agreste* is the woodland savanna transitional between the coastal evergreen rainforest (*mata*) and the xerophytic bush (*caatinga*) of the interior of northeast Brazil (KdeJ).

continuous front, behind which are the microclimatic conditions required for reproduction. Although such a contrast is uncommon, and transitional plant communities exist, it is nevertheless significant. In regions where seasonal variations are too pronounced, the plants no longer reduce them but instead adapt themselves to them. This is to a certain extent already the case of the semi-evergreen seasonal forest.

Whereas the rainforest only grows under climatic conditions strictly defined by ecologic requirements, savannas and campos cerrados are found under rather varied climates. Outside their own zonal realm, characterised by a long dry season, they are also found in climates where tropical forest could grow, but where various factors prevent it from doing so. Such extrazonal savannas offer morphodynamic conditions very different from those of the neighbouring zonal forests. We first examine their distribution and later analyse the morphogenic conditions peculiar to savannas and campos cerrados: the screening effect of the plant cover and the marked seasonal variations.

The distribution of savannas and campos cerrados

The most important restricting factor of the rainforest is the water supply. Temperature is not very important and makes itself felt more in the floristic composition than in the physiognomy of the forest, with which geomorphology is more concerned. This explains why the rainforest grows beyond the tropics, as in Atlantic Brazil or Formosa, where satisfactory hydrologic conditions are found.

The extrazonal extension of savannas can always be related to the water supply under four different conditions:

EDAPHIC ORIGIN

Soils with a small field capacity, too poor in clays, make plants more sensitive to drought. When water infiltrates too rapidly it is deficient near the surface soon after rain. For example, in Ivory Coast, according to the field work of R. Portères quoted by Lemée (1961), plants on the clayey regolith of Man react to a water shortage after 9 to 17 days, but only after 4 to 8 days on the Tertiary sands of Basse-Côte given an equal initial supply. For forest to develop on permeable soils a much greater regularity of rainfall is needed. The time limit after which a physiologic drought sets in is shorter. It must be compensated by climate.

For this reason sandy regions, including those derived from the disintegration of sandstones and quartzites, are not inviting to forests. They sometimes bear savannas in climates where more clayey soils support forests. They are also more difficult to recolonise by the forest after its destruction either by man or by a drier climatic phase. Sometimes clusters of trees forming small thickets of dense forest several hundred metres in diameter are found in certain extrazonal savannas growing on sands. Such phenomena can be seen in the savannas of the lower Amazon, the Orinoco, or the vicinity of Assinie in Ivory Coast. In the state of Pará in Brazil such groves correspond to clay lenses in the midst of sands, therefore to a greater field capacity, which compensates for rainfall deficiency (Lemée, 1961).

HYDROLOGIC ORIGIN

In regions where rainforest ought to exist, as along the lower Amazon or in the Llanos of the Orinoco, vast, completely grassy treeless savannas, except for small groves of the kind we have just mentioned, cover enormous areas. They always occupy recent or subrecent alluvial plains subject to a very contrasted hydrologic regime. Seasonal floods rise to high levels; for example, 16 to 18 m (53 to 59 ft) on the Orinoco slightly upstream of its delta. In sandy sediments such variations have repercussions on the water-table. During floods soils become waterlogged and nitrification is stopped. Only swamp forests, like the *forêts à Raphiales* of Ivory Coast, could adapt themselves to such conditions. But these forests require a permanent water-table very close to the ground surface, which is not the case here. Indeed during low stage the watertable drops appreciably with the river level and soon finds itself out of reach of the plant roots. Swamp forest is thus excluded.

Savannas adapted to highly variable watertables take over under such conditions. They are found even in the wet tropics where the watertable may fluctuate a great deal in sandy deposits, especially along exotic rivers that originate in mountains or savannas and have a highly irregular flow (e.g. the Assinie savannas of Ivory Coast).

PALAEOCLIMATIC ORIGIN

Because of their limited field capacity sandy regions are highly affected by climatic oscillations. An increased spacing of showers during a critical season may be enough to prejudice the forest, even cause its disappearance. A chain reaction is then initiated. Disappearance of the forest, after a certain climatic limit has been passed, causes the deterioration of the soil climate. Desiccation begins, temperature variations increase, the more so as sands warm up better than clayey soils. The effects are more important than the cause. Such climatic changes have certainly taken place in the Congo Basin during the Quaternary and must help explain the periods of rigorous drought that Belgian investigators have noticed.

In any case in lower Ivory Coast, where drier climates have also existed during the Pleistocene and notably during the pre-Flandrian regression, the sandy tracts of the *Continental Terminal* are still covered by patches of savanna today. To be sure, they are edaphic savannas, but as a matter of fact things are more complex. These savannas slowly lose ground as the forest front encroaches upon them, reducing them progressively. They are therefore in disequilibrium with the climate. The edaphic factor has simply retarded the reconquest by the forest when the climate has again become more humid 5 or 10 000 years ago. We can speak, in this case, of *residual savannas*.

Still other savannas occupy areas where climate permitted the growth of forest, but where the forest was destroyed by man and has not been able to reconstitute itself. This phenomenon is most widespread in Africa, with its secular practice of brush fires, the principal means to make available cultivable land to the farming tribes south of the Sahara in their migrations towards the forest margin. Deciduous seasonal forests are easily destroyed by burning during the dry season. When fire becomes recurrent they cannot reconstitute themselves, and only an impoverished plant formation, a savanna pyrosere, subsists. It is composed of rhyzomic grasses and shrubs with thick fire-resistant barks. The number of species is limited. The formation is always open due to the presence of rhyzomic plants.

In Africa north of the Equator the tropical deciduous seasonal forests have almost completely disappeared and been replaced by savanna pyroseres. Even the driest varieties of semi-evergreen seasonal forest have to a large extent been destroyed. In many places forested hills stand in the midst of the immensity of featureless savannas. They are often sacred woods spared from burning, living witnesses of the climax forest.

The edaphic factor, of course, plays an important role in this anthropic elimination of the forest. The permanent destruction of the forest is easier on sands, so that advancing savannas often follow major sandy regions, such as the Batéké Plateau in Congo (Brazzaville).

In South America shrubs often succeed the destruction of the forest, which is unable to recolonise the too impoverished or too sandy soils. This is the case of a part of the Tertiary sandy plateaus of coastal Brazil, notably in the state of Sergipe, a region of old and dense settlement. The entrenched plateaus are occupied by a characteristic secondary vegetation which gives its name, *tabuleiro*, to these landforms. But there are also anthropic savannas, those created by man to make pastures. Their dynamic geomorphology, except for the effects of the breach in the morphodynamic equilibrium, differs little from that of spontaneous savannas.

There is then in the zone of the rainforest a considerable extrazonal extension of savannas and campos cerrados, plant formations which are characteristic of the wet–dry tropics, with all the geomorphic consequences that it entails.

Defectiveness and unevenness of the plant screen

The plant formations of the alternating wet–dry tropics have a geomorphic effect that is quite different from that of the rainforest not only because of their life form but also because of the seasonal rhythm of their metabolism.

Savannas are characterised by grasses 1 to 2 m (3 to 6 ft) high, which dry up after the rainy season. They are prone to fire. Associated isolated trees and shrubs are not of geomorphic importance. Desiccation of the soil by their roots, which reach much deeper than those of the rainforest, only

favours the formation of concretions of iron hydroxides. The former existence of roots appears in this way in certain ferruginous cuirasses in the form
of tube shaped cavities with a predominant subvertical orientation. Savanna grasses, usually rhyzomic, sprout forth in bunches in the form of tufts
or tussocks. Between them bare spaces form a kind of braided network.
Tufts are more or less closely spaced depending on the luxuriance of the
vegetation. If the dry season is not too long, as in the centre of Ivory Coast,

FIG. 4.1. Savanna landscape east of Odienné, Ivory Coast
A relief thoroughly worn down through the course of a long evolution
without important tectonic deformations. Note the contrast between
rocky inselbergs (in the background) and the relief of intersecting
piedmont slopes, producing the rolling topography of the foreground.
The vegetation consists of high tufts of elephant grass, between which the
soil is poorly protected, and isolated trees. View taken in the middle of
the rainy season.

where soils are of a fair quality, elephant grass produces a dense growth
which is difficult to cross. In some places, especially on ferruginous cuirasses, rhyzomic grasses do not grow well and may even be absent, being
replaced by shrubs or bulbous plants. Patches of bare ground are large, and
one can proceed without difficulty, being hampered only by branches.
Small mud flats indicate the emplacement of old termitaries; having a
favourable soil they frequently support a small tree, and in any case a dense
grove of shrubs. All kinds of transitions exist, depending on climate, edaphic
conditions and degree of degradation.
　　The campos cerrados present a different aspect, though their geomorphic

effect is about the same. Grasses are rare, even absent. Most of the vegetation consists of shrubs, sometimes dominated by trees with xerophytic adaptations, such as the Brazilian Barriguda with a swollen trunk full of reserves, like the African baobab. The whole formation provides a dense cover in which some spiny plants make passage even more difficult. Nevertheless in spite of such thickets the soil is poorly protected. It is practically bare between stumps and stems. When ecologic conditions are poor because of infertile soils or advanced degradation, the shrubs and occasional trees give way to mere scrub 2 or 3m (6 to 10ft) high. This is the nature of the tabuleiros; passage through them is often easier as there are wide strips of bare ground winding between the bushes.

The most important physiognomic characteristic common to both savannas and campos cerrados is the great variation in the degree of plant cover, which contrasts with the rainforest. Sometimes they are quite dense, so that the plants can intercept an appreciable fraction of the rainfall, diminishing the precipitation on the ground as much as the temperate forests, or about 20 to 30 per cent. At other times it is open and comparable to the semiarid bush of Australia or the Sahelian zone. The rain then falls directly on the ground on 10 to 20 per cent of the total surface. The degree of degradation plays a role in this case, but it is not the only one. Under climatic conditions in which plants must contend with a severe seasonal drought, soil water reserves are important. Rocks that are hard to disintegrate and break up into rubble, ferruginous cuirasses and fine ferruginous gravels, are generally noted for their more open plant covers. Stones and concretions indeed do not have any water capacity and are useless parts of the soil. Rock knobs resistant to weathering are bare or half bare as only a few plants belonging to more xerophytic species are able to cling to the joints. Quartzites, with their thin soils, and compact cuirasses have the least growth. Denser vegetation is found on sandy argillaceous soils that are permeable enough to absorb even heavy rains, enabling adequate reserves to be built up. But there are numerous variations, and on certain landforms the vegetation is disposed in catenas, as those studied on Mt Nimba by Schnell (1945), in Kita (Mali) by Jaeger and Javaroy (1952) or in central Africa by Duvigneaud (1955). On the hills of the Bas-Congo, according to the last author, the following catena may be observed:

On the ridge crests, where the soil is rather sandy due to the removal of the finer particles by overland flow, there are meso-psammophytic species not too demanding in water and satisfied with light soils.
On the slopes, where soils are stony and where rillwash is more forceful and removes not only clays and silts but also sands, there is a steppelike grassy vegetation satisfied with a deficient water supply.
On the valley bottom, where colluviation supplies the finer fractions and where water is better preserved, the savannas become hygromesophytic.

In turn, the physiognomic characteristics of these divers plant associations influence the morphogenic processes: a grass steppe is not as dense as a hygromesophytic savanna and does not protect the soil as well, increasing rillwash, which precisely creates the conditions in which a grass steppe can thrive through the elimination of more demanding, competing associations. In the same way a dense hygromesophytic savanna helps increase colluviation by its comblike effect, and thus reinforces the conditions to which it owes its existence. A delicate balance between plants and landforms is thus established, which influences the morphogenic processes and their intensities over short distances. Studies of this sort remain to be made in most cases and require a collaboration between plant geographers and geomorphologists, which up to now has taken place only exceptionally.

The second characteristic common to savannas and campos cerrados is the importance of seasonal variations in metabolism. The screening effect of these plant covers varies with the seasons. During the rainy season most shrubs, which are deciduous, grow leaves. Grasses again turn green and grow new leaves, replacing the dried ones of the previous rainy season. The fullness of the plant cover rapidly increases in a few weeks. Overland flow coefficients decrease due to water absorption by the plants and increased interception of raindrops by the newly developing foliage. Stream flow is also eventually slowed down, and the lag between showers and floods is lengthened, which indicates a withholding effect of the plant cover and perhaps also an increased subterranean drainage. At least such are the results obtained by ORSTOM in a number of small experimental basins which are common to the regions of semi-evergreen seasonal forest of the Man area and the savannas of the north of Ivory Coast (Ferkessédougou) and Guinea (Kankan). There is, therefore, a seasonal screening effect, especially where the plant cover is densest.

But this screening effect changes as soon as the rainy season comes to an end. The soil dries up, lichens, mosses and algae, which sometimes cover the ground with a protective coat during the rainy season, shrivel up and disappear. Leaves fall and grasses turn yellow. Variations in temperature now become pronounced. The vegetation can no longer reduce them. Daily the soil becomes subjected to a contrasted thermal regime, the details of which are known thanks to observations made in various countries. In a bare sandy soil at a depth of 2 mm Salvador (1959) has measured mean maxima of 47 to 48°C (116·6 to 118·4°F) at the average time of 12 o'clock noon, and mean minima of 17 to 18°C (62·6 to 64·4°F) at variable hours. At a depth of 50 cm (20 in), however, the daily range drops to 1°C (1·8°F) only, and the maximum is shifted more than twelve hours, being registered at 2.00 a.m. In an analogous but more continental climate Jaeger (1952) has measured on the same day near Kita, Mali, temperatures of 53°C (127·4°F) at 11.00 a.m. and 28°C (82·4°F) at 9.00 p.m. on the surface of a bare cuirass, whereas on the floor of a gallery forest he noted 26°C (78·8°F) and 24°C (75·2°F), and in the free air 1 m above the ground 32°C (89·6°F) and 22°C (71·6°F). The bare ground, as it is frequently found between tufts, is considerably heated by radiation, especially if it consists of sands or is a cuirass. In Congo (Kinshasa), in a region of extrazonal savannas, Bernard (1945) gives the following temperatures measured at the grass roots:

DEPTH	JULY		JANUARY	
	8 a.m.	12 noon	8 a.m.	12 noon
0	24·9°C (76·8°F)	42·8°C (109°F)	24·2°C (75·5°F)	34°C (93·2°F)
−5 cm (2 in)	24·2 (75·5)	35·1 (95·2)	23·5 (74·4)	31·5 (88·6)
−10 cm (4 in)	25·3 (77·5)	35·4 (95·6)		
−50 cm (20 in)	28 (82·4)	28 (82·4)	26·1 (79)	26·1 (79)

The end of the dry season is the most critical. Indeed the completely dried up vegetation has reduced its screening effect to a minimum. Heat is now the most intense. Soil desiccation reaches a maximum. The first showers of the rainy season, usually in the form of violent thunderstorms, fall on an unprotected ground and find conditions favourable to overland flow, which is often considerable and plays an important geomorphic role.

A real seasonal morphogenic crisis occurs when the rains start to fall: the soil, dried up, minimally protected, is particularly apt to cause overland flow; gradually becoming wet it is subject to important variations of humidity.

The crisis is often aggravated by the anthropic factor of brush fires. As various investigators have demonstrated, they cause the ground surface to be heated to very high temperatures. Beadle (1940) has recorded 81 to 218°C (178 to 415°F) in ordinary grass fires, and at a depth of 30 cm (1 ft) still 67°C (153°F). With the burning of trees and shrubs the heat is even more intense. The same author mentioned temperatures of 111 to 114°C (232 to 237°F) at a depth of 30 cm and 59 to 67°C (138 to 153°F) at a depth of 1 m (3·3 ft) in blazing fires under eucalyptus. These observations agree with those recorded in West Africa. Such temperatures completely parch the soil, dehydrate the iron oxides, deflocculate and bake the clays, which lose their plastic properties. Resistance of the soil to overland flow is reduced and its permeability diminished.

A zone of transition, the wet–dry tropics are therefore characterised by morphogenic conditions that at the end of the dry season closely resemble those of the semiarid zone, as they are propitious to overland flow, and during the rainy season to those of the wet tropics, with a more protective plant cover, less overland flow, and a moistening of the lithosphere that promotes weathering.

The marked seasonal variations characteristic of this zone are also at the origin of a particular process: the formation of cuirasses.

Cuirasses

Cuirasses, especially the ferruginous, sometimes the bauxitic, are one of the characteristic features of the humid tropics. At the present time it is almost universally admitted that they form in climates with a long dry season, even though they are sometimes found under forest, as in Amazonia or in Ivory Coast. But in the forest of Ivory Coast, since a decade or so, Rougerie has patiently demonstrated that they are relict formations, inherited from epochs going back to the Tertiary, during which morphogenic conditions were different from the present. True, in certain particular cases, cuirasses can be found that are presently forming under forest, especially under semi-evergreen seasonal forest. For example, at Yangambi (near Kisangani), in the riverbank of the Congo, at low water stage, Denisoff (1957b) has reported alluvial sands consolidated into laminae 0·5 to 4cm thick (0·2 to 1·6in), cemented by a little colloidal silica on which iron oxide has been precipitated. This would have resulted from the precipitation of silica due to fluctuations in pH at the contact of the ground water and the river water. But such cases are altogether exceptional and vastly different from the cuirasses of *bowé*.

In West Africa where cuirasses are particularly common Maignien (1958) and later Daveau (1962) have outlined a 'zone of maximum cuirass development' that coincides with the savannas. Under forest cuirasses are palaeoclimatic relicts dating back to specific geologic epochs. The same is true in the Sahelian zone. Between the two the Sudanese zone has not only more of them, but they are more extensive. Some are recent: we have observed some very fine specimens in the alluvium of the Niger at Ségou and close to Mopti; they therefore must be Quaternary.

The formation of cuirasses thus appears to be a phenomenon that takes place in the alternating wet–dry tropics. But as cuirasses are resistant, there are many that are residual and were formed as the result of different palaeoclimatic conditions than obtain in the present climatic belts. The problem of cuirasses cannot be resolved without a palaeogeographic perspective, and the recent progress that has been made in their study is due to the fact that this has been understood. We now turn to study the processes at work in the genesis of cuirasses as well as their principal facies in such a manner as to account for their geomorphic importance. The palaeogeographic perspective will be examined in the next chapter, for it extends beyond the limits of the presently wet–dry tropics.

The cuirassing mechanisms

Cuirasses result from a more or less localised enrichment of certain soil layers in iron or alumina or in a mixture of both, which causes their induration. This enrichment can be realised in two different ways as d'Hoore (1954a) has shown: either by leaching, removing the other elements (*cuirasses of relative accumulation*), or by precipitation of elements brought in solution by lateral influx (*cuirasses of absolute accumulation*). More geograph-ically, one could speak of *authigenic* and of *allogenic cuirasses*.

AUTHIGENIC CUIRASSES (OF RELATIVE ACCUMULATION)

Authigenic cuirasses are formed through an *in situ* concentration of iron or aluminium hydroxides. This implies, first of all, the elaboration of these elements and later their mobilisation, but under such circumstances that they remain in the soil profile and are not distantly removed; this is followed by precipitation in the form of stable bodies that are not remobilised and thus accumulate progressively.

The elaboration of the metallic elements is linked to the action of living organisms. Indeed the formation of cuirasses stops at a certain elevation in the highlands of the humid tropics. In Madagascar, for example, Isnard (1955) notes that 'laterites' are never found above 2 000 m (6 000 ft). They also disappear to the southwest of a line connecting Tulear and Fort-Dauphin, in drier regions (less than 800 mm: 32 in of rain). They form, therefore, only in regions that are sufficiently hot and where there is enough seasonal moisture. Bacteria with ferruginous envelopes seem to play an important role in the mobilisation of iron. Savanna grasses have high con-centrations of silica, ten to twenty times higher than that of perennial plants such as trees. In Congo (Kinshasa) 1 to 8 per cent silica is found in their ashes, which make up to 50 per cent of their weight. This silica extracted at a certain depth by the roots is abandoned on the ground surface every dry season when the leaves die and are decomposed. Each year the flux amounts to several quintals per hectare (several pounds per acre). Part of the silica is reincorporated into the soil and circulates in a closed circuit. Another part, variable in size depending on the case, is exported. For example, ashes produced by brush fires are submitted to considerable aeolian deflation, obscuring the atmosphere for months (dry haze). The first showers wash them down slope: a downslope migration of free silica with increasing con-centrations toward the base has indeed been recorded by d'Hoore's measurements. If the relief is sufficiently steep, an appreciable removal of silica takes place owing to this biogeomorphic relationship. On the crests of interfluves silica extracted by savanna grasses is removed in greater quan-tities by wind and rillwash. An impoverishment in silica on rocks poor in silica could in this way lead to the formation of bauxites. It seems that silica is in part extracted by plant roots from clays, especially kaolinite. In any case this desilication favours the concentration of gibbsite. It is clear, then, why certain bauxitic cuirasses are found on ridge crests, which are usually old residual reliefs.

The iron hydroxides are furnished by dark minerals: biotite mica, amphi-bole and pyroxene, by ferruginous sandstones or ferruginous quartzites, and even more by certain rocks forming iron ores, such as the itabirite of the Brazilian Ferriferous Quadrilateral. Their elaboration is much easier than that of gibbsite, which is more stable.

Mobilisation primarily concerns the iron hydroxides. Indeed gibbsite, which is even stabler than silica, is not mobilised and becomes elaborated *in situ* through the extraction of silica. It practically does not participate in the migrations within the weathered profile, in contradistinction to iron. Of all the mechanisms involved in the formation of cuirasses it is the most authigenic.

The iron hydroxides are mobilised by plants and bacteria. As we have seen, soluble organic compounds of iron only form in an acidic environment. Iron carbonate, which forms in the presence of CO_2, is not stable in the presence of oxygen but is precipitated by oxidation. The mobilisation of iron by bacteria, which enables more distant transport, requires a certain temperature. All these conditions are only realised in the rainy season when savanna soils are replete with water, and organic acids freed by the decomposition of plants penetrate into the soils. Hydroxide solutions then form and migrate. Down slopes they move through subterranean drainage, reappearing in springlets at the foot of the hills or broaching the surface at outcrops of unweathered rockbars. On level surfaces solutions stagnate, the more so if drainage is poor. Iron thus remains in the profile and there is no *lateral drainage* as on the slopes.

Accumulation is determined by the climate and the seasonal rhythm of plant life. Desiccation of the soil in the dry season allows the penetration of air, and consequently phenomena of oxidation. Carbonates as well as certain humic compounds are decomposed. Evaporation causes the dehydration of hydroxides and their transformation into more stable sesquioxides: iron is precipitated. In savannas there is also the influence of silica, which is electro-negative and combines with electro-positive bivalent iron to form very stable compounds (e.g. ferruginous quartzite). They precipitate as soon as they form. But, as we have seen, savanna grasses free large quantities of tiny particles of silica that are more soluble than quartz crystals. Part of them, however, are rendered insoluble by brush fires. During the rainy season the water dissolves the silica abandoned by the grasses as soon as it penetrates into the ground; the solutions then mix with solutions of iron hydroxide, provided by the weathering of ferruginous minerals, and as they combine, precipitate.

Ferruginous solutions do not migrate as far in savanna soils as they do in forest soils. There are three reasons: the intense seasonal desiccation that causes oxidation, the lesser abundance of organic acids (especially oxalic and tartaric) and the greater abundance of dissolved silica, due to the grasses. Everywhere where drainage is not excellent, for example on gentle slopes, iron precipitates in very stable forms frequently combined with silica. Particularly favourable to this kind of precipitation are the surroundings of roots, where desiccation is more intense due to the extraction of water by the plant; this is especially true of trees which, adapted to drought, have a high osmotic pressure. Thus concretions are formed, and sometimes the emplacements of former roots are detectable in cuirasses in tube-shaped structures. Such casts, however, seem to be more the result of desiccation caused by air circulation along the mould of the vanished root rather than by extraction of water by the live root itself.

The essential difference between savanna and forest lies in the mobility of the iron oxides. Under forest a swifter weathering produces more of them, but part is removed out of the profile, and those that remain isolate themselves imperfectly in the form of concretions, fixing themselves in the clays, which they colour.

The iron oxides that have been removed from the soil profile will be found back in the absolute accumulations.

ALLOGENIC CUIRASSES (OF ABSOLUTE ACCUMULATION)

Allogenic cuirasses result from the induration of products that are not provided by the soil profile, but which on the contrary come from more or less distant areas. They are therefore forms of accumulation related to more or less distant chemical weathering, exactly as the travertine deposits of karstic springs and streams or as certain calcareous crusts of dry regions.

This origin restricts the formation of allogenic cuirasses to products that are sufficiently mobile to leave the pedogenic profile. As a matter of fact, cementation is always due to iron oxide, never to alumina. This origin is directly related to the topographic position of the cuirasses: they can only form in depressions that are overfilled with iron oxides from higher areas. Cuirasses of absolute accumulation are really the result of a kind of chemical colluviation or alluviation. They are found in the zone of colluviation at the base of slopes, on seasonally flooded bottomlands, or on fluvial floodplains.

For absolute accumulations to take place solutions of hydroxides must be brought in and be precipitated *in situ*. These two conditions are partly contradictory. The arrival of solutions, sometimes from quite distant sources, indeed supposes that they are stable, otherwise precipitation would have occurred earlier. Precipitation, however, implies that the solutions cease to be stable where they become precipitated. Some kind of change must therefore take place.

The mechanism of absolute accumulation, as d'Hoore (1954a) has emphasised, implies the coexistence of three environments more or less distant from one another:

1. An environment of *exportation* where the solutions are formed, especially solutions of iron, and subsidiarily solutions of manganese. Forested regions with acidic soils and a deep subsoil are particularly favourable. The waters of the streams that drain them export a high tonnage of iron oxides, especially in the form of bacteria with ferruginous envelopes. The solutions thus formed, even if they are not very concentrated, play an important role, for they are relatively stable. Slopes of savanna regions also contribute solutions during the rainy season when the soil is rich in CO_2, the water is abundant and when reducing conditions prevail, in short, when circumstances are similar to those that permanently reign in the rainforest.

2. An environment of *transportation* during which the solutions remain stable. The original conditions that enabled their elaboration should therefore remain approximately unchanged. River floodplains on which vegetation provides both CO_2 and rotting organic matter are favourable. Rapids, on which the waters become mixed with air, may produce a certain amount of iron precipitation, which results in a patina and coatings of which we have already spoken.

3. An environment of *accumulation* that must not only enable deposition but also fixation. Several physicochemical processes may take place in such circumstances: adsorption by active surfaces, such as clay laminae, which thoroughly fix iron oxides; oxidation of bivalent ions into trivalent ions in an environment rich in oxygen, i.e., practically, at the contact of air; destruction of organic compounds; flocculation of colloidal hydroxides; and dehydration of hydroxides followed by crystallisation. At the foot of slopes the water of springlets loses part of its carbon dioxide and some oxidation takes place, causing the precipitation of iron. Maignien (1958) thus explains the frequent basal thickening of scarp producing cuirasses with springlets at the contact of the less permeable underlying regolith, restoring the water that infiltrated the cuirass.

Cuirassing conditions are particularly favourable on lowlands subject to alternating floods and low watertables, which are common in savanna regions. At high water the waters saturate the regolith in which a reducing, anaerobic environment is formed. Gley phenomena appear. Biologic activity produces carbon dioxide and organic acids. Solutions are stable and easily penetrate the soil. During the dry season conditions are completely reversed. Water evaporates and air enters the soil. Oxidation takes place. Organic acids as well as organic compounds decompose. Carbon dioxide becomes rare and its degree of concentration approaches that of the air. The various processes enumerated by d'Hoore come into action and precipitate the iron oxides. The gleys that form in the ground water during the dry season grade upward into cuirassing phenomena in the zone of the fluctuating watertable, saturated at high water and dry at low water. Here forms what is called a *ground water cuirass*.

The two types of cuirasses each have their own geomorphic significance:

The authigenic cuirasses of relative accumulation correspond to landforms where chemical weathering makes possible the *in situ* concentration of the less soluble elements, principally gibbsite. They sometimes are caused by an on the spot immobilisation of solutions due to an interruption of the bioclimatic equilibrium. We will mention them again in connection with palaeoclimatic influences. They always develop on reliefs that are sufficiently high so that they are well drained. They are found on the interfluves of dissected regions.

The allogenic cuirasses of absolute accumulation owe their existence to mechanisms of chemical accumulation and result from the influx of hydroxides of iron and manganese in an environment where they cannot remain in solution. They develop in low regions either in a colluvial form at the foot of slopes or in an alluvial form on floodplains and marshy bottomlands. They represent in a chemical form the dynamics of a depositional relief. Although they may appear in embryonic form in the forested zone, they normally develop under the influence of a contrasted seasonal rainfall and are thus characteristic of the savannas. In drier environments the humidity no longer permits a sufficient mobility of the iron for its elaboration, and in the Sahelian zone they give way to patinas or desert varnish.

Cuirasses thus have through their genesis an invaluable geomorphic significance. But they also constitute a material that is attacked by erosion. For this reason it is necessary to study their characteristics from this double point of view.

Characteristics of cuirasses

The concentration of minerals such as alumina or the oxides of iron and manganese in the superficial layers of the regolith is, to be sure, an indispensable condition for the formation of a cuirass. However, it is not the only condition as d'Hoore (1954a) and Maignien (1958), the two authors who have made the most important recent contributions relative to this question, have pointed out. The induration of the material must indeed still take place.

As Maignien (1958) has shown, ferrallitisation and cuirassing do not necessarily derive the one from the other; it is necessary to separate the two phenomena. Every time sesquioxides are isolated, migrate, and become concentrated cuirassing may result. The ferrallitic environment is favourable to cuirassing only in so far as it liberates sufficient quantities of sesquioxides. All ferrallitic soils are not necessarily cuirassed, and some cuirasses form outside ferrallitic soils, notably under hydromorphic conditions (periodically inundated valley bottoms). We must, therefore, now study the conditions of induration, which play an essential role in the facies of cuirasses.

THE ROLE OF CLAYS

Clays constitute an environment unfavourable to the cuirassing mechanism although they exert a great attraction on hydroxide molecules, which tend to fix themselves in films on the argillaceous crystals (adsorption). In a rich clayey soil iron oxides are easily retained, adsorbed by the clays, which they colour. Only a small proportion of iron, 3 to 4 per cent for example, is sufficient to cause bright colours. But proportions of 10 to 15 per cent are possible in argillaceous soils. When this happens the deflocculated clays lose certain of their properties, notably their plasticity. But there is no induration. A high proportion of iron oxides is typical of the red clays of kaolinitic profiles under forest.

Because of local influences scattered accumulations of iron oxides form in certain places in a clayey soil where solutions are more common or where a more intense desiccation causes their precipitation. When the environment is perhumid the positions of old roots are often marked by bleaching due to dissolution of iron by organic acids produced during the putrefaction of the roots. When, on the contrary, the environment is alternating wet and dry, living roots cause a desiccation through water extraction, while dead roots leave behind holes in which air circulates. Sometimes concretions form in sandy spots where there is more permeability, at other times around roots, and are then tubular. The latter are commonly composed of successive coatings, which demonstrate a genesis in the confined atmosphere of a cavity, such as the laminated coatings on cave walls. Microstalactites, microfolds and draperies formed by migrating drops of water that seep through the soil during the rainy season and then evaporate, often between two showers, have occasionally been observed. The surface of such coatings is shiny. Sometimes the centre of the tube-shaped structures is filled with a less consolidated, sandlike, heterometric material. Fissures, especially desiccation cracks, which are never very wide because of the nature of the predominantly kaolinitic clays, are also avenues through which solutions and air penetrate. Their walls become coated and indurated with iron oxides, which prevent them from being closed and facilitate their reopening at the same place, more or less as in the formation of periglacial ice wedges. Fissures are the starting point of epigenic phenomena through the action of hydroxides.

Depending on the abundance of hydroxides and the type of climate and vegetation, clays include a series of epigenic phenomena:

(*a*) A simple deflocculation by hydroxides, which dehydrate and become reddish. The mechanical properties of the clays are then modified: there is a decrease in plasticity and resistance to splash erosion. This type normally forms under rainforest.

(*b*) The formation of concretions and ferruginous pisolitic pellets with a core commonly formed by a few grains of quartz that have served as a nucleus to the concretionary iron (high affinity of silica for iron). Such concretions, which are very resistant, are found back in alluvial formations or accumulate on the ground surface as a result of sheet erosion caused by overland flow. In France they can be found as relicts in many regions, some dating from the Tertiary (iron ores of Barrois and Alsace), others from the early Quarternary. Tropical deciduous seasonal forests and savannas are excellent environments for their formation.

(*c*) The formation of an indurated network of veins and tubular structures in a mass of deflocculated clays with concretions along fissures, roots and abandoned termite galleries (Tessier, 1959b). Clay cores, owing to their low permeability, can subsist quite a long time in the midst of the tangle, which is more or less dense depending on the stage of evolution. Such a facies requires an abundant allogenic influx of hydroxides and stability of the geomorphic relationships. As far as we are concerned, we have encountered such conditions only on level surfaces, notably on old argillaceous alluvial formations now forming terraces and which must still have been periodically flooded during the induration of the network.

The research of Australians (Turton *et al.*, 1962) has demonstrated that ferruginous concretions sometimes contain unweathered minerals and ions of phosphorus, molybdenum, vanadium, and gallium that normally should have been leached. It seems to us that this implies a relatively small degree of weathering in the hostal environment and constitutes a solid argument for an allogenic origin of the concretionised iron oxides.

Sometimes a soil rich in kaolinite with more or less gibbsite remains unconsolidated as long as it does not dry up. Dug up and exposed to the air it indurates and can serve to make poor building stones. Such were certain laterites studied by Buchanan in India. This kind of induration results from precipitation of solutions due to evaporation of interstitial water, the salts forming a continuous but more or less consolidated framework. Such observations have accredited the legend of rapid 'laterisation' of deforested regions. In nature a rapid elimination of the forest can cause a hardening of the soil, which becomes more impermeable and increases erosion, but on the human time scale it cannot involve the formation of a cuirass.

THE ROLE OF SANDS AND GRAVELS

Surface deposits that are more permeable than clays—homometric, such as fluvial sands and gravels; heterometric, such as slope deposits with rubble in a loamy matrix—are often cuirassed, and always by iron oxides. The higher the permeability, the more abundant must be the supply of ferruginous cement. In the rubble of some terraces, as those of Ségou, and in some ferruginous sandstones its proportion approaches 30 per cent. An abundant supply is therefore necessary. It occurs in depressions that benefit from appreciable inflow of allogenic solutions, or at the foot of highlands that provide substances extremely rich in iron, such as the itabirites of the Brazilian Ferriferous Quadrilateral (near Belo Horizonte). The time factor tied to a great geomorphic stability can to a certain degree compensate for abundance of supply, as seems to apply at the base of Mt Nimba (Guinea).

The importance of allogenic inflow of iron oxides is demonstrated by the high frequency of cuirassing phenomena in detrital formations not capable of liberating much iron themselves. Typical in this regard are certain cuirasses observed along the Niger (Tricart, 1959a) and which are the result of the cementation of pure siliceous gravels (quartz and quartzites of Ségou) or quartz sands (of Bourem and Gao).

The great affinity of silica for iron explains certain aspects of the cuirassing mechanism in a coarse clastic environment, which may be compared to the formation of concretions in an argillaceous environment. Often, moreover, cuirassing takes place in a regolith of intermediate texture, containing

182

gravel and sand, as well as clay, as in the case of certain colluvial deposits, or a mixture of sand and clay only, as commonly occurs in lowlands subject to seasonal flooding.

In sandy loams iron oxides affix themselves to quartz grains. These become ferruginised, first where fissured and in re-entrants that result from the corrosion of impurities or pockets. For example, quartz fragments originating in schists in which the quartz was injected in the form of irregular veins, branching and even almost completely engulfing them (forming pockets), become ferruginised first in the re-entrants, which then increase in size. Such ferrugination may result in a fragmentation of the quartz, whether pebble or sand grain, in any event, in its debilitation. A weak renewal of dissolution of iron, for example in the presence of organic solutions, is all that is necessary to clean up and enlarge the ferruginised fissures and re-entrants and transform them into points of rupture. A binocular examination of these phenomena enables the observer to identify current or past actions: intact or more or less corroded coatings and impregnations.

In pure sands where iron oxides are few, plants produce a soil-forming process that remobilises the iron and at the same time causes:

(*a*) the formation of silt-size quartz fragments, especially close to the ground surface. They fill most of the voids between the larger grains and thus decrease permeability, permitting overland flow. This is the case, for example, of the palaeosols on the oldest dunes of lower Senegal. A slight colluviation softens their forms when showers are violent. We have even observed gullies on the confines of the Sahelian and the subarid zones near Nouakchott and Boutilimit, Mauritania. When this fine dust is rather abundant, sands have a notably red colour.

(*b*) the fixation of a red, somewhat hydrated iron oxide on the quartz grains. It forms an irregular coating, thicker in re-entrants, thinner on salients and sometimes interrupted in places. It colours the sand mass. As the thickness of the coating does not vary much in relation to grain size, it is all the more abundant as the grains are small; for this reason finer sands are more highly coloured, sometimes salmon-red rather than orange-red or pink on coarser sands.

In the Senegal delta such red sands, which we have studied in detail, are palaeosols, 2 to 3 m (7 to 10 ft) thick, formed under a more humid climate of Sudanese type. Recent research carried on further south by Maignien (1961) has confirmed our views, and this author proposes a forest environment to account for the origin of the red soils to the south of the Gambia, which are a southern extension of the red sands of the Senegal delta. Under savannas iron fixation sometimes proceeds under a lighter, brown or brownish horizon where the organic matter causes a slight leaching.

When iron oxides become more abundant ferruginous sandstone concretions start to form. In loamy soils they preferably form in small sandy accumulations that are more permeable and in which microphreatic lenses develop in the rainy season. On breakage one often discovers that the centre of the concretions consists of pure ferruginous sandstone, frequently marked by a violet tint caused by manganese oxide. Their outer surfaces are smooth and shiny, sometimes with a few large quartz grains sticking through, clean as if they had been polished. In that case the outer surface is composed of films of iron oxide formed in a more argillaceous, almost homogeneous, environment. The mixing of alluvial sands and clays often results in isolated sandy accumulations, which are at the origin of such concretions. A few centimetres in diameter, they constitute fine or medium textured *ferruginous gravels* larger than the pisolites that form in a more purely argillaceous environment. In purer sandy soils ferruginous sandstone occurs in the form of more ordinary aggregates whose surfaces are rough, hardening progressively with the development of a kind of patina under the effect of exposure to air. Such aggregates are often irregular with a scoriaceous look (*scoriaceous aggregates*). Sand grains can still be distinguished clearly although sometimes corroded by the ferruginous environment; from the petrographic point of view they are in fact ferruginous quartzites. Near Bourem, Mali, such aggregates have a diameter of 1 to 2 cm (0·4 to 0·8 in).

If the iron oxides are even more abundant the aggregates coalesce, forming a *scoriaceous* or *nodular cuirass* if the accretionary bodies are nodular. Such cuirasses remain very permeable in their initial stages. Concretions coalesce, leaving the interstices between them partly filled with loose sand or smaller, free concretions. Later, as concretions continue to form, the remaining voids are gradually filled with smooth and shiny coats of iron oxide, recalling those of caverns, and which form superimposed laminae sometimes 1 cm (0·4 in) thick. The sandy or loamy pockets are also partially transformed into ferruginous sandstones. But some always persist, holding loose materials in cavities that have been entirely closed through the consolidation of their surroundings. In this way is formed a *vesicular scoriaceous cuirass*. Generally, such scoriaceous cuirasses either form layers a few decimetres thick (several inches to a few feet), seldom 1 to 3 m (3 to 10 ft), or the degrees of consolidation vary from place to place. At other times one only finds isolated blocks, half a metre to several metres in length (2 to 10 ft or more), sometimes closely spaced, in groups, separated by non-indurated veins, a few centimetres or 1 to 2 decimetres wide (1 to 8 in). If such blocks occasionally result from a breaking up of old cuirasses, they also frequently are perfectly in place and mark a small area where cuirassing has been more favourable for divers reasons. In this last case their margins are generally less frank.

Such cuirasses may not only cement scoriaceous concretions formed *in situ* but also fragments of former cuirasses. They are then *secondary cuirasses*, as distinct from *primary cuirasses* formed from locally developed concretions. They may also be composed of angular rubble, forming *brecciaceous cuirasses*, which are common at the foot of slopes that have a thin regolith or none at all. If they contain pebbles or cobbles they are *conglomeratic cuirasses*. In the same way one can also speak of *sandstone cuirasses* if the original sand can

FIG. 4.2. Nubbly conglomeratic cuirass on the calvaire de Popenguine, Senegal

Ferruginous pebbles, less than 2 cm in diameter, sometimes enrobed in concentric hydroxide coatings; ferruginous concretions and pisolites embedded in a ferruginous cement, sometimes producing the effect of nubbly crusts.

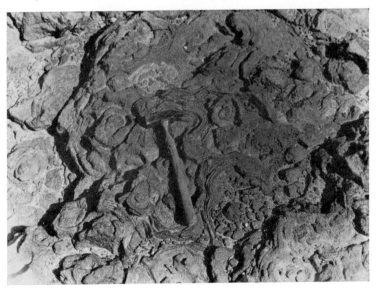

FIG. 4.3. Ferruginous cuirass on bedrock (sandstone) at Tosaye Rapids on the Niger, Mali

Series of laminae in nodular arrangement forming bundles moulded on top of one another. Products of present disintegration, accumulating in the centre of some depressions, may be recemented to form a breccia.

still be recognised but is associated with iron oxide concretions. If concretions are absent it is better to speak of *ferruginous sandstones* in order not to deviate from the habitual terminology used to describe detrital rocks.

In favourable circumstances cuirasses may reach advanced stages of diagenesis. For example, at Ségou on the Niger a Pleistocene terrace entirely composed of siliceous gravel, including cobbles not larger than 5 cm (2 in), is cemented together by iron oxide through a thickness of at least 5 m (16 ft). The surface, down to a depth of 2 or 3 m (7 to 10 ft), is composed of a very compact scoriaceous facies with such small voids that it is analogous to poorly leavened bread. Below, the conglomerate reveals a dissolution of cer-

Fig. 4.4. Very epigenised alluvial cuirass on early Quaternary terrace of the Niger River lowland near San, Mali

Detrital particles can still be seen, especially the moulds of quartz pebbles, but the whole material has been epigenised by hydroxides.
Traces of bedding planes emphasised by harder and more brilliant crusts remain visible. Such a material is extremely resistant.

tain quartzose or quartzitic pebbles, creating voids lined with knobby siliceous coatings occasionally covered by a few 'pearls' and partly filled with a very hard ferruginous quartzite. The pebbles are progressively dissolved and the silica precipitated in association with the iron. They lose their original characteristics, but the primitive detrital silica can still be recognised in the stratified material (lenses, cross-bedding, etc.). Such very advanced epigenic growths correspond to the zone where the fluctuating watertable now only reaches during major floods.

Although diagenesis occasionally destroys the original characteristics of the component parts, ferruginous induration in this way preserves the detrital formations enriched in allogenic solutions as a result of their location in flood zones and occasionally on footslopes.

INDURATIONS IN THE BEDROCK

Sometimes ferruginous solutions penetrate direct into the more or less weathered bedrock. Indurations through epigenesis may then occur. But a large quantity of oxides is necessary, a condition usually not satisfied in the case of cuirasses of relative accumulation in which authigenic solutions precipitate in the form of concretions in the soil itself or are removed by subterranean drainage. The epigenisation of slightly weathered bedrock by iron oxides requires a periodic waterlogging, which is only realised in flood zones. And concretions should not develop in the alluvial or colluvial deposits of such zones, which means that the transported mantle must not be more than a few metres thick, allowing the watertable to drop into the bedrock during the dry season. In this way the bedrock is alternately subjected to weathering, which enables the penetration of solutions, and to epigenisation by iron oxides.

Such conditions are best realised in shales or schists. Commonly deformed, their truncated, weathered laminations facilitate the penetration of solutions. These rocks weather into clays and readily fix the oxides. Epigenesis proceeds into the rock from the bedding or foliation planes, and eventually the rock becomes transformed into a more or less amalgamated iron ore in which the iron oxides are combined to the clay, but in which the rock fabric remains recognisable. Such ferruginised schists transformed into *epigenic cuirasses*, buried under the remains of thin detrital cuirasses, are found locally in West Africa, on Precambrian Birrimian micaschists. Daveau *et al.* (1962) have drawn attention to them and photographed an excellent example in Upper Volta.

In certain formations very rich in iron, such as the quartzites of Mt Nimba and the itabirites of the Ferriferous Quadrilateral, such epigenic cuirasses are common. They constitute very high grade iron ores. Epigenic cuirasses may form anywhere as long as highly ferruginous rocks are present: ordinary pedogenesis is all that is required. It produces enough iron oxides to tint seepages. In these circumstances the evaporation of subsurface water during the dry season causes the precipitation of the oxides, which cement together the rocks. This remobilisation of the iron is here the elementary and initial form of weathering. It may even cuirass abrupt slopes (50 to 60° in certain areas of the Ferriferous Quadrilateral). Such cuirasses are usually composed of two superposed horizons: at the top a few decimetres (4 to 20 in) of a brecciaceous kind of cuirass, in which the more or less epigenised slope rubble can still be recognised, and below, the bedrock epigenised by iron oxide. These are, of course, exceptional cases.

Ferruginations that have some connection with cuirassing may occasionally be observed in sandstones, as, for example, in the *Continental Terminal* in the area of Bourem, Mali. But conditions are less favourable to the pene-

FIG. 4.5. Piedmont slope cuirass on the back slope of the Serra do Curral del Rey homocline, near Belo Horizonte, Brazil

Brecciaceous cuirass containing slightly rounded itabirite boulders. Very high degree of cementation due to abundance of iron hydroxides, becoming irregular towards the base and reaching a thickness of 2 to 3 m (7 to 10 ft). It tends to form pockets due to the irregular penetration of the solutions. Not to be confused with pockets due to geliturbation.

tration of solutions than in schists. Usually one only finds laminae of ferruginous quartzite on the rock surface or along joints. As similar phenomena may occur in a marine environment (hard grounds), their palaeogeographic interpretation may present some risks.

Only authigenic cuirasses are really a part of the pedogenic process. The other cuirasses, exactly like the calcareous crusts of dry regions, belong to the geomorphic realm. They have only distant connections with pedogenesis: the origin of the iron-bearing solutions. Their precipitation enters into the framework of landforms of chemical accumulation associated, in nearly all cases, with the accumulation of products of mechanical erosion. Cuirassing at the foot of slopes is part of the process of colluviation, representing its chemical component. For this reason iron oxides cement together the rubble or loams coming from the same hillside. The cuirassing of alluvial formations fits into the framework of the evolution of valleys. Because of it iron oxides cement together an alluvium of varied texture in valley bottoms: sands, gravels and mixtures resulting from the mixing of sands and clays on floodplains covered with vegetation. For historical reasons pedologists were the first to undertake the study of cuirasses. Pedologic concepts, however, have often misled them and caused them to believe in upward movements of solutions under the effect of evaporation, an error which Maignien (1958a) has discussed fully. Cuirasses must be studied with petrographic methods, which have enabled the understanding of the genesis of the detrital deposits, and with geomorphological methods of reasoning, as cuirasses are integrated with the landforms, a fact well understood by Maignien in spite of an initial training as a pedologist. Geomorphologists were the first to insist on the importance of the detrital materials of cuirasses: Rougerie discovered their significance in the period of 1946 to 1951, and since 1953 Lamotte and Rougerie have clearly expressed this idea; we also have adopted it, beginning with our first visits to West Africa, and it has been subsequently developed by Vogt and Michel. Cuirasses, therefore, are extremely important milestones in the evolution of landforms. After having studied their geomorphic role, we return to them in connection with palaeoclimatic inheritances (Chapter 5).

In order to understand the lithologic role of cuirasses in the evolution of landforms we must first examine the interplay of other geomorphic processes operating in the alternating wet–dry tropics.

Other morphogenic processes

If the cuirassing mechanism is the most important specific aspect of the morphogenic processes of the alternating wet–dry tropics, it is not the only one. Cuirasses also evolve, change and break up. Runoff plays an important role in the fashioning of slopes; it is not only dependent on climate but on other morphogenic processes. While eroding cuirasses it also influences their evolution. Finally termites, which have only a limited morphogenic influence in forests, reach their greatest development in this type of tropical climate with a marked dry season. The African savannas are the domain of giant termitaries and of the most evolved and active termites. They mix the soil energetically and in this way intervene in the geomorphic evolution.

We begin our analysis with the termites, as they determine the other processes more than they are dependent on them; we later examine the relationship between soils and sheet erosion and, lastly, study the evolution of cuirasses.

The geomorphic importance of termites

Although termites are well known to naturalists, who have written voluminously about them, their geomorphic influence has hardly been studied. A pioneer work, that of de Heinzelin (1955), was necessary to attract attention to them, followed by a note of Taltasse (1957); summations have been published in the *Revue de Géomorphologie dynamique* by Grassé and Noirot (1959) and by Boyer (1959).

In the forest termites are satisfied to consume plant remains and do not build large earth mounds. In the savannas and campos cerrados conditions are different. In Africa cathedral-like termitaries reach heights of 2 to 4 m (7 to 13 ft) and volumes of 2 to 4 cu m (35 to 70 cu ft). In Brazil shapeless termite mounds scattered through the deciduous seasonal forest like ordinary heaps of earth have identical volumes. In some regions, like northeast of Andarai, they have an average spacing of 15 m (50 ft). They create a chaotic relief of hummocks in quincuncial order, 1 to 2 m (3 to 6 ft) high and with a diameter of 4 to 6 m (13 to 20 ft), in which all stages of earth spreading due to rainwash after abandonment of the nests may be observed. Smaller termitaries, also in mounds about 1 m (3 ft) high and with diameters of 2 to 3 m (7 to 10 ft) and an average spacing of 7 m (23 ft), giving the impression of a public dumping ground where trucks have carelessly unloaded their loads of dirt, are found in the thorn scrub of Jequié and Palmeiras (Bahia), which have a rainfall of 700 to 800 mm (21 to 24 in). Their volume is about 2 to 3 cu m (35 to 52 cu ft) in the average. Near Lumumbashi they cover up to 7·8 per cent of the ground surface (Sys, 1955). In Kivu de Heinzelin (1955) has estimated at 1 m (3 ft) per 1 000 years the thickness of earth mixed by termites, which is a considerable average in comparison to other morphogenic processes, especially on level surfaces. According to our own observations the distribution of large termite mounds is very irregular, especially in Brazil. In some regions they may be densely spaced for several square kilometres, whereas in others there are few or none at all. Sometimes there is some relation with the nature of the underlying rock. The huge area of mounds located to the northeast and east of Andarai, Bahia, which measures some 30 by 40 km (19 by 25 miles) coincides with partly silicified Palaeozoic limestones overlain by thick, weathered red silty clays. The termitaries on them are both very large and densely spaced. At Palmeiras they are found on proglacial Eocambrian siltstones sometimes ferruginised into practically coalescing scoriaceous nodules, whereas at Jequé they are found on arenised gneiss.

In West Africa termitaries do not reach the Sahelian zone: their farthest advance is near Diré, Mali, on old silty levees of the Niger. At Ségou, San and Bla they are densely spaced. Aeolian sands and floodplains do not have any. Large mounds occur where surface deposits are sufficiently thick, silty, not too argillaceous, and rich in fine sands to provide the materials for their construction. They may even be found on a regolith hardened by a cuirass on condition that the latter is neither too compact nor too thick. The Guinean *bowé* are inhospitable to them. They are also absent from the cuirassed terrace of Ségou but common on the surrounding plateaus, where vesicular cuirasses are only $\frac{1}{2}$ to 1 m thick. One can find one about every

FIG. 4.6. Cathedral-like termitary near
Youkounkoun, Guinea (drawn by Pierre
Tricart from photograph by E. de
Chetelat, 1938)
Height 6 m (20 ft). Woodland savanna
vegetation; a magnificent tree grows in
the termitary.

20 to 50 m (50 to 150 ft) (Tricart, 1957). As Boyer (1959) has shown, termites need clay for construction and a good supply of water to keep humid
the confined atmosphere of their nests. Besides food resources these factors
determine their distribution.

The work of Sys (1955) and later that of Grassé (1957, 1959) and his co-
workers have demonstrated that termites affect the soil-forming processes
locally. The clays they extract, sometimes from deep layers, to construct
their nests (which then have a different colour from that of the ground
surface) are softened with saliva, mixed with dejecta and sand grains. They
are thus stabilised, and the mound becomes resistant. Several years of abandonment are needed once the organic matter is decomposed to destroy the
aggregates and for rainwash to spread the materials, as we have been able
to observe in Mali and in the state of Bahia, Brazil. The deposits of organic

matter and the humid atmosphere in the mound normally impede cuirassing as long as the nest is inhabited. Waterlogging on a microscopic scale may even occur at the base with nodulation of calcium carbonate, according to Boyer (1959). But once the nest is abandoned the greater permeability of the soil promotes increased desiccation and consequently the development of cuirassing phenomena. If as a consequence of local relief a submergence with waters containing hydroxides occurs, the unfilled and uncollapsed cavities favour infiltration and may become filled with coatings in the same way as the openings of old roots. The networks then become

FIG. 4.7. Termitary on ferruginous gravels north of Bamako, Mali
Level surface on which rainwash and aeolian deflation have left behind ferruginous gravels formed by soil concretions. They form a pavement on top of a soil which is very poorly protected by a sparse vegetation whose wide spacing is in large part due to poor edaphic conditions. The small termitary is composed of fine materials brought up from under the pavement. Once abandoned, it will be destroyed and its products scattered. The pavement will be reinforced in this case.

permanently marked in the structure of the cuirass. This, according to Tessier, is the origin of certain cavities in cuirasses of the neighbourhood of Dakar and, according to Taltasse (1957), of the cavities he has described in the cuirasses of the northeast of Brazil. In both cases these phenomena date back at least to the early Pleistocene, probably to the Tertiary. Termite specialists find these cavities different from the ones made by known species and attribute them to other phenomena, for example, burrows of other animals. But one may very well accept that termites have evolved, like other animals, and that the Tertiary species were not necessarily the same as the present ones and produced a different architecture. However that may be, it is not important. For us the consequences are the same, and the actions of various burrowing animals add themselves to those of the termites.

Termites and other insects play an important role in the evolution of cuirasses, as we have observed in Mali. Where the crusts are not too great an obstacle, the insects raise clay and grains of sand through their openings. The cuirass may thus be undermined, producing some subsidence; there is perhaps also a little corrosion because of water percolation through the structure. The accumulation of fine materials with good structure after the abandonment of the termitaries is favourable to plant life; shrubs commonly colonise old termite mound sites. Moreover, they are always of the same species, sometimes evergreen. Their roots penetrate through the cuirass and probably contribute to its breakup. Such patches of two or three shrubs often contrast with the bare surface of the surrounding cuirass. At the base trunks commonly have a diameter of 20 cm (8 in), which should effectively help the dismantling. The shrubs protect the tumulus from splash erosion and prevent the materials from spreading by surface wash.

But shrubs do not always occupy the sites of old termitaries. For example in Brazil we have observed bare tumuli washed by the rain, which slowly spreads the debris and progressively produces flatter cones. They are finally recolonised by shrubs before disappearing completely. The same may occur in Africa. After a few years all that is left of some large cathedral-like mounds are cones 30 cm (1 ft) high with a radius of 3 to 4 m (10 to 13 ft).

De Heinzelin (1955) maintains that termitaries play an important role in the formation of superficial sheets of fine, yellowish soil which cover certain gravel beds in Kivu. Mesolithic artifacts found at their base indicate a recent genesis. Their age and the depth at which they are found coincide with the rate of upward movement of the earth by the termites. Such an evolution may, moreover, very well be placed within the framework of a climatic oscillation; the fine soil cover corresponding to the present climate, and the deposition of the underlying gravel to a drier climate, the gravel representing the riverbed of an intermittent stream. Near Andarai there is an obvious decolorising of the original yellowish material, which is explained by the dissolution of iron owing to the conveyance of organic matter by the termites. The activities of termites thus produce a convergence with the formation of yellow soils under more humid climates. This causes an added difficulty in palaeoclimatic interpretations and demonstrates the importance of interdisciplinary co-operation between termite specialists, pedologists and geomorphologists.

The role, then, of burrowing insects and more particularly of termites, which are the most active of them, is considerable in the alternating wet–dry tropics. It is characterised by impressive constructions made by the upward conveyance of fine earth, which may modify the evolution of cuirasses, and, later, during the destructive phase facilitate the downward migration of material and colluviation.

Pedogenesis and sheet erosion

The particular conditions of pedogenesis in the wet–dry tropics favour overland flow. For example, in Senegal Maignien (1961) has calculated that the percentage of clay at the ground surface essential for the stability of aggregates is only 3 to 10 per cent in beige tropical soils as against 12·5 per cent in red palaeosols under forest. The beige soils, which are typical of savannas and have a tendency to become cuirassed, clog faster on the surface. Red soils are practically never waterlogged during the wet season. This difference of behaviour, which is lessened by the violence of the showers falling on a dried forest soil at the end of a short dry season, has repercussions on the landforms. Maignien notes that the present beige soils with ferruginous pellets develop on long piedmont slopes, whereas a relief of hills and ridges is preserved in regions dissected in red palaeosols. The piedmont slopes, of course, correspond to more intense rillwash, in keeping with the properties of the soils.

The action of man, which has made itself felt for centuries in the African savannas, favours overland flow by making the soil harder through increased desiccation (depleted plant cover) and recurrent brush fires. Aeolian deflation, which affects only the rather scarce silty particles, must be added to these in areas swept by fire and where the plant cover is sufficiently open. It has as yet been insufficiently studied but reflected in a kind of dry haze that does not only consist of smoke. Some rains precipitate mineral dusts that originate in red tropical soils. Due to the vegetative cycle, overland flow is especially active at the beginning of the wet season. It depends to a great extent on the nature of the regolith, but the rapid decomposition of the humus and the poor structure of the savanna soils increase it. It always forms during violent showers. Measurements of the ORSTOM on experimental basins have demonstrated at Tiémoro near Kankan, Guinea, in a slightly dissected cuirassed basin with intensively cultivated argillaceous slopes, a volume of overland flow about equal to that of subterranean drainage. But this runoff occurs only during showers that drop more than 20 mm (0·8 in), even on saturated soil. In the basin of the Flakoho River (50 sq km : 20 sq miles) near Ferkessédougou in the north of Ivory Coast, 20 to 25 per cent of the rainfall is drained as overland flow during downpours of 90 to 100 mm (3·6 to 4 in), which occur nearly every year. During August, in the midst of the rainy season, this percentage goes up to 30 per cent. During average rains, however, overland flow is only 2 to 10 per cent of the total water dropped. It is true that dissected cuirasses cover a large area and feed springs that flow even during the dry season. But the concentration of floods in savannas is about three times as swift as in forests. This is an important fact, for it determines the intensity of the mechanical morphogenic processes. Larger rills are stronger and their greater competence enables them to attack the soil, to remove clays and, on fairly long slopes, even sand and ferruginous pellets, which indicate sheet erosion if they are concentrated on the ground surface. This happens frequently: the heavier pellets, which are removed with greater difficulty, are left behind by the finer fractions and accumulate in a layer that gives some idea of the amount of soil removed. In *marigots* near Siguiri, Guinea, Pelissier and Rougerie (1953) have shown that in the midst of the Sudanese zone the high water alluvium forming the levees is silty, whereas the river bedload is composed of coarse sands, 1 to 2 mm in diameter. On steeper slopes, 10 to 20° or more, debris of ferruginous rubble is common when cuirassing processes have affected the interfluves. At the foot of cuirass margins ferruginous gravel and rubble up to 3 and 6 cm (1·2 and 2·4 in) in diameter commonly forms a layer approximately 10 cm (4 in) thick, washed clean of its finer fractions and exposed to alternations of wetting and drying, which hardens the stones and give them a violaceous, blackish tint and a varnished appearance. They slide slowly through undermining. When their density is sufficient they finally block runoff and the slope then evolves as a humid scree, together with some undermining and sliding lubricated by showers. The gradient increases towards 30° and the slope then retreats slowly, parallel to itself. Ferruginous sandstones cause analogous forms, which can be observed in the much degraded anthropic savannas of Camaçari near Salvador, Bahia.

All authors who have described savanna regions insist on the morphogenic importance of overland flow, which of course is most important on medium and steep slopes (10° and more) where soils are thin and stony. Here the accumulation of coarse debris is general, as on inselbergs. Ruxton (1958) has demonstrated in the Sudan near the Ethiopian border, at the margin of the Sahelian zone, the relationship between overland flow and the weathering of biotite, microcline and other feldspars. These minerals drop from a content of 72 per cent in the bedrock to 15 per cent on the surface of the residual mantle. Inselbergs owe their forms to the combined actions of weathering and runoff. Waters that flow down steep hillsides partly infiltrate at the base, favouring weathering at this level, as Rougerie (1961) has shown for the region of Beyla, Guinea. When the waters spread on piedmont slopes they evaporate, enabling the precipitation of iron and in some cases the formation of laminated cuirasses composed of smooth, undulated and superimposed laminae. It is also the action of runoff that gives savanna amphitheatres a different form from those developed under forest, as on the sands of the Batéké Plateau. Here Sautter (1951) has shown that their floor is occupied by a very flat but rather steeply inclined sandy cone (9 to 18°) extending progressively backward as the walls, which provide the sands, retreat, eventually eliminating the amphitheatre itself. The amphitheatre apparently can develop properly only under forest, thanks to the undermining effect of underground drainage. Under a savanna runoff would gradually transform it into a regular valley.

In Madagascar certain red soils are particularly impermeable. Segalen (1948) has described some that are grass covered and probably were degraded a long time ago. Below a superficial crust 1 to 2 cm thick, occasionally humic, they have a compact red horizon rich in clay hardened by desiccation and the precipitation of gels of iron and silica. After several days of rain these soils are wet down to a few centimetres only. In Brazil Ruellan (1953) has given an excellent analysis of the effects of heavy showers in the interior, which receives 500 to 1 100 mm (20 to 44 in) of rain from November to March and only 25 to 150 mm (1 to 6 in) from July to September, and which is covered by campo limpo merging into caatinga in the driest parts. The rills that form where the ground is bare at the beginning of the torrential rains coalesce and form a sheet flood at the heaviest of the downpour, after which it breaks up again into rills at the waning of the storm. The ground is thus thoroughly swept and smoothed, the fines washed away and the depressions filled with standing water. Cuirasses or an impermeable substratum favour such phenomena. But the runoff remains unconcentrated only on gentle or stony slopes. On steeper slopes the vegetation is no longer able to prevent its concentration and gullies appear.

The campos cerrados and deciduous seasonal forests are much less favourable to overland flow, which restricts its effects to colluviation. Such is the Andarai region of Bahia with a rainfall close to 1 000 mm (40 in). On argillaceous soils abandoned termite mounds supply most of the material. In a region of sandstones in the west of Bahia with a rainfall of 927 mm (37 in) a study by Bramao and Black (1955) emphasises the contrast between plateaus and valley sides. On the plateaus yellow, orange and reddish sandy ferrallitic soils attain a thickness of 5 to 7 m (16 to 23 ft). A little humus on the surface facilitates infiltration. It causes the leaching of clays, which increase downward and occasionally produce a watertable at a depth of about 4 m (13 ft). This provides the plants with a fairly good water supply, ensuring the development of the campo cerrado. The morphogenic processes are slowed down, giving an extra lease of life to the plateaus. On the slopes, on the other hand, soils are thin, at the most 1 m (3 ft), reddish or red-brown and rather clayey. They can only support a caatinga as a result of their restricted water supply and are subject to rillwash.

If overland flow plays a more important role than under forest, it none the less remains subordinate to weathering, which provides it in very unequal quantities with the materials that it transports. Lithology and steepness of slope intervene and sometimes produce great contrasts, as in the area of Barreiras (Bahia) or in the Katanga, studied by Lefèvre (1953).

On cuirasses runoff is more intense and, because of it, influences their evolution.

The evolution of cuirasses

Morphogenic processes continue to alter cuirasses after their formation. Termites and other burrowing animals help to indurate or to dismantle them, depending on the circumstances. Not very permeable, they are rather easily bared, like genuine structural surfaces or the calcareous crusts of dry regions, under the combined effect of a certain amount of aeolian deflation and runoff, which are very intense under a sparse plant cover. The more massive the cuirass and the closer it is to the surface, the more unfavourable the environment is for plant growth, which helps erosion. There is then, here again, a chain reaction that tends to produce cuirassed plateaus scantily occupied by clustered bushes or shrubs, often growing on the sites of former termitaries, or covered by sparse and poor grasses during a later stage of evolution. They are the so-called *bowé* (singular: *bowal*). By definition a bowal is a plateau whose margins dominate entrenched canyons or neighbouring depressions. We now turn to the evolution of these two fundamental elements of tropical relief: the bowé surfaces and their margins.

Evolution of bowé surfaces

The very concept of a bowal is tied to its geomorphic evolution. It implies first of all the genesis of a cuirass and later its dissection through lowering of the base-level. In this way the hydromorphic phenomena that play such an important role in the cuirassing process are eliminated and the regolith is well drained. Two series of changes then take place simultaneously:

1. Sheet erosion through overland flow, which progressively removes the friable superficial horizons. The best cuirassing conditions are obtained, as we have seen, in an argillaceous, sandy overburden that is sufficiently but not too permeable. The surface of such materials becomes puddled and, insufficiently protected by vegetation, is subject to severe splash erosion during heavy showers. The bowal surface reveals rills, even on very gentle slopes. Its small irregularities are exploited, protuberances reduced to rock knobs, while mud deposits, frequently hardened by desiccation, accumulate in hollows. There is thus a slow degradation accompanied by a gradual

Fig. 4.8. Bowal surface near Négaré, Guinea (drawing by P. Tricart after a photograph by E. de Chetelat, 1938)
Smooth almost bare surface, slightly rolling, covered with scattered blocks. The concentration of water in the vale in the foreground permits the incision of trenches.

197

infiltration and entrainment of the finer fractions into the cavities underneath the mud deposits. As cuirasses often have appreciable slopes, this degradation prepares their denudation together with a regressive evolution of the vegetation, which accelerates the process. The resulting landform is a cuirassed pavement, like a genuine structural or stripped surface, spotted with residual deposits of fine materials accumulated in occasionally undrained hollows (a few decimetres deep—a foot or so) and covered by sparse vegetation.

2. A hardening of the cuirass through rearrangement and diagenesis. Pedologists, notably d'Hoore (1954) and Maignien (1958), have shown that cuirasses are sensitive to alternations of wetting and drying. They permit a very limited remobilisation of iron oxides, which results in their hardening. Exposed knobs are intensely heated by the sun as a result of their dark colour, and evaporation is high. There is intense desiccation, patinas form rapidly. As d'Hoore has shown, the immobilisation of iron goes through the stage of a highly hydrated and colloidal oxide. Through desiccation and crystallisation it forms coherent films that form a whole weft of impregnation through superposition. The protruding knobs of cuirasses long exposed to the air become darker, a brilliant violet-black. When too salient they disintegrate, perhaps partly due to temperature variations but certainly due to alternations of wetting and drying. There is also a certain amount of corrosion. When they are not too highly cemented, they produce a little rubble. Waters on the surface carry in solution a few hydroxides owing to the presence of a little organic matter chemically provided by the vegetation. But such solutions are very unstable and subject to rapid evaporation caused by considerable temperature variations. The iron, therefore, cannot be moved very far. Penetrating together with the clay into the cavities of

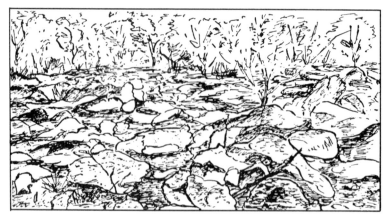

FIG. 4.9. Bowal rubble near Bové, Guinea (drawing by D. Tricart after a photograph by E. de Chetelat, 1938)
A thick cuirass is broken up into blocks, which form a chaotic assemblage. Sparse shrub vegetation.

the cuirass, notably along the tube-shaped openings, it precipitates at a depth of a few decimetres (10 to 20 in). Dissected cuirasses therefore undergo a slow evolution. If there is no outside source of iron, the surface disintegrates into ferruginous rubble, forming a scattering of loose stones. In the mass of the cuirass there is, on the contrary, increasing consolidation through recementation, secondary filling of cavities, hardening, encrusting along conduits, especially in the form of superposed films, or a diffuse impregnation by soil products entrained from above. Living organisms of course intervene in the evolution. Termites help the disintegration and the remobilisation of iron oxides as long as their colony is alive. Once abandoned the termitary instead facilitates oxidation and, dismantled, its galleries allow the penetration of solutions originating in rainwash. Shrubs which grow on bowé usually have strong roots which penetrate and widen the fissures in which they grow. Sometimes the wind blows them down, causing cuirass blocks to be dislodged. It is infrequent, however, because of firm, deep rootage, though reported from Guinea by de Chetelat (1938).

Summital bowé therefore evolve; they are never finished but in perpetual genesis. This evolution is characterised by:

(*a*) a progressive hardening, resulting in highly cemented forms, genuine massive iron ores with a permeability practically limited to fissures. The oldest cuirasses are the hardest. Such old sheets, perched for a long time, have an ability to persist that increases with time. Here is a paradox that is peculiar to the wet–dry tropics; it plays a primordial role in the evolution of landscapes.

(*b*) a slight *in situ* lowering: the cuirass being dismantled on the surface gains in depth to the detriment of the subjacent regolith. Under a detrital cuirass the regolith may become consolidated through ferrugination by the downward penetration of solutions.

(*c*) the acquisition of a gently rolling topography, determined by the degree of compaction of the initial cuirass. The more massive areas, where there is but little infiltration, progressively stand out in relief, forming protruding knobs covered by a scattering of loose rubble. The more permeable zones are transformed into mud flats indurated at a certain depth.

When cuirasses are not in summital locations evolution is different. They are then subject to overland flow originating in uplands such as inselbergs or plateaus, or even the upper part of a cuirassed piedmont slope.[1] If there

---

[1] The term *piedmont slope* in this work refers to all gently sloping lowlands found at the foot of uplands. The origin may be erosional or aggradational or both. When erosional and truncating crystalline rocks, as in certain semiarid climates, such piedmont slopes are called *pediments*. When aggradational they are called *bajadas* or alluvial aprons. When truncating the regolith, as commonly happens in the wet–dry tropics, they are called *pediplains*. English geomorphological terminology does not have an equivalent term for the French *glacis*, which, *sensu lato*, is a piedmont slope. A *glacis d'érosion*, however, is a 'pediment' truncating *sedimentary* rocks (as in certain semiarid regions) or the regolith of any kind of rock but

are enough iron oxides as, for example, at the foot of uplands composed of dark coloured rocks such as basalts, diabases, ferruginous schists, or of highly cuirassed landforms, then the waters as they spread and evaporate after showers precipitate their hydroxides, which cause a superficial induration, however irregular. Certain piedmont slopes thus become covered with laminated cuirasses composed of successive layers. A good plant cover reflecting a sufficiently humid climate on the uplands helps the mechanism, as it favours the dissolution of iron. An evolution of this kind is common at the foot of cuirasses of relative accumulation, especially bauxitic.

FORMS AND EVOLUTION OF BOWÉ MARGINS

Bowé generally terminate in conspicuous escarpments several metres high, 2 to 10 generally (7 to 33 ft), often dominating gentler slopes with an overall concave profile. These slopes cut through the regolith that has formed underneath the bowé whenever the rocks are favourable, such as acidic igneous rocks. In the neighbourhood of Tsaratanana, Madagascar, Brenon (1952) has described such escarpments with a maximum height of 10 m (30 ft) above slopes composed of weathered kaolinitic material, 15 to 50 m (50 to 160 ft) high. Similar forms are also common in the north of Ivory Coast. The lower concave slopes are realised every time the substratum produces a sufficiently thick regolith and therefore are missing on sandstone and, especially, on quartzites, where the bedrock is found immediately below the cuirass. On the margin of the Sahelian zone, in the area of Bandiagara, Mali, Daveau (1959) has described cuirasses that are slightly set back from the escarpment composed of more resistant quartzitic sandstones.

A typical bowal margin is realised every time a cuirass overlies a weathered mantle, which as a matter of fact is a common phenomenon. The escarpment is determined by tunnelling processes (rather than by 'piping', a misnomer).[1] This tunnelling peculiar to cuirasses is caused by differences in permeability between cuirass and friable regolith. Water that infiltrates the cuirass is partially blocked during the rainy season by impermeable layers. Whereas there is a fissural permeability in the cuirass, there is only an irregular textural permeability, or micropermeability in the friable regolith. Rather high in an acidic igneous environment, the textural permeability is much lower in a basic igneous environment, as for example in the clayey regolith on the diabases of the Fouta Djallon. A seasonal waterlogging commonly occurs at the base of the cuirass in the rainy season, forming a ground water mass that is fed by the entire bowal surface and slowly flows toward the margins, obeying the gentle grade of the plateau. The water reappears in springs, which play an important hydrologic role and are a godsend to many villages. In the Flakoho basin, studied by the

usually crystalline in the wet–dry tropics. As the exact nature of such gently sloping plains is not always immediately obvious unless cross-sections are exposed, it is more prudent to speak of piedmont slopes until more details are known (KdeJ).

[1] A pipe is a long hollow cylinder whereas a tunnel is an underground passageway (KdeJ).

FIG. 4.10. Bowal scarp, Bassari Plateau, Guinea (drawing by P. Tricart after a photograph by E. de Chetelat, 1938)

Woodland savanna. Abrupt massive scarp breaking up into blocks, which slide on the subjacent clays. A number of gullies have been cut by the waters of springlets. There appears to have been a succession of climatic phases starting with a dry period with rockfalls and development of the present piedmont slope, followed by a more humid phase during which springlets were more active than they are now.

ORSTOM (1957), near Ferkessédougou, Ivory Coast, they still yield late in the dry season when their waters wane and evaporate at the foot of the cuirassed plateaus.

FIG. 4.11. Cuirass margin breaking up; bowal at Foumbaya, Guinea (drawing by Pierre Tricart after a photograph by E. de Chetelat, 1938)

As a result of settling due to undermining at the foot of the cuirass scarp, the margin of the cuirass is breaking up into quadrangular blocks, which will eventually be reduced to scree. In the background a block is already dislodged.

Springs have a great geomorphic importance and play an important role in the dismantling of bowé, as de Chetelat (1938) was the first to show. The mechanisms are as follows.

The waterlogging of the regolith underneath the cuirass causes, as on some cuesta fronts but without the dip being here unfavourable, a sliding that dismantles the cuirass, breaking it up into blocks, which end by falling and then slowly slide on the subjacent slope, which they clutter. Cleaned by rainwash they pile up and usually form an irregular cover. Under the influence of palaeoclimatic oscillations they are sometimes recemented into very coarse breccias, for their dimensions usually exceed one metre (3 ft). The waterlogging of the regolith is most intense where the flow of water is greatest. The basal slope is therefore unequally affected, resulting in the formation of spring recesses that gradually widen by chain reaction. Bowé margins of a certain age are thus marked by small arcs of circles convex towards the interior of the plateau, intersecting like so many fingernail strokes and corresponding each to a spring.

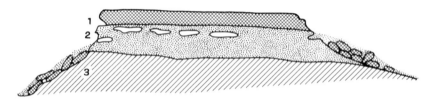

FIG. 4.12. Cuirassed butte of the Bassari Plateau, Guinea (after E. de Chetelat, 1938)

Old cuirassed, slightly inclined piedmont slope in homoclinal attitude. As a result springs are more active on the lower margin where they cause a more frequent sliding of cuirass blocks. In white: caves. 1. Cuirass; 2. Friable regolith with caves; 3. Schists.

Tunnelling helps waterlogging, especially in materials with a high plastic limit. The effect of tunnelling is to produce a subterranean drainage system with tunnels several metres, even 100 m (330 ft) long below the cuirass, with their roofs against it, their floors dug into the regolith, and springs emerging generally at the outlet of caves. The tunnels commonly follow fractures or joints and sometimes form right-angle patterns. Water quickly infiltrates the fissures, becomes concentrated into a current capable of evacuating the mantlerock, and the tunnel thus grows essentially by mechanical means. Draining the groundwater at the base of the cuirass, the water must of course emerge into the open and counterslopes be excluded. Cave-ins may occur, as with ordinary tunnelling. The spring recess in the escarpment then extends itself as a narrow canyon into the bowal margin.

Closed depressions, occupied by mud flats occasionally waterlogged or forming ponds during the rainy season, together with canyons, caves and spring recesses on the margins of bowé are peculiar forms that are found associated with the underground water circulation of cuirasses, whether ferruginous or bauxitic. They offer some analogies with karst topography, but there is a fundamental difference: the processes are essentially mechanical in a rather insoluble mass.

The marginal escarpments of bowé separate two distinct drainage systems: that of rillwash combined with infiltration on top of the bowé, which produces the hardening process and tunnelling, and that of spring fed streamlets on the flanking slopes. The line of demarcation is generally sharp, as on the edges of the Causses: outside of small canyons, always few in number, the escarpment isolates the bowal from channelled runoff that permits the resumption of erosion. This favours the evolution of bowé, as we have analysed it above.

Frequently the margin of the bowal is the locus of modifications in the cuirassing process that, also, help to block regressive erosion and hinder the dissection of the plateau. On the one hand the escarpment, as all bare cuirass exposures, undergoes a certain amount of mechanical disintegration, which is variable with climate, a point to which we will return. Ferruginous rubble is detached from it and slowly slides down the subjacent slopes where it sometimes forms a scree that prevents overland runoff. But under different climatic conditions the waters that seep in the area between spring recesses evaporate in the scree, consolidating it into a hillside breccia, which helps to block the retreat of the bowal margin. In other cases, also dependent on climate, there is simply a thickening and a hardening of the escarpment itself. The ground water in the upper zone of the regolith holds oxides in solution and in the dry season is subjected to intense evaporation, which dries up the springs. This happens in the immediate vicinity of the escarpment. The iron oxides are precipitated, extending the cuirass at the base to the detriment of the underlying regolith, as Maignien (1958) has found out. In this way cuirasses that are only 2 or 3 m (6 to 10 ft) thick under the bowal surface may be as thick as 10 m (33 ft) on its margin. This, of course, blocks regressive erosion all the more effectively.

The morphogenic mechanisms peculiar to the evolution of cuirasses undergoing resumed erosion thus have the effect of isolating bowal and valley. The bowal is subjected to a very slow degradation that gives its surface a typical undulating form with flat, closed depressions and an increased resistance due to the diagenesis of the cuirass. Valleys open up between marginal escarpments that gradually become more and more indurated. Regressive erosion comes to a stop at their base due to sliding cuirass rubble or scree, and never penetrates more than several tens of metres (50 to 200 ft) into the plateau, being limited to canyons formed by tunnelling. Bowé surfaces therefore have a considerable geomorphic endurance. This fact plays an essential role in the evolution of the relief of the alternating wet–dry tropics.

Evolution of the relief of the wet–dry tropics

The important role played by cuirasses is a significant characteristic of the wet–dry tropics. Even more than in the dry realm with calcareous crusts, tropical cuirasses determine the evolution of the relief. Indeed calcium carbonate can be more easily mobilised than iron or alumina. The permanency of calcareous crusts is therefore less and their effect on the evolution of the relief is not as long continued. Such caliches are moreover tied to Quaternary climatic oscillations. The oldest are only a million years old. Cuirasses, on the other hand, are characteristic of epochs well before the Quaternary, even if the climatic oscillations of this period have had repercussions on them. They sometimes date back to the Eocene or even to the Cretaceous. Their influence has therefore been felt for a very long period of time.

Because of the importance of cuirasses we must therefore distinguish between cuirassed and non-cuirassed regions when considering the evolution of the relief of the wet–dry tropics.

Evolution of the relief in cuirassed regions

Having studied the genesis and evolution of cuirasses, we must now put them back into their geomorphic framework. Only then can we work out what is the evolution of regions in which they play a predominant geomorphic role.

THE GEOMORPHIC FRAMEWORK OF THE CUIRASSING PROCESS

The formation of a cuirass is a complex phenomenon that can be well understood only if replaced in its geomorphic framework. But the cuirasses themselves materially influence the land-forming processes. There is therefore interaction, and we separate the two aspects of the interaction only for the sake of clarity of exposition.

As we have already indicated, there are two theoretical types of cuirasses: authigenic cuirasses (of relative accumulation) and allogenic cuirasses (of absolute accumulation). Their importance is very unequal, which is not always well emphasised in pedologic studies, which mostly treat them equally. Authigenic cuirasses are, as a matter of fact, much less common than allogenic cuirasses, and the latter make up the enormous majority of bowé. This error in appreciation results from a tradition that consists of viewing the cuirassing process as a pedogenic phenomenon rather than a sedimentological phenomenon. It is only the distant echo of Harrassovitz's point of view, which made of the cuirass the inevitable end product of tropical pedogenesis.

It often happens that cuirasses of relative accumulation develop at the expense of cuirasses of absolute accumulation placed in the dominant topographic position of bowé. Indeed, we have seen that under such circumstances the cuirass undergoes internal rearrangements with leaching in the upper horizons and accumulation in the lower horizons. If leaching is in an advanced stage it may result in a modification of the cuirass, as d'Hoore (1954a) has indicated. The more a geomorphological point of view is applied to this kind of research, the more often it appears that bowé cuirasses are in large measure allogenic cuirasses developed in detrital formations, which can still be recognised in spite of diagenesis. Their dominant position results, therefore, from an inversion of relief.

Distinction between authigenic and allogenic cuirasses is therefore not only of pedologic interest. It is fundamental in geomorphology. We must therefore verify what are the required conditions for the formation of each kind of cuirass.

1. *Authigenic cuirasses* can develop only under particular circumstances, permitting an *in situ* induration of a sufficient quantity of bauxitic and ferruginous products, the first less soluble than the second. This can take place in two different ways:

(a) On rocks very poor in quartz in which the very stable mineral kaolinite does not form. This is the case of basic igneous rocks and probably also of certain very aluminous clays. The lack of silica allows the alumina to remain isolated (instead of forming kaolinite as it combines with dissolved silica) and to become proportionally more abundant as the other mobile elements are leached away. The result is a ferruginous bauxite that becomes progressively more bauxitic as it loses its iron hydroxides with the good drainage of a bowal. But evolution is long and for this reason bauxitic cuirasses often indicate very old surfaces. The concentration of alumina depends, of course, on the initial proportions of iron hydroxides, which are all the more easily removed as they are less abundant. For this reason the concentration of alumina is slower on basic igneous rocks. The bauxitic ores covering the diabases (initially rich in iron) of the Fouta Djallon date back at least to the Eocene, perhaps even to the Cretaceous. It is only thanks to the ancient and continuous uplift of the Guinea Arch, which has maintained geomorphic conditions favourable to leaching, that they have been able to form. Such geographic conditions, however, present the beds with a great danger of erosion before the elaboration of the ore is completed. Indeed, the culminating position necessary for a good elimination, through leaching, of the iron hydroxides implies a high potential erosion. In the case of the Fouta Djallon the beds have escaped erosion only thanks to the substratum of sandstones and quartzites, which are the most resistant rocks under tropical climates. It is thus, in the last analysis, the existence of basic intrusions in an environment of sandstones and quartzites that has enabled the unlatching of a series of favourable circumstances necessary to arrive at this always more or less exceptional result: an ore mine.

The geomorphic lesson to draw from this example is the following: first, there was formed a very uniform relief, consisting of an undulated surface of erosion truncating the sandstone series and their intrusions. This relief

205

was so minor that mechanical erosion was very slight. Instead an intense pedogenic evolution set in, initiating a relative accumulation of alumina and iron. Next, the peneplain was uplifted but thanks to the substratum and the peculiar conditions of fluvial erosion in the tropics has undergone too little erosion for all the superficial mantle to be swept away. The formation of a bauxitic cuirass went hand in hand, under such conditions of excellent drainage, with a leaching of iron that has resulted in the formation of commercial bauxites. The importance of the leaching of iron implies an old age for the deformed peneplain, arched from the Los Islands to the crest of the Fouta Djallon. The genesis of the cuirass, which can be reconstructed through petrographic studies, allows one to retrace the geomorphic evolution, as can be done with any detrital formation.

In environments poorer in iron, bauxites can form more rapidly. This is the case, for example, of the bauxites formed from the decalcified clays of limestones poor in silica and iron. In Jamaica the mined beds mark the surface of an eroded plateau truncating Oligo-Miocene limestones, tectonically deformed and karstified, according to Hose (1961). They can therefore have been formed only during the last 20 million years at the most. Karstic drainage permits, it is true, an excellent leaching of pockets in which bauxite occurs, but the same is true in the Fouta Djallon. The rapidity of the elaboration of the ore is, according to us, due to the paucity of iron in the bedrock.

In spite of such differences in speed essential to geomorphological comprehension, the genesis of bauxitic cuirasses implies, on the one hand, a long persistence in the position of a dominant residual erosion surface and, on the other hand, a considerable leaching of iron, which in the wet tropics may be distantly removed, but in the wet–dry tropics can only travel short distances to lower elevations. In the north of Guinea and in the west of Mali, where it is abundant, iron has formed cuirasses on piedmont slopes at the foot of the high plateaus covered with bauxitic cuirasses (Michel 1960 and Vogt 1956).

(*b*) On rocks very rich in iron ferruginous cuirasses of relative accumulation may form. But their importance has been highly exaggerated by pedologists influenced by old and erroneous notions. This phenomenon is indeed altogether exceptional and deserves to be compared with the ironcap that covers the regolith above certain mineralised veins. A very great abundance of iron is necessary for summital ferruginous cuirasses to persist. Otherwise, as Maignien (1958) has pointed out, they are progressively destroyed by disintegration and leaching, especially in the wet tropics. The thickening of cuirasses at bowé margins has as a corollary a thinning and a weakening of the central parts that supply the required iron. So far as we are concerned, we have only seldom observed authigenic ferruginous cuirasses and only on rocks exceptionally rich in iron that by themselves constitute iron ores of a high content. Such are the itabirites in the area of Belo Horizonte, Brazil. On the crest of the residual Precambrian homoclinal ridge of the Serra do Curral del Rey (Tricart, 1961), completely formed of ore, there persists a compact ferruginous cuirass, 2 to 4 m (6 to 12 ft) thick, enclosing small angular fragments of itabirite. It extends to the south along the backslope of the homocline. It probably dates from the Eogene. In the rest of the Ferriferous Quadrilateral certain flanks of itabirite ridges, whose slopes are 50° and even 60°, are draped with thin cuirasses ($\frac{1}{2}$ to 1 m), which can be dated back to the Pliocene. They sometimes reach the ridge crests, linking up with what is left of the Eogene cuirass. They are located on sandstones and graywackes and are brecciaceous and discontinuous, never extending more than 2 to 3 km (2 miles) from the iron bearing itabirite schists. They are therefore allogenic, the more so as their topographic position, which permits a considerable runoff, necessarily implies an influx of iron from the higher parts of the slopes and the crest of the ridge.

Analogous observations have been made on the Cerro Bolivar in Venezuela, where a Precambrian 'appalachian' ridge composed of metamorphics very rich in iron, like itabirite, forms a belt in the midst of gneiss. The ridge was formed during the dissection of a surface of erosion dated as late Eocene or Oligocene. The good drainage resulting from the gradual exhumation of the ridge produced an intense leaching, especially of the silica, concentrating the iron into an excellent commercial ore. This superficial ore is up to 10 m (33 ft) thick on the flanks of the mountain and is absent near the top. The crusts sometimes affect the regolith, at other times the bedrock made permeable by the leaching of silica. Evolution is therefore complex and combines, sometimes in the same locality, characteristics of an authigenic as well as of an allogenic cuirass. The migration of iron has affected smaller distances than in Africa, probably for palaeoclimatic reasons. Allogenic piedmont slope cuirasses are small and relatively poorly developed at the foot of the Cerro Bolivar. The most important seems to indurate a system of piedmont slopes, now dissected, dating back to the uppermost Pliocene.

An evolution such as in the Ferriferous Quadrilateral and of the Cerro Bolivar is of great practical importance as it permits the formation of high grade iron ore, poor in silica and of much higher quality than that of the parent formation. It also seems to have played an important part in the formation of the iron ores of Liberia.

Very exceptional cases of ridges with a high iron content are therefore necessary for ferruginous cuirasses of relative accumulation to develop and to persist through a rather long geomorphic evolution. For even in the Ferriferous Quadrilateral most ferruginous cuirasses are allogenic and mould the surfaces of piedmont slopes. Furthermore, one can recognise periods of cuirassing and periods of dissection with partial dismantling of the cuirasses and diagenesis of those that persist. Neither phenomenon is marked by continuity in time, which poses problems of palaeoclimatology. They are discussed in Chapter 5.

Only bauxitic, more or less ferruginous, cuirasses are therefore authigenic. Ferruginous cuirasses are, with certain exceptions, allogenic.

2. *Allogenic cuirasses* result from the impregnation of certain horizons in the regolith by ferruginous solutions brought by surface or subsurface drainage.

Just as in the case of the formation and evolution of authigenic cuirasses, geomorphic stability is necessary, and this is what explains differences in cuirassing from region to region, differences that are connected to the influence of tectonics and the progression of erosion. Regions subject to uplift and dissection are unfavourable to the formation of cuirasses. In sufficiently humid regions iron hydroxides are distantly exported from them to feed cuirasses on alluvial piedmont slopes. In drier regions part of them remain in place, forming only ferruginous concretions, the size of pellets or fine gravel, which may be removed and transported to neighbouring depressions during a period of more violent runoff. These cases exist on the Atlantic Arch of Brazil and the Guinea Arch of West Africa. On the Atlantic Arch of Brazil there are no cuirasses on the perhumid seaboard in spite of the occasional presence of syenites and micaschists relatively rich in iron, whereas on the drier interior flank they are found only in connection with rocks that are exceptionally rich in iron. On the Guinea Arch the leached iron of the basic igneous rocks has to a large extent accumulated in the Sudanese zone, where it has been incorporated in the alluvium of regions that had inland or partial inland drainage for a long period of time and have experienced climates with a marked dry season.

3. *Alluvial cuirasses* develop when three conditions are met simultaneously. The first, stability, is characterised by a slow accumulation of sediments, thus providing enough time for a good impregnation of iron. Rapid subsidence with a too massive detrital accumulation is unfavourable, it causes only ferruginous stains, as in the Llanos of the Orinoco. On the other hand extensive alluvial spreads as those of the western Sudan offer excellent conditions (e.g. the *Continental Terminal* of Senegal, Siguiri Basin, the region of Bamako). Cuirassed alluvial deposits usually reveal an alluvial fan type of facies, as the *Continental Terminal* of Ivory Coast (Tricart, 1962b), and are nearly always thin, sometimes accompanied by a cuirassing of the subjacent bedrock (Daveau *et al.*, 1962). The second condition is climatic: a marked dry season is indispensable, otherwise there would be only concretions or isolated blocks, but nothing that would perpetuate itself in the ulterior geomorphic evolution. Lastly, a good supply of iron is necessary, which depends on the lithology and the climate. Basins whose peripheries are very humid and overgrown with forests, as on part of the Guinea Arch, are in a favourable situation. Indeed, the wetter the climate, the more intensive is the leaching of iron oxides. If there is no marked seasonality, iron concretions do not form, and although the mottled soils have rutilant colours they contain only minimal amounts of iron. It goes without saying that if the rocks are moreover rich in iron, such as the diabase intrusions of the Fouta Djallon or the biotite rich Birrimian micaschists of Ivory Coast, conditions are even more favourable. This is why the western Sudan is an

208

ideal region from this point of view. The distribution of alluvial cuirasses is therefore an excellent instrument for palaeoclimatic reconstructions. For example, on the interior flank of the Atlantic Arch of Brazil or in the dry Northeast, there are important remains of alluvial cuirasses dating back to the Tertiary, as in the Borborema or in the interior of the state of Bahia. They imply a climate that was already semihumid and a lack of serious erosion in a region that was subjected to a regime of alluviation. Further in the interior the large *canga*[1] plateaus, once thought to be Cretaceous, reveal a Sudanese type of regime at the end of the Mesozoic or during the Eocene.

4. *Colluvial cuirasses* develop at the foot of uplands capable of supplying sufficient quantities of iron. Climate and lithology again meet in this case. No colluvial cuirass develops at the foot of a sandstone relief except when the sandstone is very ferruginous, as at Mt Nimba (Guinea Highlands). For example, in the area of Boa Vista in the Rio Branco territory of northern Brazil no cuirass develops in alluvial aprons composed of 6 to 8 m (20 to 26 ft) of yellow or orange argillaceous sands resting on a planated gneissic substratum, here and there exposed (Teixeira Guerra, 1957). Although the lithologic environment has a favourable texture and an impermeable bedrock blocks the ground water the cuirassing process does not go beyond the formation of concretions and a few blocks within which a ferruginous crust only encloses a poorly epigenised argillaceous mass. And yet the climate is that of typical savannas. But the marginal uplands are either gneissic hills, hardly supplying any iron, or iron poor sandstone plateaus. It is only in the immediate vicinity of the Rio Branco, which comes from the more humid Guiana Highlands where there is some iron, that cuirasses begin to form. They are alluvial, groundwater cuirasses dependent on the presence of a river bringing enough iron.

On the other hand, at the base of uplands rich in iron, cuirasses develop quite well, especially where runoff is slowed down and spreads in the form of rillwash. Slightly concave alluvial piedmont slopes or alluvial fans are excellent sites for colluvial cuirasses to form. Thanks to the dry season evaporation is high. Moreover, the spreading of the runoff itself is unfavourable to an excessive detrital sedimentation that would not allow enough time for the cuirassing process to operate.

Alluvial fans and alluvial piedmont slopes situated at the base of uplands supplying iron are ideal sites for cuirasses to develop. In the area of Mopti, Mali, Pleistocene fans coming from the Bandiagara Plateau are generally transformed into compact, brecciaceous or conglomeratic cuirasses (Tricart, 1959a). The same is true of the stepped piedmont slopes of the upper Gambia Basin (Michel, 1960), of the Siguiri Basin (Pelissier and Rougerie, 1953), of the Kéniéba area in Mali (Vogt, 1956), of eastern Guinea (Rougerie, 1961), or of the Ferriferous Quadrilateral of Brazil (Tricart, 1961b). It is a very general phenomenon whose particularities only depend on the

---

[1] Popular name for a ferruginous cuirass in Brazil.

lithology within an adequate genetic and climatic framework. When iron hydroxides are very abundant they are mobilised by the least runoff. The cuirass then extends far upslope. This is the case, which is extreme and pathologic, in the Ferriferous Quadrilateral. When the iron hydroxides are less mobile, only the piedmont slope becomes cuirassed, principally farthest down where there is a permanent watertable, as the water must be subject to high evaporation for the very dilute iron to precipitate.

The selective character of the cuirassing process materially influences the geomorphic evolution.

EVOLUTION OF CUIRASSED REGIONS

Cuirasses play a considerable stabilising role, which is all the more important as they are well indurated and, also, discontinuous, that is, connected to certain outcrops in the case of authigenic cuirasses, and to certain conditions of sedimentation in the case of allogenic cuirasses. When their life-span is sufficient, they thus cause inversions of relief that are characteristic of tropical landscapes.

Cuirasses are always local phenomena in a moderately dissected region sufficiently rich in iron. The interfluves, composed of ridges that are the source of the iron, do not become cuirassed except in the case of authigenic cuirasses. On the other hand readily flooded depressions become the locus of iron precipitation, especially if they are not subjected to a resumption of erosion, as for example due to the local protection of a quartzitic rockbar. Whereas on the ridges the unweathered rock is covered by a regolith that becomes friable through leaching and is readily mobilised, in the depressions the precipitation of iron indurates the loose detrital deposits, cementing them into a cuirass. Once the cuirassing process is sufficiently advanced a moderate and slow resumption of erosion facilitates its diagenesis and further induration.

At this stage of evolution there normally appears a depression that separates the cuirassed piedmont slope or cuirassed alluvial fan from the upland or valley that was its parent. A similar phenomenon is often observed in old calcareous crusts and in the siliceous gravelly fans of the perhumid mid and low latitudes. Rougerie (1961) has studied this problem in the savannas of northern Guinea. The formation of such trenches that interpose themselves between the old cuirassed piedmont slope and the upland seems to be the result of a combination of two factors:

1. During the formation of the cuirass there is less infiltration on the upper than on the lower alluvial piedmont slope (or alluvial fan) because of its greater steepness. On the other hand the material of the upper slope is coarser and generally has a greater porosity, but this also means that there is less (ferruginous) cementation. Most cementation takes place on the lower slope because of the presence of ground water and a fluctuating watertable. The cuirass therefore does not develop as well on the upper slope, except when the iron content of the running water is exceptionally high.

2. During resumption of erosion the upper part of the piedmont slope is raised higher above the watertable, which improves subterranean drainage while a decreasing slope favours infiltration, as Rougerie has emphasised. The circumstances are thus favourable for a certain amount of leaching, which is not the case on the lower slope, which receives less water and is less porous because it is more indurated. Here, especially, the local waters percolate and produce the diagenesis of the cuirass. On the upper slope the already precipitated iron can easily be remobilised and leached, especially if the climate becomes simultaneously more humid. Weathering also proceeds more rapidly and more deeply in the underlying rocks. Indeed it benefits from the influx of runoff from the upland, exactly as happens at the base of inselbergs. The more friable and thinner cuirass and a more intensely weathered and thicker subjacent regolith therefore constitute a zone of weakness. Before long it will be exploited by rivulets that gather the runoff from the upland and undermine and disrupt the cuirass through erosion of the subjacent regolith. The trench then develops and becomes continuous as a col is formed at the head of streams running in opposite directions.

The trenches at the foot of the upland thus isolate plateaus of colluvial cuirasses from the uplands that have supplied the iron.

The cuirassed plateaus then continue to evolve in the manner already described. They further indurate through recementation as long as the climate does not become too humid, which would lead to a regressive evolution and the degradation of the cuirasses. They are attacked only at their margins, resulting in a slow, parallel retreat of the plateau flanks, cut in the regolith and capped by the cuirass. The iron that they set free contributes to the cuirassing of new slopes that form below the base of the escarpment and merge with the bottom of the depression. In this way, because of the low mobility of iron in regions of marked dry season, stepped piedmont slopes become the rule. The iron leaching from the cuirassed plateaus permits both the reinforcement of the marginal escarpment, which protects it from dissection, and the supply of material that indurates the surrounding depressions. A rather intense leaching causes some loss of material in the central part of the cuirassed plateau, barely visible however in the topography in the form of closed depressions resembling dolines. Exactly as in the case of a karst plateau, the relief opens out without otherwise much altering its appearance. These processes produce stepped or multiple level cuirasses, and this particular evolution explains why the induration of multiple level cuirasses decreases downwards. Only part of the iron is remobilised, and each younger surface is less affected by diagenesis. Possible influences of palaeoclimatic oscillations should, of course, be added.

In the area of Kéniéba, Mali, Vogt (1956) has described two piedmont slope levels at the foot of a large Ordovician cuesta composed of sandstones and diabase intrusions, which supplied the iron. The oldest level, which is cuirassed, ends in a marked escarpment and is almost detached from the cuesta, whose face, however, has hardly retreated due to its sandstone composition. The lower level, formed after the dissection of the upper level and merging into the incipient trenches, is formed by incompletely cuirassed heterogeneous material composed of blocks and isolated indurated masses. A similar disposition in a shale depression at the foot of the same Ordovician cuesta in the Siguiri area (Guinea) has been described by Pélissier and Rougerie (1953), but evolution is not as far advanced (because the upper Niger drained inland until the end of the middle Quaternary, whereas the Falémé, at the foot of the Ordovician cuesta, is an affluent of the Senegal, which has always had exterior drainage). Near Beyla, Guinea, in a crystalline basin surrounded by diabase uplands Rougerie (1961) has described a system of three piedmont slopes converging towards the centre of the depression but increasingly stepped towards the margin. The oldest is cut from the peripheral highlands by a trench. In the Ferriferous Quadrilateral Tricart (1961) has studied two sets of cuirasses, the oldest of which, probably Eogene, moulds piedmont slopes frequently eroded into inverted relief at the elevation of 1 200 m (4 000 ft). Here there are no trenches when the ridges that supplied the iron were not eroded at a later date, for the rock, which is itabirite, is not subject to chemical weathering. Moreover itabirite supplies so much iron that not only the upper parts of the piedmont slopes but the footslopes of the hills themselves are thoroughly cuirassed. This counterproof confirms the correctness of Rougerie's (1961) point of view on the origin of the trenches.

After the trenches become well developed cuirassed piedmont slopes behave like cuestas. The cuirass indeed has the same gradient as the surface of the piedmont slope, which has a dip of 1 to 5–10° in its highest part. Overlying a less resistant regolith, the conditions necessary for the formation of a homoclinal ridge facing the upland that was the source of the iron are perfectly realised. Some highlands looking like inselbergs and dominating old piedmont slopes detached from them by trenches give an entirely erroneous impression of an anticlinal core forming a ridge surrounded by an annular homoclinal depression. A number of petroleum photogeologists have fallen into this trap . . .

The cuestas of cuirassed piedmont slopes retreat rather rapidly in the beginning. This is, first, because the cuirass generally thins out at the foot of the upland where the trench develops. Here the steeper inclination of the cuirass helps infiltration waters to migrate towards the outskirt of the old piedmont slope, causing leaching and only a partial precipitation of hydroxides in the higher end of the cuirass. Secondly, a thickening of the escarpment through lateral migration of iron solutions does not take place, and inversely, there are only a very few spring recesses. Instead there is a continuous scree of ferruginous rubble supplied by the escarpment, which gives a rectilinear outline to the cuesta front.

The formation of trenches and cuestas from cuirassed piedmont slopes is only the first stage of an evolution that sometimes ends in inversions of

relief. Given enough time and, especially, when climatic oscillations interrupt the cuirassing processes, it often happens that only the lowest parts of
the relief, for example old depressions or basins indurated by cuirasses, resist
and are transformed into summital plateaus, after the old uplands have
been completely eroded away and replaced by lowlands. Such a new relief
composed of a second generation of piedmont slopes and rolling hills occasionally dominated by slightly inclined cuirassed plateaus, locally preserving
the shape of cuestas, is frequent in regions of calm tectonic evolution and very
slow erosion. Good examples exist in the northeast of Ivory Coast and the
west of Upper Volta (Daveau *et al.*, 1962). At Bouré (90 km : 60 miles west
of Ouagadougou) an old bauxitic cuirass is in a summital position describing the figure of a horseshoe around a basin. It forms narrow, disconnected
tablelands, representing what are probably the remains of a highly dissected
piedmont slope or alluvial apron. The cuirass is partly epigenised on a
detrital formation. Below, other, ferruginous cuirasses with reworked blocks
of bauxite form less dissected, larger cuesta-shaped surfaces, usually detached from the dominant uplands crowned with the bauxitic cuirass (see
Fig. 4.13). At Vouriba the faceslope of a cuesta-shaped mesa is itself

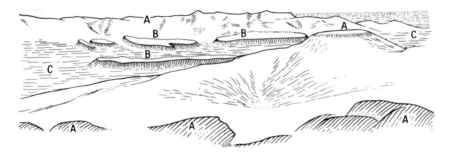

FIG. 4.13. Relief of multiple level cuirasses at Bouré, Upper Volta (from S. Daveau,
M. Lamotte and G. Rougerie, 1962)

In the foreground and in the background, remains of an old bauxitic cuirass
reduced to the shape of a horseshoe around an amphitheatre (A). In the middle,
remains of cuirassed piedmont slopes occupying the centre of the amphitheatre (B)
above the present non-cuirassed surface (C).

cuirassed, and at its foot an equally cuirassed piedmont slope is inclined in
the opposite direction of the capping cuirass (Fig. 4.14). The lowest cuirass
came into being as a result of a complete inversion of relief at the very place
of the upland against which the highest piedmont slope and cuirass were
constructed. A similar situation exists in the Ferriferous Quadrilateral
(Tricart, 1961). Here are preserved a number of Eogene cuirassed piedmont
slopes that have become transformed into summital plateaus as the uplands
against which they leaned have eventually been eroded away. It presupposes
that all the iron oxides of the uplands were exported far afield instead of

cuirassing the new piedmont slopes in the same way as the Eogene cuirass was formed, which implies climatic oscillations.

Such inversions of relief are aided by rock diversity. In the southeast of Ivory Coast, presently forested but having known periods of drier climate,

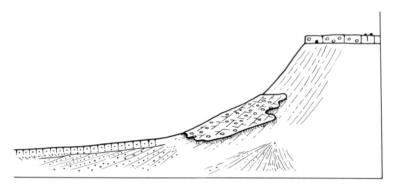

FIG. 4.14. Antithetic cuirassed piedmont slopes at Vouriba, Upper Volta (after Daveau *et al.*, 1962)

An old cuirassed piedmont slope inclined toward the right has been put into inverted relief and transformed into a cuesta. At the foot of the face of the cuesta, a new piedmont slope, in obsequent position, has been formed and cuirassed in turn. A conglomeratic cuirass covering moulds part of the cuesta inface.

Rougerie (1950) has observed a northeast–southwest trending appalachian type of relief in which the ridges correspond to basic rocks truncated at an elevation of about 200 m (660 ft) by an erosion surface, which now is cuirassed. Between the ridges wide marshy depressions obstructed by colluvium have been cut from schists or granitoid rocks. On their sides there are the remains of another cuirassed level at about 130 m (430 ft).

Cuirasses, therefore, produce a particular type of inverted relief in the wet–dry tropics as a result of their mechanical as well as their chemical resistance. They form especially in depressions that become indurated because of important influxes of allogenic iron. In this way depressions can be put in inverted relief and transformed into bowé during a sufficiently long evolution. The reworking of their iron feeds the cuirasses of new piedmont slopes at their feet, often on the location of the upland that once dominated the old cuirasses. Such an evolution, which is common, can however only take place within a definite palaeoclimatic framework. It supposes a certain permanence of iron within the region, therefore the absence of too much exportation that would cause the destruction of the cuirasses. To finish with the discussion of this problem, we must, therefore, place ourselves in a palaeoclimatic perspective, which we do in Chapter 5. Before doing so we turn to the other types of relief found in the alternating wet–dry tropics.

Evolution of non-cuirassed landforms

In addition to cuirassed surfaces the wet–dry tropics are characterised by a trilogy of landforms, some elements of which are more akin to the landforms of the wet tropics (bottomlands and embayments, inselbergs), whereas others are more akin to those of the semiarid realm (plains and pediplains). It is due to the transitional character of the savannas and the campos cerrados.

VALES, BOTTOMLANDS AND EMBAYMENTS (AKIN TO THE WET TROPICS)

Valleys are much like those of the wet tropics, especially when drainage is poor. They are seasonally flooded, which causes waterlogging of the soil and considerable weathering. During the dry season there is no weathering, instead cuirassing processes may occasionally be active.

Under certain conditions, which will be specified below, the form of the valleys is in sharp contrast with that of the interfluves. On these a rather intense rillwash and rapid evaporation after showers permit only a very limited amount of weathering. Intermittent streams carry coarse sands, often rich in muscovite, and ferruginous pisolites, which create permeable deposits capable of retaining abundant seasonal ground water and of assuring a good waterlogging of the substratum. The following succession of runoff and landforms, starting at the divides, is normally observed in the area of Bouna, Ivory Coast, and in the northwest of Ghana:

1. A zone of unconcentrated runoff (rillwash) beginning with small coalescing pools formed during torrential showers on the divide. There is generally a concentration on the surface of quartz pebbles and fine ferruginous gravels that are not easily entrained. The slopes have a wide summit convexity merging downslope into a concave profile. Their amplitude varies considerably with the spacing of the stream courses and the abundance of coarse debris, which can maintain gradients of 25 to 30° on short slopes, such as below escarpments formed by cuirasses. But generally the gradient is less, or about 10 to 20°, on the rectilinear sector joining the summit convexity to the basal concavity, and 5° near the base. Micaschists and shales, which break up easily and weather into relatively argillaceous products, deflocculated by iron oxides and much subject to splash erosion, generally form rather gentle slopes, 7 to 12°, with a long median rectilinear sector. Sands and sandstones, on which rillwash occurs only as a result of poor soil structure and where the heavy load impedes the flow of water, form steeper slopes, often close to 20°. Daveau (1959) has noted that in an area close to Bandiagara, Mali, 900 mm (36 in) of annual rain, or more, are necessary for weathered sandstone products to include fractions finer than sand. Because of such silty sand rillwash becomes possible. Being more coherent the material provides a lesser load, and the slopes develop gentler gradients.

2. A zone of concentrated runoff where slopes converge at the extreme head of the vales. Here there are distinct watercourses, genuine small washes with

sandy beds and marked banks, sapped by a sporadic, short but forceful stream flow. The banks, often vertical, are frequently cut into sandy colluvium and crumble in places. The force of the concentrated runoff succeeding the unconcentrated runoff depends on the degradation of the vegetation and the general steepness of the relief. In regions subjected to a resumption of erosion and where vales are sufficiently incised, it may be important. At the foot of the Bandiagara Plateau, on the margin of the Sahelian zone, Daveau (1959) mentions gullies deeply incised in silty sands that pass upward into amphitheatres in the zone of rillwash. In Congo-Kinshasa Delhaye and Borgniez (1948) report genuine gullies cutting up the slopes. In Katanga Lefèvre (1953) described dead-end torrential gullies eating their way into the steep slopes of amphitheatres. It happens sometimes that gullies debouch directly into important rivers, as along the Baoulé (Mali) or in the north of Dahomey, where they have been described by Vogt (1961). Here, a terrace composed of thick silts overlying 'underbank gravels' flanks the rivers. Its sides are cut by narrow gullies that pass through the gallery forest and seem to have been caused by the footpaths of large animals on their way to drink. Back of the gallery forest they ramify, their branches spreading fanwise and opening up into an amphitheatre, whose floor is on the bedrock or on rubble washed by rills. The gully sides are abrupt but may become less steep through stabilisation, but then they are cut by new gullies, thus renewing the mechanism.

3. A zone of sheet flooding in flat floored vales with steep sides (20 to 30°) joining the vale bottom at a clear break-of-slope. There is no clearly marked channel. The vale floors, periodically flooded, are covered by tall, dense, swamp grasses. Their decomposition supplies enough organic matter to cause a complete discoloration of the upper soil horizons, and sometimes a little, very black acid humus. Rillwash from the hillsides often concentrates quartz pebbles and fine ferruginous gravel at their margins, especially if cuirass remnants persist on the interfluves. Similar forms have been described in the campos of the territory of Amapá in Brazilian Amazonia by Teixeira Guerra (1954). In Zambia Ackermann (1936) has described slightly different forms under the name *dambos*. Dambos are U-shaped vales at the head of perennial streams; they are moist only during the rainy season but lack a stream bed and concentrated runoff, exactly as the flat-floored vales just described. Again, like the flat-floored vales they form strips of herbaceous vegetation, often marshy, in the midst of the deciduous seasonal forest or thorn woodland and scrub. The only difference is their U-shaped rather than their flat-bottomed form. Ackermann suggests that they represent a form of headward erosion attacking the interfluves. It may be the reason for the difference. In our region (Ivory Coast and Ghana) there is no regressive erosion and the forms are ancient. The flat bottom is explained by weathering of the foot of the vale sides, causing a certain amount of chemical undermining, sometimes reflected in slight settling movements. Evolution is mainly chemical and, has been for a very long time. In dambos,

on the other hand, there is colluviation due to rillwash and creep on the vale sides, and if there is a tendency towards deepening of the vale a flat bottom becomes impossible. The difference would thus be due to successive stages of evolution.

4. A zone of subperennial runoff in flat-floored valleys or bottomlands that dry up slowly during the dry season. Runoff being very unequal, there are distinct low water channels and high water floodplains although the former are sometimes discontinuous (like marigots). The floodplain is overgrown with marshy vegetation, whereas the low water channel is marked by steep banks in graded reaches that locally fade away in the floodplain and circumscribe poorly delimited depressions in which water holes persist. The alluvium is finer and richer in silt and mud than in tributary bottomlands. This seems to be due to sorting, which stops the coarser sands higher up, and partly to more advanced weathering. Such bottomlands always have steep sideslopes usually marked by a clear break-of-slope. They evolve by weathering at the foot of the slope, aided by the moisture of the flood zone and by chemical undermining. There is some rillwash and creep. Seepages appear at their foot during the rainy season. Groundwater cuirasses may form in such bottomlands, where weathering is considerable and wells often go down to 10 m (33 ft) or more before reaching bedrock. These flat-floored valleys are generally poorly graded and widened by embayments due to retreat of the valley sides, as in the wet tropics. The finest of them are located upstream from rockbars that have for a long time provided stable base levels, as for example above Itaberaba, Bahia, Brazil, in gneisses traversed by a dike of granulite. Such forms are much better developed in eastern Brazil than in West Africa as a result of lesser climatic oscillations.

This type of drainage and land sculpturing, typical of regions where erosion is moderate, therefore indicates, from the hills to the bottomlands, the transition of landforms related to those of semiarid regions, sculptured essentially by overland flow, to landforms similar to those of the wet tropics, with bottomlands and predominant chemical weathering.

In more rapidly eroded regions the landforms are different. In schistose regions of Cuba, for example, Massip and S. de Massip (1949) have described a relief with knife edge ridges (*cuchillos*). Schists, slates and quartzites are easily fragmented and dislodged along foliation planes by alternate wetting and drying. Runoff is strong during heavy showers and dissection is very severe, with a high density of stream courses separating sharp and sinuous ridges. The bedrock is periodically bared by mudflows and landslides. The main streams are often obstructed by waste coming from the sides. Similar landforms may be observed in Venezuela's Cordillera de la Costa, formed by varied and slightly metamorphic schists. Between December 1950 and February 1951 landslides displaced no less than 570 000 cu m (20 million cu ft) of material at La Guaira.

PLAINS AND PEDIPLAINS (AKIN TO SEMIARID REGIONS)

When evolution is more advanced, overland flow (rillwash and sheet floods) makes possible the formation of remarkable planations similar to the pediments of semiarid regions. One should not relate this to a simple application of the cyclic concept, however. These savanna plains are in large part due to climatic oscillations, and in the case of the western Sudan to an evolution of inland drainage that has ceased to exist only recently.

Savanna plains generally develop in regions of relatively weak rocks, whose weathered waste is removed by overland flow. Frequently the general morphology is that of a depression surrounded by crystalline inselbergs or marginal uplands formed by sandstones, diabases or caps of resistant cuirasses. The depression itself may be occupied by a number of piedmont slopes, cuirassed or not, often stepped, reflecting the importance of palaeoclimatic oscillations in the genesis of the relief. In West Africa, for instance, high Ordovician quartzitic sandstone cuestas frequently dominate over depressions eroded into the subjacent crystalline basement, as is the case with the Mandingo Plateau overlooking the Falémé (Tambouara Cuesta) and Siguiri basins. In an analogous fashion the Bandiagara-Hombori Plateau, also capped by Ordovician quartzitic sandstone, dominates the Gondo Basin filled with terrestrial Tertiary deposits (*Continental Terminal*). In Amazonia a similar disposition exists in the depression of the Rio Branco dug from gneisses and bordered in the north by the Triassic sandstone plateaus of the Guiana Highlands. Sandstones and quartzites, which have a tendency to recement themselves under the effect of alternations of wetting and drying, in fact always stand out as residual reliefs in the wet–dry as well as in the wet tropics. They are more resistant than gneiss and granite, as Domingues and Keller (1956) have emphasised in connection with the geomorphology of the state of Bahia.

In Brazil Ruellan (1953) has insisted on the role of sheet floods downstream from V-shaped vales. During heavy showers they sweep flat-floored vales and valleys like a rising tide, progressively transforming them into very smooth erosional piedmont slopes. The mud they carry off fills the pores of the more permeable formations, making them impermeable and permitting a more distant water flow. It results in very flat, gently inclined surfaces with a few projecting rock knobs and minor depressions where mud deposition occurs during flood recession. Granite blockpiles (or bedrock boulders) as well as unjointed bedrock masses in the shape of turtlebacks (ruwares) thus come into being. Residual quartz pebbles lie scattered about, oxidising under the effect of wetting and drying, and migrating very slowly as they are undermined and pushed along. Their sharp edges become blunted by chemical weathering more than by mechanical wear. At a certain depth the infiltrated water aids the formation of ferruginous concretions or scoriaceous masses, which are the beginning of a cuirass. Due to the lack of nearby uplands true cuirasses are the exception. They are confined to the surroundings of inselbergs that supply enough iron. The finest planations of the campos cerrados are found on micaceous rocks: micaschists or schistose gneisses rich in biotite, both of which disintegrate readily and provide clayey and sandy products favourable to overland flow. In such regions inselbergs are usually composed of highly quartzose gneisses, granulites, quartzitic schists, or slaty phyllites.

At the Rio Branco a vast piedmont slope, a true pediplain, covered by an alluvial veneer not more than 6 to 8m (20 to 26ft) thick, masking the irregularities of the gneisses truncated by erosion, has thus been formed. It is composed of argillaceous sands with debris of reworked ferruginous concretions and fragments of quartz and quartzite derived from the marginal reliefs. Similar pediplains are found in the territory of Amapá, also described by Teixeira Guerra (1954), and forming a savanna covered plateau at about the 100m (330ft) contour. Concentrations of ferruginous concretions sometimes consolidated into a secondary cuirass are found on its surface. Underneath, pisolitic clays cover a more compact cuirass. Here too there must have been climatic oscillations. Again, similar landscapes are found in the Sudan, from the middle Nile Basin to the Atlantic, as referred to by Dresch (1953).

As in drier regions, the present geomorphic processes operating in savannas are capable of reducing the relief by pediplanation. The more the relief is already reduced, the greater is the water lost by evaporation on ever larger pediplains. The water supply of streams is thereby diminished by an equal amount as well as their potential for regressive erosion. Once sufficiently developed, pediplains can continue their evolution in the climates of the tropical savannas in spite of moderate climatic oscillations.

INSELBERGS

Inselbergs, like pediplains, are not landforms restricted to the wet–dry tropics. They are found, with minor differences, in dry regions and in the wet tropics as well. Again we are in a zone of transition.

The reduced intensity of chemical weathering in comparison to the wet tropics results in a higher rock resistance and a multiplication of inselbergs. Rocks that would be incapable of forming monolithic domes in a more humid climate can form them here. Less marked differences in lithology are reflected by differential erosion and therefore produce a greater variety of inselbergs. In crystalline regions, for example, slightly less jointed granite or gneiss is all that is necessary to produce a turtleback or blockpiles. A smaller proportion of biotite, very sensitive to variations of humidity, or even a finer texture is all that is needed. On the Rio Branco pediplain, inselbergs are formed by granites, gneisses and diabase dikes. Blockpiles and turtlebacks are numerous. In Tikar, India, they are formed by syenites in the midst of granites, according to Thorbecke (1921). In the Fort-Lamy area of Chad, Barbeau and Gèze (1957) describe some very regular monolithic domes formed at their base by granite with curved joints parallel to the surface, and, at the crest, after a sudden transition of only a few metres,

FIG. 4.15. Residual blockpiles in the savanna area of Boundiali, Ivory Coast

Weather resistant compact granitoid gneiss, but jointed. Differential weathering has made possible a slight hill. The thin regolith has been denuded and Zones III and IV exposed. Typical savanna in which rillwash has helped clear the corestones.

by a granite with vertical prismatic jointing followed by fractures filled first with microgranite and then with rhyolite. These fractures become wider and wider in such a way that the proportion of rhyolite increases until it forms the entire rock. It takes only 10m (33ft) to go through the whole transition. The domes are caused by the exhumation of the true granite by removal of the more friable rocks that cover it. In Uganda Ollier (1960) describes granitic and gneissic inselbergs formed by rocks that are simply fresher than those which surround them and which, more easily weathered, have been eroded away and truncated to form piedmont slopes, the resistant hard cores having been progressively put into relief by wastage of the surrounding material.

The genesis of inselbergs is therefore identical to that of the monolithic domes of the wet tropics. The cause is differential weathering. But because of its lesser intensity weathering is more selective, and only a slight difference in rock resistance is necessary for inselbergs to appear. This explains the greater lithologic variety of inselbergs in the wet–dry tropics. The importance of overland flow also plays a role by emphasising the contrast between inselbergs and the weaker rocks that are denuded into ridges with gentler slopes than are found in the wet tropics, or are even completely truncated by pediplains.

Such genetic differences are reflected in the detailed topography of inselbergs. Some rocks, such as the granites of the Fort-Lamy area, produce

FIG. 4.16. Monolithic dome in the savanna area of Boudiali, Ivory Coast

The granite is particularly massive, with typical curved joints. Characteristic structural exfoliation and flutings. Because of lithologic conditions the relief is similar to that of forested regions. At the base a hilly topography with gentle slopes sculptured by rillwash in the jointed and weathered enclosing rocks. The hills are covered by savanna, whereas a deciduous seasonal forest surrounds the monolithic dome, taking advantage of deeper weathering thanks to the rainwater that runs off the bare slopes.

inselbergs that are true monolithic domes. But whereas in the wet tropics only a limited number of rocks are sufficiently resistant to form monolithic domes or sugar loaves, here there are a great variety. Many inselbergs are composed of jointed rocks, which have their influence on topography. Cavities develop along fissures. Blocks are detached and rubble is formed. Debris appears, sliding slowly, in part due to slope wash. At Balos on the Sudanese side of the Ethiopian border, on the margin of the Sahelian zone, Ruxton (1958) describes a granitic inselberg littered with boulders 2 to 4 m (6 to 12 ft) in diameter and between which rubble and cobbles over 10 cm (3 in) increase in quantity towards the base, causing an increase in slope. A petrological study reveals the part played by chemical weathering in the fashioning of the slope. The feldspars drop from 72 per cent in the bedrock to 15 per cent in the most weathered rubble at the base of the inselberg. It decreases even further beyond this point, whereas the percentage of clay increases. Biotite is the first mineral to be attacked, cleaving as a result of differences in humidity. Rainwash at once removes the fine weathered products: sands, silts and clays. They are spread at the foot of the inselberg where they form a permeable soil, which, because of waterlogging due to a high influx of water supplied by rillwash on the inselberg, facilitates the weathering of the subjacent rock. The regolith thus thickens in an annular fashion around the upland while the reduction of the inselberg itself proceeds together with the parallel retreat of its slopes. The rocks being less resistant to weathering than those of a monolithic dome of the wet tropics, evolution is rather rapid. In a perhumid climate it never would have formed a monolithic dome. The multiplicity of inselbergs is often related to a relatively rapid evolution of their forms.

It seems that some inselbergs of the wet–dry tropics cannot be explained by lithologic factors but must simply be explained by their position on interfluves. Having thin or completely eroded soils, lines of weakness such as fissured zones are rapidly exploited. Runoff, which becomes easily channelled on steep slopes, cuts gullies that gradually dissect the interfluves, as Louis (1959) has shown in connection with an area west of Bangkok, subject to a monsoonal climate.

With a resumption of erosion the rings of weathered material surrounding inselbergs are eroded into depressions, as Clayton (1956) has shown in Ghana and Cameroon. They insert themselves between the foot of the inselberg and the piedmont slopes formed as a function of the previous base-level. Evolution is analogous to that which ends in the formation of trenches between uplands and cuirassed piedmont slopes.

Inselbergs of the wet–dry tropics are therefore composite and display varied topographies. Some are composed of very resistant rocks capable of forming monolithic domes in a perhumid climate, such as the granitic domes of the Fort-Lamy area. The denudation of the more fragile materials in which the resistant core is enclosed is the cause of such forms. Flutings may have the time to develop if the rock is resistant enough. Other inselbergs are simply formed by rocks that are more resistant than the enclosing rocks but not resistant enough to form monolithic domes as those of the wet tropics. Even though they may form rather high uplands, they are relatively rapidly eroded as chemical weathering does take place and is helped by active denudation caused by rillwash. Such inselbergs are littered with loose boulders, rubble and debris in transit. The most friable zones are the most exploited. In the most advanced cases all there are left are blockpiles. Yet other inselbergs owe their existence only to their position on drainage divides. Bestriding piedmont slopes, they have steep slopes, from 20 to 30°, but, not very resistant, they are covered with thin stony soils. Runoff becomes channelled and takes advantage of weak structural zones, eventually producing a number of isolated reliefs.

Besides these various types of inselbergs corresponding to present dynamic factors, we still have to take into consideration the palaeoclimatic legacies that are also reflected in the landscape. To distinguish them is often a delicate matter. The existence of monolithic, domal inselbergs does not necessarily mean, as Czajka (1957) maintains, a perhumid palaeoclimatic origin.

They may simply be due to lithologic influences, as those described by Barbeau and Gèze (1957).

Being a zone of transition between the wet tropics and the semiarid zone, the alternating wet–dry tropics combine original landforms with landforms that are intermediate between those of its neighbouring zones. Characteristic and original are the landforms related to the cuirassing process, which finds its optimum climate here. By themselves these landforms introduce a noteworthy originality in the geomorphic evolution by producing such landforms as bowé, peculiar kinds of cuestas, and particular inversions of relief. The other landforms, on the contrary, assure a transition, for example by associating hilly topographies that are akin to the wet tropics with erosional piedmont slopes and pediplains that are akin to the semiarid zone, and inselbergs that are akin to both. Much remains to be done to work out all the landforms in detail and to relate them more precisely to climatic conditions. The task is complex because of the existence of palaeoclimatic legacies, which are more difficult to detect in such transitional regions than they are in the heart of the more franc morphoclimatic zones.

Bibliographic orientation

Biogeographic and climatic factors affecting morphogenesis

Essential references concerning climatic conditions at the ground surface and about the vegetation of the wet–dry tropics are listed:

Colloquium: 'Le problème des savanes et campos dans les régions tropicales', *XVIIIth Int. Geogr. Congr.*, Rio de Janeiro, 1956, vol. I, 299–348.

BEADLE, N. (1940) 'Soil temperatures during forest fires and their effect on the survival of vegetation', *J. Ecology*, **28**, 180–92.
 Includes many useful measurements.

BERNARD, E. (1945) *Le Climat écologique de la cuvette congolaise*, Brussels, Publ. INEAC, 240 p.
 Includes valuable data about forests and savannas.

CHEVALIER, A. (1929) 'Sur la dégradation des sols tropicaux causée par les feux de brousse et sur les formations végétales régressives qui en sont la conséquence', *C.R. Acad. Sc.* **188**, 84–6.

DOMMERGUES, Y. (1956) 'Etude sur la biologie des sols des forêts tropicales sèches et de leur évolution après défrichement', *Rept. IVth Int. Congr. Soil Sc.*, Paris, v, 605–10.

DUVIGNEAUD, P. (1955) 'Etudes écologiques de la végétation en Afrique tropicale', *Année Biol.*, 3rd ser., **31**, pp. 375–92.
 Shows relationships between types of savannas and morphodynamic conditions in the Bas Congo.

GUILLOTEAU, J. (1958) 'Le problème des feux de brousse et des brûlis dans la mise en valeur et la conservation des sols en Afrique au Sud du Sahara' (part I), *Sols Africans*, **4**, no. 2, 65–104.
 Soil and ground surface temperatures, effects on vegetation, numerous data, and important bibliography to which we refer.

HUMBERT, H. (1952) 'Le problème du recours aux feux courants', *Int. Union Prot. Nat.*, 3rd meeting, Caracas, mimeographed, 9 p.
 Excellent summation.

JAEGER, P. (1952) 'Mission au Soudan occidental', *Encyl. Mens. d'Outre-Mer*, Suppl. to no. 24, August, 7 p.
 Data on soil climate.

JAEGER, P. and JAVAROY, M. (1952) 'Les grès de Kita (Soudan occidental); leur influence sur la répartition du peuplement végétal', *Bull. IFAN*, sér. A, **14**, 1–18.
   Ecologic monograph of a sandstone massif in a region of Sudanese climate. Persistence of gallery forests in refuge areas.

MAACK, R. (1950) 'Notas preliminares sôbre clima, solos e vegetação do Estado do Paraná', *Bol. Geogr.*, Rio de Janeiro, **7**, no. 84, 1401–87.

MIÈGE, J. (1953) 'Relations entre savanes et forêts en Basse Côte-d'Ivoire', *CR Vᵉ Réunion Afric. Ouest*, Abidjan, pp. 27–9.

PITOT, A. and MASSON, H. (1951) 'Quelques données sur la température au cours des feux de brousse aux environs de Dakar', *Bull. IFAN*, sér. A, **13**, no. 3.

ROBYNS, W. (1952) 'Les feux courants et la végétation', *Int. Union Prot. Nat.*, 3rd meeting, Caracas, mimeographed, 4 p.

SALVADOR, O. (1959) 'La température et le flux de la chaleur dans le sol à Dakar', *Ann. Fac. Sc. Univ. Dakar*, **4**, 47–54.

SCHNELL, R. (1945) 'Structure et évolution de la végétation des Monts Nimba (A.O.F.) en fonction du modelé et du sol', *Bull. IFAN*, sér. A, **7**, 80–100.

WALTER, H. (1939) 'Grasland, Savanna und Busch der ariden Teile Afrikas in ihren eoko-logischen Bedingtheit', *Jahrb. für wiss. Bot.* **87**, no. 750.

Pedogenesis and morphogenic processes, excluding cuirasses

TERMITES AND THEIR ACTIONS:

BOYER, P. (1955) 'Premières études pédologiques et bactériologiques des termitières', *C.R. Acad. Sc.* **240**, 569–71.

BOYER, P. (1959) 'De l'influence des termites de la zone intertropicale sur la configuration de certain sols', *Rev. Géomorph. Dyn.* **10**, 41–4.
   Good summation by an entomologist. Detailed bibliography to which we refer.

DE PLOEY, J. (1964) 'Nappes de gravats et couvertures argilo-sableuses au Bas-Congo. Leur genèse et l'action des termites', in A. Bouillon, *Etudes sur les termites africains*, Edit. Univ., Leopoldville, pp. 399–415, 6 fig., 2 phot., bibl.

ERHART, H. (1951) 'Sur le rôle des cuirasses termitiques dans la géographie des régions tropicales', *C.R. Acad. Sc.* **233**, 804–6.
   Interpretation different from that of P. Boyer. Termites would facilitate the cuirassing process.

GRASSÉ, P. and NOIROT, C. (1957) 'La genèse et l'évolution des termitières géantes en Afrique Equatoriale Française', *C.R. Acad. Sc.* **244**, 974–9.

GRASSÉ, P. and NOIROT, C. (1959) 'Rapports des termitières avec les sols tropicaux', *Rev. Géomorph. Dyn.* **10**, 35–40.
   Summation by one of the main specialists on termites. Bibliography.

HEINZELIN, J. DE (1955) *Observations sur la genèse des nappes de gravats dans les sols tropicaux*, Brussels, Publ. INEAC, sér. Sc., no. 64, 37 p.
   Attributes the stone-line to a fossilisation under fine earth brought up by termites. Essential.

SYS, C. (1955) 'L'importance des termites sur la constitution des latosols de la région d'Elizabethville', *Sols Afr.* **3**, 392–5.
   Substantial although rapid analysis.

TALTASSE, P. (1957) 'Les cabeças de jacaré et le rôle des termites', *Rev. Géomorph. Dyn.* **6**, 166–70.

TRICART, J. (1957c) 'Observations sur le rôle ameublisseur des termites', *Rev. Géomorph. Dyn.* **8**, 170–2, 179.

SOILS AND MORPHOGENIC PROCESSES, INCLUDING PLUVIAL FACTORS OF OVERLAND FLOW:

ALEXANDRE, J. (1966) 'L'action des animaux fouisseurs et des feux de brousse sur l'efficacité érosive du ruissellement dans une région de savane boisée', *Congr. and Coll. Univ. Liège*, **40**, 1967, pp. 43–9.

BAKKER, J. P. (1951) 'Bodem en bodemprofielen van Suriname, in het bijzonder van de noordelijke savannenstrook', *Landbouwkundig Tijdschr.* **63**, 379–91.
Podzolisation occasionally ending in the formation of a completely impermeable concretionary horizon.

BASU, J. K. and PURANIK, N. B. (1951) 'Land erosion with particular reference to agricultural lands in the dry areas of the Bombay State', *Int. Union Geod. Geophys., Int. Ass. Sci. Hydrol., Brussels Meeting,* vol. 2, 57–63.

BERLIER, Y., DABIN, B., and LENEUF, N. (1956) 'Comparaison physique, chimique et microbiologique entre les sols de forêt et de savane sur les sables tertiaires de la Basse Côte-d'Ivoire', *Rept. VIth Int. Congr. Soil Sc.,* Paris, vol. 5, 499–502.

BIRCH, H. F. and FRIEND, M. T. (1956) 'Humus decomposition in East African soils', *Nature, Lond.* **178**, 500–1.
Alternations of desiccation and humidification accelerate the decomposition of humus.

BRAMAO, L. and BLACK, C. (1955) 'Nota preliminar sobre o estudo solo-vegetação de Barreiras (Bahia)', *Bol. Serv. Nac. Pesquisas Agron.,* Rio de Janeiro, **9**.

COMBEAU, A. (1960) 'Quelques facteurs de la variation de l'indice d'instabilité structurale dans certains sols ferrallitiques', *C.R. Ac. Agric.* (France) **46**, 109–15.
Shows rapid increase in soil erosion with decreasing amounts of humus.

DABIN, B., LENEUF, N., and RIOU, G. (1960) *Carte pédologique de la Côte-d'Ivoire,* scale 1:2,000,000. Explanatory text, Adiopodioumé, Office de la Recherche Scientifique et Technique pour les Pays d'Outre-Mer, 31 p.

DAVEAU, S. (1959) *Recherches morphologiques sur la région de Bandiagara,* Mem. IFAN, no. 56, 120 p., 19 fig., 30 pl.
Excellent monograph on a sandstone region on the margin of the Sahel and the Sudan. Description of processes, especially of overland flow. Should also be referred to in connection with the study of landforms.

FAUCK, F. (1954) 'Les facteurs et les intensités de l'érosion en Moyenne Casamance', *Proc. & Trans. Vth Int. Congr. Soil Sc.,* Leopoldville, vol. 3, 376–9.

FOURNIER, F. (1955) 'Les facteurs de l'érosion du sol par l'eau en Afrique Occidentale Française', *C.R. Acad. Agric.* (France) **41**, no. 15, 660–5.
Excellent analysis of the particular conditions of overland flow in savannas.

MAIGNIEN, R. (1961) 'Le passage des sols ferrugineux tropicaux aux sols ferrallitiques dans les régions Sud-Ouest du Sénégal', *Sols Afr.* **6**, 113–72.

MATON, G. (1960) *Introduction à un important programme de construction de barrages en Haute-Volta,* Ouagadougou, Min. of Agric., mimeographed, 136 p.
Data on rain intensity.

ORSTOM (1957) *Etudes hydrologiques des petits bassins-versants d'Afrique Occidentale Française. Rapport préliminaire pour la campagne de 1957.* Mimeographed.

QUANTIN, P. and COMBEAU, A. (1962) *Erosion et stabilité structurale du sol,* Int. Ass. Sci. Hydrol., Publ. no. 59, 124–30.

RADWANSKI, S. A. and OLLIER, C. D. (1959) 'A study of an East African catena', *J. Soil Sc.* **10**, no. 2, pp. 149–68.

RODIER, J. (1961) 'Transports solides en Afrique Noire à l'Ouest du Congo', *Bull. Int. Ass. Sci. Hydrol.* **6**, 32–4.

RODRIGUES DA SILVA, R., LIMA ALMEIDA, G. C. and SOUTO MAIOR, R. (1957) 'Identificação microscópica dos componentes minerais dos solos', *Inst. Pesquisas Agron. Pernambuco, n.s.,* no. 3, 51 p.
Study of a soil profile in the agreste, with granulometric and petrographic analyses.

RUELLAN, F. (1953) 'O papel das enxurradas no modelado do relévo brasileiro', *Bol. Paulista de Geogr.* **13**, 5–18 and **14**, 3–25.
Excellent study, fundamental. Should also be consulted for types of landforms.

SECK, A. (1955) 'La Moyenne Casamance, étude de géographie physique', *Rev. Géogr. Alpine,* **43**, 707–55.

SEGALEN, P. (1948) 'L'érosion des sols de Madagascar', *Conf. Afr. Sols*, Goma, 1127–37.

TEIXEIRA GUERRA, A. (1955) 'Os lateritos dos campos do Rio Branco e sua importáncia para a geomorfologia', *Rev. Brasil. Geogr.* **17**, 220–4.
Baring by sheet erosion of cuirasses and gravel beds.

TRICART, J. and CARDOSO DA SILVA, T. (1958) 'Algumas observações concernentes as possibilidades do planejamento hidráulico no estado da Bahia', *Est. de Geogr. da Bahia*, Salvador, 51–110.

VERSTAPPEN, H. (1955) 'Geomorphologischen Notizen aus Indonesien', *Erdkunde*, **9**, 134–44.

Cuirasses, their genesis and their relief

From the genetic point of view the essential publications are those of Betremieux and d'Hoore.

ALIA MEDINA, M. (1951) 'Datos geomorfológicos de la Guinea continental española', *Cons. Sup. Invest. Cient.*, Madrid, 63 p., 9 pl.

AUBERT, G. (1948) 'Observations sur le rôle de l'érosion dans la formation de la cuirasse latéritique', *Bull. Agron. Congo Belge*, **40**, no. 2, 1383–6.

AUBERT, G. (1953) 'Quelques problèmes pédologiques de mise en valeur des sols du Delta Central Nigérien (Soudan Français)', *Desert Res.*, Jerusalem, 392–9.
Influence of the lowering of the water table on the precipitation of iron.

AUBREVILLE, A. (1947) 'Erosion et bowalisation en Afrique noire française', *Agron. Trop.* **7/8**, 339–57.

AUFRÈRE, L. (1936) 'La géographie de la latérite', *C.R. Soc. Biogéogr.* **13**, 3–11.
Old but of historic interest.

BETREMIEUX, R. (1951) 'Etude expérimentale de l'évolution du fer et du manganèse dans les sols', *Ann. Agron.* no. 3, 193–295.

BIROT, P. (1955) *Les Méthodes de la morphologie*, Paris, PUF, Coll. Orbis, 177 p.
Useful summation, pp. 63–70.

CARVALHO, G. SOARES DE (1961) *Nota sobre as formações superficiais de pediplanicie da região da Quibala-Catofe (Angola)*', Lisbon, Garcia de Orta, vol. 9, 779–90.

CHETELAT, E. DE (1938) 'Le modelé latéritique de l'ouest de la Guinée française', *Rev. Géogr. Phys. Géol. Dyn.* **11**, 5–120.
Old but good descriptions and perspicacious views.

CRAENE, A. DE (1954) 'De la néoformation du quartz dans le sol', *Sols Afr.* **3**, 298–300.

DAVEAU, S., LAMOTTE, M., and ROUGERIE, G. (1962) 'Cuirasses et chaînes birrimiennes en Haute-Volta', *Ann. Géogr.* **71**, 460–82.

DENISOFF, I. (1957a) 'Contribution à l'étude des formations latéritiques du Parc National de la Garamaba (Congo Belge)', *Pédologie* (Ghent) **7**, 124–32.

DENISOFF, I. (1957b) 'Un type particulier de concrétionnement en cuvette centrale congolaise', *Pédologie* (Ghent) **7**, 119–23.

DRESCH, J. (1952) 'Dépots de couverture et relief en Afrique Occidentale française', *Proc. XVIIth Int. Geogr. Congr.*, Washington, 323–6.
Demonstrates the allogenic nature of certain cuirasses and inversions of relief.

ERHART, H. (1951a) 'Sur l'importance des phénomènes biologiques dans la formation des cuirasses ferrugineuses en zone tropicale', *C.R. Acad. Sc.* **233**, 804–6.

ERHART, H. (1951b) 'Sur le rôle des cuirasses termitiques dans la géographie des régions tropicales', *C.R. Acad. Sc.* **233**, 966–8.

ERHART, H. (1953a) 'Sur la nature minéralogique et la genèse des sédiments dans la cuvette tchadienne', *C.R. Acad. Sc.* **237**, 401–3.

ERHART, H. (1953b) 'Sur les cuirasses termitiques fossiles dans la vallée du Niari et dans le massif du Chaillu (Moyen Congo, A.E.F.)', *C.R. Acad. Sc.* **237**, 431–3.

FREISE, F. W. (1936) 'Bodenverkrüstung in Brasilien', *Z. Geomorph.* **9**, 233–48.

HARMS, J. E. and MORGAN, B. D. (1964) 'Pisolitic limonite deposits in North-West Australia', *Proc. Australasian Inst. Mining Metall.* **212**, 91–124, 2 maps, 2 cross-sections, 5 phot., 2 tables.

HINGSTON, F. J. and MULCAHY, M. J. (1961) 'Laboratory examination of the soils of the York-Quairading area, W.A.', *Commonw. Scientif. Industr. Res. Org., Div. Soils Rept.*, Adelaide, 9/61, 76 p.
Analysis of soils derived from Tertiary ferruginous cuirasses.

HOORE, J. D' (1954a) *L'accumulation des sesquioxydes libres dans les sols tropicaux*, Brussels Publ. INEAC, *Sér. Sc.*, no. 62, 132 p.
Basic.

HOORE, J. D' (1954b) 'Essai de classification des zones d'accumulation des sesquioxydes libres sur des bases génétiques', *Sols Afr.* **3**, 66–81.
Basic, completed by the article cited above.

ISNARD, H. (1955) *Madagascar*, Paris, Colin, 219 p.
Descriptions, some useful data.

JAEGER, P. (1956) 'Contribution à l'étude des forêts reliques du Soudan occidental', *Bull. IFAN*, sér. A, **18**, 993–1053.
Geomorphic data and cuirassing of sandstones.

LAMOTTE, M. and ROUGERIE, G. (1952) 'Coexistence de trois types de modelé dans les chaînes quartzitiques du Nimba et du Simandou (Haute Guinée Française)', *Ann. Géogr.* **61**, 432–42.

LAMOTTE, M. and ROUGERIE, G. (1953) 'Les cuirasses ferrugineuses allochtones. Signification paléoclimatique et rapports avec la végétation', *C.R. IVe Réunion Afric. de l'Ouest*, Abidjan, pp. 89–90.

LAMOTTE, M. and ROUGERIE, G. (1962) 'Les apports allochtones dans la genèse des cuirasses ferrugineuses', *Rev. Géomorph. Dyn.* **13**, 145–60.
Basic. Excellent discussion of the problem. Important bibliography.

LAPLANTE, A. and BACHELIER, G. (1954) 'Un processus pédologique de la formation des cuirasses latéritiques dans l'Adamaoua (Nord Cameroun)', *Rev. Géomorph. Dyn.* **6**, 214–19.

LAPLANTE, A. and ROUGERIE, G. (1950) 'Etude pédologique du bassin de la Bia', *Bull. IFAN*, sér. A, **12**, 883–904.

LAPPARENT, J. DE (1939) 'La décomposition latéritique du granite dans la région de Macenta (Guinée Française). L'arénisation prétropicale et prédésertique en Afrique Occidentale Française et au Sahara', *C.R. Acad. Sc.* **208**, 1767–9; **209**, 7–9.

MAIGNIEN, R. (1954a) 'Formation de cuirasses de plateau, région de Labé (Guinée Française)', *Proc. Vth Int. Congr. Soil Sc.* iv, 13–18.

MAIGNIEN, R. (1954b) 'Cuirassement de sols de plaine, Bally (Guinée Française)', *Proc. Vth Int. Congr. Soil Sc.* iv, 19–22.

MAIGNIEN, R. (1954c) 'Différents processus de cuirassement en A.O.F.', *C.R. IIe Conf. Interafr. Sols*, doc. 116, 1469–86.
These three articles contain Maignien's first conclusions regarding the cuirassing process, conclusions which were reworked and elaborated in his doctoral dissertation (1958a).

MAIGNIEN, R. (1956) 'De l'importance du lessivage oblique dans le cuirassement des sols en A.O.F.', *Rept. VIth Int. Congr. Soil Sc.*, Paris, vol. 5, pp. 463–7.

MAIGNIEN, R. (1958a) *Le cuirassement des sols en Guinée (Afrique Occidentale)*, Mem. Serv. Carte Géol. Alsace-Lorraine, no. 16, 239 p.
Basic work; essential geomorphological point of view although occasionally insufficiently stressed.

MAIGNIEN, R. (1958b) 'Le cuirassement des sols en Afrique tropicale de l'Ouest', *Sols Afr.* **4**, no. 4, pp. 5–41.

MAIGNIEN, R. (1966) *Review of Research on Laterites*, UNESCO, ser. Research on natural resources, no. 4, 148 p.

MICHEL, P. (1958) *Rapport de mission dans le Nord du Fouta-Djalon et dans le pays bassaris (Guinée)*, Dakar, SGPM, 50 p., mimeographed, 31 phot., 20 pl. Reviewed in *Rev. Géomorph. Dyn.* **11** (1960), 72–4.

Firsthand very precious observations made in a geomorphological and palaeogeographical spirit.

MICHEL, P. (1959) 'L'évolution géomorphologique des bassins du Sénégal et de la Haute-Gambie, ses rapports avec la prospection minière', *Rev. Géomorph. Dyn.* **10**, 117–43.
Basic work. Excellent monograph about the morphogenic processes of a savanna region. Practical applications.

MICHEL, P. (1960) *Notes sur les formations cuirassées de la région de Kédougou* (March–April 1960), Bur. Rech. Géol. Min., Paris, mimeographed, 23 p.

MICHEL, P. (1962) *Etude géomorphologique des sondages dans les formations cuirassées de la région de Kédougou (Sénégal)*, Bur. Rech. Géol. Min., Dakar, A5, mimeographed, 56 p., 20 pl.

MULCAHY, M. (1960) 'Laterites and lateritic soils in south-western Australia', *J. Soil Sc.* **11**, no. 2, 206–26.

MULCAHY, M. and HINGSTON, F. (1961) 'The development and distribution of soils of the York-Quairading area, Western Australia, in relation to landscape evolution', *Commonw. Scientif. & Industr. Res. Organiz., Soil Publ.* no. 17, 43 p., 3 pl., 1 map.
Excellent monograph carefully analysing the relationships between the geomorphic evolution and pedology in a region of a dissected Tertiary cuirass.

NEIVA, J. M. COTELO and NEVES, J. M. CORREIA (1957) 'Latérites de l'Ile du Prince', *Mem. e Noticias, Mus. e Lab. Miner. e Geol. Univ. Coimbra*, **44**, 1–9.
Good description and mineralogic study.

PALLISTA, J. W. (1953) 'Erosion levels and laterite in Buganda province, Uganda', *Proc. XIXth Int. Geol. Congr.*, Algiers, **21**, 193–9.
Describes the evolution of a cuirassed plateau.

PELISSIER, P. (1960) 'Un point de vocabulaire relatif à la morphologie tropicale', *Inf. Géogr.* **24**, 113–14.

PELISSIER, P. and ROUGERIE, G. (1953) 'Problèmes morphologiques dans le bassin de Siguiri (Haut Niger)', *Bull. IFAN*, sér. A, **15**, 1–47.
Interesting regional monograph.

PITOT, A. (1954) 'Végétation et sols et leurs problèmes en A.O.F.', *Ann. Ec. Sup. Sc.*, Dakar, **1**, 128–39.
Accepts that a cuirassing process is presently going on in the north of the forest zone. Useful bibliography.

PLAYFORD, P. E. (1954) 'Observations on laterite in Western Australia', *Austr. J. Sc.* **17**, no. 1, 11–14.

POUQUET, J. (1954) 'Altération de dolérites de la presqu'île du Cap Vert (Sénégal) et du plateau du Labé (Fouta-Djalon, Guinée Française)', *Bull. Assoc. Géogr. Franç.*, 173–82.
Notes on the cuirassing process.

POUQUET, J. (1956) 'Aspects morphologiques du Fouta Djalon, régions de Kindia et de Labé, Guinée Française, A.O.F. Aspects alarmants des phénomènes d'érosion des sols déclenchés par les activités humaines', *Rev. Géogr. Alpine*, 231–46.

POUQUET, J. (1956) 'Méthodes d'études des versants et principaux résultats obtenus sur le Labé. Guinée Française, A.O.F.', *1er Rapport Comm. Et. des Versants*, Amsterdam, 85–95.

PREEZ, J. W. DU (1954) *Notes sur la présence d'oolithes et de pisolithes dans les latérites de Nigeria*, Bur. Interafricain Sols, mimeographed, 7 p.

PRESCOTT, S. A. and PENDLETON, R. L. (1952) 'Laterite and lateritic soils', *Commonwealth Bur. Soil Sc. Tech. Comm.* no. 47, 51 p.
Already a somewhat old summation, useful however to show the evolution of ideas.

REFORMATSKY, N. (1935) 'Quelques observations sur les latérites et les roches ferrugineuses de l'Ouest de la colonie du Niger', *Bull. Soc. Géol. Fr.* **5**, 575–89.

RIQUIER, J. (1954) 'Formation d'une cuirasse ferrugineuse et manganésifère ou latéritique', *Proc. & Trans. Vth Int. Congr. Soil Sc.*, iv, 229–36.

ROUGERIE, J. and LAMOTTE, M. (1952) 'Le Mont Nimba (Guinée, Côte-d'Ivoire). Etude de morphologie tropicale', *Bull. Assoc. Géogr. Franç.* **226/8**, 113–20.
Useful monograph. Describes cuirasses on steep slopes.

ROUGERIE, G. (1959) 'Latéritisation et pédogenèse intertropicales', *Inf. Géogr.* **23**, 199–206.
Useful and rapid summation.

SABOT, J. (1952) 'Les latérites', *Proc. XIXth Int. Geol. Congr.*, Algiers, **21**, 181–92.
Useful summation but has become inaccurate on certain points.

SAURIN, E. and CARBONNEL, J. P. (1964) 'Les latérites sédimentaires du Cambodge oriental', *Rev. Géogr. Phys. Geol. Dyn.*, n.s., **16**, 241–56.

SCAETTA, H. (1938) 'Sur la genèse et l'évolution des cuirasses latéritiques. Rôle des cuirasses latéritiques dans l'évolution ultérieure des sols sous-jacents', *C.R. Soc. Biogéogr.* **15**, no. 125, 14–18; no. 126, 26–7.
Pioneer work.

SCAETTA, H. (1941) 'Limites boréales de la latéritisation actuelle en Afrique occidentale. L'évolution des sols et de la végétation dans la zone des latérites en Afrique occidentale', *C.R. Acad. Sc.* **212**, 129–30, 169–71.

SCHOFFIELD, A. N. (1957) 'Nyasaland laterites and their indication on aerial photographs', *Overseas Bull.*, Road. Res. Lab., 5 p.
Only notes allogenic cuirasses.

SECK, A. (1955) 'La Moyenne Casamance, étude de géographie physique', *Rev. Géogr. Alpine*, **43**, 707–55.
Describes an old dissected cuirass.

SHERMAN, G. and KANEHIRO, Y. (1954) 'Origin and development of ferruginous concretions in Hawaiian latosols', *Soil Sc.* **77**, 1–8.

TESSIER, F. (1954) 'Oolithes ferrugineuses et fausses latérites dans l'Est de l'Afrique Occidentale Française', *Ann. Ec. Sup. Sc.*, Dakar, **1**, 113–28.
A lacustrine origin is attributed to oolitic horizons.

TESSIER, F. (1959) 'La latérite du Cap Manuel à Dakar et ses termitières fossiles', *C.R. Acad. Sc.* **248**, 3320–2.
This interpretation was arrived at simultaneously by P. Grassé.

TESSIER, F. (1959) 'Termitières fossiles dans la latérite de Dakar (Sénégal). Remarques sur les structures latéritiques', *Ann. Fac. Sc. Univ. Dakar*, **4**, 91–132.
Detailed description.

TEIXEIRA GUERRA, A. (1957) 'Estudio geográfico do territorio do Rio Branco', *Cons. Nac. Geogr.*, Rio de Janeiro, **9**, 669–72.
Old concepts.

TEIXEIRA GUERRA, A. (1953) 'Formação de lateritas sol a floresta equatorial amazônica (Territorio Federal do Guaporé)', *Rev. Brasil. Geogr.* **14**, 407–26.
Good example of misapprehensions often made about ancient laterites.

TEIXEIRA GUERRA, A. (1955) 'Ocorréencia de lateritas na bacia do Alto Purús', *Rev. Brasil. Geogr.* **17**, 107–14.

TEIXEIRA GUERRA, A. (1957) 'Estudio geográfico do territorio do Rio Branco', *Cons. Nac. Geogr.*, Rio de Janeiro, 252 p., 150 fig.
Good descriptions. From now on considers cuirasses as being ancient.

TOTHILL, J. D. (1952) 'A note on the origin of the soils of the Sudan from the point of view of the man in the field', *Agriculture in the Sudan*, 2nd edn, Oxford University Press, pp. 129–43.

TRICART, J. (1959) 'Géomorphologie dynamique de la moyenne vallée du Niger, Soudan', *Ann. Géogr.* **68**, 333–43.

TRICART, J. (1961) 'Le modelé du Quadrilatero Ferrifero, au Sud de Belo Horizonte (Brésil)', *Ann. Géogr.* **70**, 255–72.

TRICART, J. (1962) 'Quelques éléments de l'évolution géomorphologique de l'Ouest de la Côte-d'Ivoire', *Rech. Afr.*, Conakry, **1**, 30–9.

TURTON, G. A., MARSH, N. L., MCKENZIE, R. M. and MULCAHY, M. J. (1962) 'The chemistry and mineralogy of laterite soils in the southwest of Western Australia', *Commonw. Scient. & Industr. Res. Organ.*, *Soil Public.* no. 20, 40 p., 5 fig.
Detailed mineralogic study of an old cuirassed surface and the soils derived from it.

VANN, J. H. (1963) 'Developmental processes in laterite terrain in Amapá', *Geogr. Rev.* **53**, 406–17.

WAEGEMANS, G. (1951) 'Introduction à l'étude de la latéritisation et des latérites du Centre Africain', *Bull. Agr. Congo Belge*, **42**, no. 1, 13–56.
Historical review of theories on laterisation and cuirassing; summation of concepts at that time; the point of departure of present concepts.

WAEGEMANS, G. (1954) *Les Latérites de Gimbi (Bas-Congo)*, Brussels Publ. INEAC, Sér. Sc., no. 60, 27 p.
Very good monograph about ferruginous concretions gradually grading into a scoriaceous cuirass on the margin of a plateau. Accepts the idea of a ground water cuirass before dissection. Analyses and photographs.

WEISSE, G. DE (1952) 'Note sur quelques types de latérites de la Guinée Portugaise', *Proc. XIXth Int. Geol. Congr.*, Algiers, **21**, 171–9.
Demonstrates an allogenic origin of most cuirasses.

Landforms and geomorphic evolution

One should refer back to a large number of publications cited above, in particular to de Chetelat (1938), Daveau (1959), Daveau, Lamotte and Rougerie (1962), Ruellan (1953), Tricart (1959, 1961), etc. First on the list will be a series of references concerning common landforms, to be followed by sources on inselbergs, in order to facilitate comparative studies dealing with the wet, wet–dry, as well as with the dry tropics.

COMMON EROSIONAL LANDFORMS

ACKERMANN, E. (1936) 'Dambos in Nordrhodesien', *Wiss. Veröffentl. Dtschen Mus. Länderkunde zu Leipzig*, n.s., **4**, 149–57.

BATTISTINI, R. and DOUMENGE, F. (1966) 'La morphologie de l'escarpement de l'Isalo et de son revers dans la région de Ranohira (Sud-Ouest de Madagascar)', *Madagascar, Rev. de Géogr.*, no. 8, 67–92.

BÜDEL, J. (1965) *Die Relieftypen der Flächenspülzone Süd-Indiens am Ostabfall Dekans gegen Madras*, Colloquium geographicum, Bd. 8, Bonn Univ., F. Dümmler, 100 p.

CLAYTON, R. W. (1956) 'Linear depressions (Bergfussniederungen) in savannah landscapes', *Geogr. Studies*, **3**, 102–26.

COTTON, C. A. (1961) 'The theory of savanna planation', *Geography*, **46**, 89–101.

DAVEAU, S. (1960) 'Les plateaux du Sud-Ouest de la Haute-Volta, étude géomorphologique', *Trav. Dépt. Géogr. Univ. Dakar*, no. 7, 65 p., 11 fig.

DAVEAU, S. (1962) 'Principaux types de paysages morphologiques des plaines et plateaux soudanais dans l'Afrique de l'Ouest', *Inf. Géogr.* **26**, 61–72.
Rapid summation.

DELHAYE, F. and BORGNIEZ, G. (1947) 'Contribution à la connaissance de la géographie et de la géologie de la région de la Lukénie et de la Tshuapa supérieures', *Bull. Soc. Belge Géol. Paléont. Hydrol.* **56**, 349–71.

DIXEY, F. (1943) 'The morphology of the Congo-Zambezi watershed', *South Afr. Geogr. J.* **25**, 20–41.

DOMINGUES, A. PORTO and KELLER, E. (1956) *Guidebook* no. 6, Bahia, XVIIIth Int. Geogr. Congr., Brazil, 254 p., 35 phot.

DRESCH, J. (1952) 'Dépots superficiels et relief du sol au Dahomey septentrional', *C.R. Acad. Sc.* **234**, 1566–8.

DRESCH, J. (1953) 'Plaines soudanaises', *Rev. Géomorph. Dyn.* **4**, 39–44.

FAIR, T. J. (1944) 'The geomorphology of the Ladysmith basin, Natal', *South Afr. Geogr. J.* **26**, 35–43.

GOUROU, P. (1956) 'Milieu local et colonisation réunionaise sur les plateaux de la Sakay (Centre-Ouest de Madagascar)', *Cahiers d'Outre-Mer*, **9**, 36–57.
Notes on runoff.

HANDLEY, J. R. (1952) 'The geomorphology of the Nzega area, Tanganyika, with special reference to the formation of granite tors', *Proc. XIXth Int. Geol. Gongr.*, Algiers, pp. 201–10.

JAEGER, P. (1953) 'Contribution à l'étude du modelé de la dorsale guinéenne: les Monts Loma (Sierra Leone)', *Rev. Géomorph. Dyn.* **3**, 105–13.

KREBS, N. (1933) 'Morphologische Beobachtungen in Südindien', *Sitz. Ber. Preuss. Akad. Wiss., Phys.-Math. Kl.*, no. 23.

LEFEVRE, M. (1953) 'Notes sur la morphologie du Katanga', *Bull. Soc. Belge Et. Géogr.* **22**, 407–31.

LOUIS, H. (1964) 'Uber Rumpflächen- und Talbildung in den wechselfeuchten Tropen besonders nach Studien in Tanganyika', *Z. Geomorph.*, n.s., **8**, Sonderheft, 43–70.

MASSIP, S. and MASSIP, S. DE (1949) 'La cordillera de los Organos en la porción occidental de Cuba', *Proc. Int. Geogr. Congr.*, Lisbon, **2**, 734–47.

MATHIEU, F. (1931) 'Notes sur la géographie physique des bassins de la Dungu et de la Duru (Haut Uele)', *Ann. Soc. Géol. Belgique*, **53**, 663–8.

MORTELMANS, G. (1951) 'Observations sur la morphologie de la région Mituaka-Haute Kalumengongo (Monts Kilara, Katanga)', *Bull. Soc. Belge Géol. Paléont. Hydrol.* **59**, 383–99.

POLINARD, E. (1948) 'Les grands traits de la géographie physique et les particularités des formations du plateau dans le Nord-est de la Lunda (Angola). Interprétations des observations des premières missions de recherches', *Bull. Soc. Belge Géol. Paléont. Hydrol.* **57**, 541–53.

RADWANSKI, S. A. and OLLIER, C. D. (1959) 'A study of an East African catena', *J. Soil Sc.*, **10**, 149–68.

RATHJENS, C. (1957) 'Physische-geographische Beobachtungen im nordwestindischen Trockengebiet (Ein erster Forschungsbericht)', *Erdkunde*, **11**, 49–58.
Changes in relief on the margin of the dry zone.

SAVIGEAR, R. (1960) 'Slopes and hills in West Africa', *Z. Geomorph.*, Suppl. i, 156–71.
Confusing and theoretical.

TEIXEIRA GUERRA, A. (1954) *Estudio geografico do territorio do Amapá*, IBGE, 366 p., 246 fig.

THORP, M. (1967) 'Closed basins in younger granite massifs, northern Nigeria', *Z. Geomorph.*, n.s., **11**, 459–80.

VOGT, J. (1956) *Rapport provisoire de mission à Kéniéba, Soudan*, DFMG.

VOGT, J. (1961) 'Badlands du Nord-Dahomey', *Actes 85ᵉ Congr. Soc. Sav.*, 1960, Geogr., 227–39.

INSELBERGS

BARBEAU, J. and GÈZE, B. (1957) 'Les coupoles granitiques et rhyolitiques de la région de Fort-Lamy (Tchad)', *Bull. Soc. Géol. Fr.*, 6ᵉ sér. **7**, 341–51.

BEHREND, F. (1919) 'Über die Entstehung der Inselberge und Steilstufen besonders in Afrika und die Erhaltung ihrer Formen', *Z. Dtschen Geol. Ges.* **70**, 154–67.

BERGER, H. (1959/60) 'Beobachtungen zur Inselbergbildung in Nordost Uganda', *Geogr. Jahresver. aus Oesterr.* **28**, 72–9.

BORNHARDT, W. (1900) *Zur Oberflächengestaltung und Geologie Deutsch Ostafrikas*, Berlin, 595 p., 28 pl., atlas of 8 maps.

CZAJKA, W. (1957) 'Das Inselbergproblem auf Grund von Beobachtungen in Nordost-brasilien', *Matchatschek Festschrift*, 321–33.

FREISE, F. (1938) 'Inselberge und Inselberg-Landschaften im Granit- und Gneissgebiete Brasiliens', *Z. Geomorph.* **10**, no. 4/5, 137–68.

HANNEMAN, M. (1951/2) 'Eine Inselberglandschaft in Zentraltexas', *Erde*, 354–65.

KING, L. C. (1948) 'A theory of bornhardts', *Geogr. J.* **112**, 83–7.

KREBS, N. (1942) 'Über Wesen und Verbreitung der tropischen Inselberge', *Abh. Preuss. Akad. Wiss., Math.-Nat. Kl.*, no. 6.

LOUIS, H. (1959) 'Beobachtungen über die Inselberge bei Hua-Hin am Golf von Siam', *Erdkunde*, **13**, 314–19.

MORTENSEN, H. (1929) 'Inselberglandschaft im Nordchile', *Z. Geomorph.* **4**.

NOWACK, E. (1936) 'Zur Erklärung der Inselberglandschaften Ostafrikas', *Z. Ges. Erdkunde*, Berlin, **7/8**.

OLLIER, C. D. (1960) 'The Inselbergs of Uganda', *Z. Geomorph.*, n.s., **4**, 43–52.

OLLIER, C. D. and TUDDENHAM, W. G. (1961) 'Inselbergs of Central Australia', *Z. Geomorph.*, n.s., **5**, 257–76.

PASSARGE, S. (1904) 'Die Inselberglandschaften im tropischen Afrika', *Natur. Wochenschr.*, Iena, **19**, 657–65.

PASSARGE, S. (1904) 'Rumpfflächen und Inselberge', *Z. Dtschen Geol. Ges.* **56**, 193–215.

PASSARGE, S. (1912) 'Inselberglandschaften der Massaisteppen', *Pet. Geogr. Mitt.*

PASSARGE, S. (1924) 'Das Problem der Skulptur-Inselberglandschaften', *Pet. Geogr. Mitt.*, 66–70, 117–20.

RIBEIRO, O. (1954) 'Contribution à l'étude géographique des pays de l'Océan Indien', *Pan-Indian Oc. Sc. Ass.*, March 8, mimeographed.
Problems posed by inselbergs; study suggestions.

ROCH, E. (1952) 'Les reliefs résiduels ou Inselberge du bassin de la Bénoué (Nord Cameroun)', *C.R. Acad. Sc.* **234**, 117–18.

ROCH, E. (1953) 'Itinéraires géologiques dans le Nord de Cameroun et le Sud-Ouest du Territoire du Tchad', *Bull. Serv. Mines Cameroun*, no. 1, 110 p.
Useful descriptions. Precise geologic definitions.

ROUGERIE, G. (1955) 'Un mode de dégagement probable de certains dômes granitiques', *C.R. Acad. Sc.* **240**, 327–9.

RUXTON, B. P. (1958) 'Weathering and subsurface erosion in granite at the piedmont angle, Balos, Sudan', *Geol. Mag.* **95**, 353–77.

THORBECKE, F. (1921) 'Die Inselberglandschaft von Nord-Tikar', *Festschr. A. Hettner*, Breslau, 215–42.

WAIBEL, L. (1925) 'Gebirgsbau und Oberflächengestalt der Karrasberge in Südwestafrika', *Mitt. Dtschen Schutz.* **33**.

5

Interruptions in the bioclimatic equilibrium

We have seen that the climactic landforms of the humid tropics vary considerably according to the humidity of the climate, which determines the nature of the vegetation. Under rainforest weathering is far more important than erosion, which results in varying proportions depending on the kind of weathering from the combined effects of runoff and creep, to which must be added occasional landslides. When erosion prevails over weathering the forest is interrupted and monolithic domes appear. Under savanna, on the contrary, less intense weathering produces a more irregular regolith, often broached by rock outcrops. Erosion, which is more rapid, is mainly due to the effects of runoff. Pediplains are cut from the regolith and colluvium accumulates at the base of hillsides. Lastly and specifically, indurations occur in the form of cuirasses, which play an important geomorphic role.

But the forest is fragile, for it has important ecological requirements. Its destruction causes a profound change in geomorphic conditions, even if it is not immediately followed by 'laterisation', as was formerly believed. An interruption of the bioclimatic equilibrium in such a case has important repercussions on the relief. It occurs in two different cases, which are not of the same order of magnitude:

(*a*) Through the impact of man, who destroys the forest to extend his fields or his pastures. It has resulted, notably in Africa, in widespread replacement of the deciduous seasonal forests by burnt-over savannas.

(*b*) Through the influence of palaeoclimatic oscillations, which have affected the tropics not only during the Quaternary but also during the Tertiary and earlier, and have reshaped the landforms more owing to fluctuations in humidity than in temperature.

We must now study their effects.

The impact of man

This is not the place to review the study of the influences of the most ancient and most permanent of the actions of man, such as those of brush fires: they have lasted long enough to create a real artificial climax, that of the burnt-over savannas, which are only a more extreme form of the morphoclimatic

233

environment with a long dry season. Our attention will be drawn to the consequences of a more recent destruction of the vegetation. The destruction of vegetation in the tropics produces a much greater morphogenic disturbance than in other morphoclimatic zones, the temperate zone for example. Before studying the peculiar landforms that result from this destruction, we must determine its cause.

Factors contributing to the magnitude of man-induced morphogenic processes peculiar to the tropics

Several causes contribute to make an interruption in the climactic equilibrium of the vegetation of the tropics particularly serious. Some result from the peculiar conditions of the environment, others from the particularities of the soil-forming processes.

FACTORS DUE TO THE NATURE OF THE ENVIRONMENT

As the factors of the environment have already been mentioned, we will only recall them briefly here, emphasising their relationships.

On the one hand the climate is marked by a great deal of aggressiveness. In many places the proportion of violent downpours is high due to the convective nature of the rainfall. The first rains following the dry season are usually the most intense; for example, in monsoonal regions or the African savannas, as have among others shown the studies of Maton (1960) in Upper Volta. The first rains fall on hardened ground that is vulnerable to degradation due to the loss of superficial permeability. Indeed, the essentially kaolinitic intertropical clays have a small contraction coefficient. They cleave little. Wide cracks open to running water do not exist but only narrow fissures, which absorb a small proportion of the precipitation. These characteristics do not prevent kaolinitic soils from playing an important morphogenic role however, as we see later.

The natural vegetation plays a protective role even in savannas. If it does not prevent desiccation in savannas, at least it protects part of the ground against splash erosion. The soil is not struck too violently, it is not made too impermeable by the impact of raindrops. Pools that form on bare mud flats and migrate with slowly shifting gradients lose part of their volume by peripheral infiltration. The critical limit beyond which runoff becomes channelled is not easily attained. The obstacles that are plants slow down the overland flow and help delay its concentration. Once the vegetation is destroyed and the soil becomes degraded, the waters start to concentrate and gullies to form. Only the denudation of a cuirass or a considerable cover of rubble or ferruginous concretions is capable of maintaining an unconcentrated type of runoff. Sheet erosion is then arrested for want of material to remove. The worse the ecological conditions are, the greater is the danger of sheet erosion, for the recolonisation by plants of the cleared area is slower and not as complete. *Bowé* offer excellent examples.

The protection provided by vegetation is much greater under forest than under savanna. As we have seen, the forest creates a particular soil climate. The vegetation causes an effective, albeit incomplete, interception of the precipitation, especially in the rainforest. But above all the soil is protected against variations in temperature and humidity, and its surface never dries up completely. When the forest is destroyed, the protection against runoff that it provided and its role as a screen against splash erosion are simultaneously eliminated, and the soil climate is completely modified. The soil rapidly dries in the sun, hardening in a few rainless days. Its surface becomes similar to that of a savanna at the end of the dry season and identical conditions favourable to intensive runoff then exist, except that they are also aggravated by the thickness of the regolith, the very nature of the soils, and the relief, especially the steeper hillsides.

It is therefore under forest that the disequilibrium is greatest, which is easily understood, for it is here that the soil climate, protected by the vegetation, is most different from that of the free air.

But the rapid degradation of fragile soils exposed to the free air further aggravates the situation by making them increasingly vulnerable to attack.

THE RAPID DEGRADATION OF SOILS

Tropical soils are fragile. The destruction of vegetation affects them very rapidly and in several ways concurrently. The supply of organic matter is interrupted, as the cropping systems, with certain exceptions, do not provide any manure. In the banana plantations of Guinea one metre of mulch must be added yearly to maintain the humus content of the soil! The very rapidity of the mineralisation of the humus aggravates the consequences of the interruption in the supply of organic matter, making its effect felt all the more rapidly. The cleared soil is already degraded at the end of several months.

The decomposition of the humus left by the cleared forest is itself accelerated, as Birch and Friend (1956) have shown. According to these authors the decomposition is activated by alternations of wetting and drying, which are precisely more numerous and intense when the plant cover is suppressed, especially in the case of a forest. The mechanism would be as follows: the organic matter in process of decomposition in the soil becomes covered with an argillaceous film that protects it, slowing down the decomposition of the organic matter by keeping out the air. However, the film cracks under the effect of desiccation, enabling oxidation to rapidly dispose of the organic matter.

In turn, the swift disappearance of the humus has grave consequences.

Under natural conditions, as Quantin and Combeau (1962) have shown on experimental plots at Grimari, Central African Republic, slightly ferrallitic red soils on gneiss have a good structural stability under seasonal deciduous forest with intense dry season and an annual rainfall of 1 800mm

(72 in). Henin's (1952) index of erodibility[1] is low (0·24). Under cultivation without manure it reaches 1·08 after a short period of time. On slopes of 1 to 3° erosion is intense and runoff reaches 8 to 20 per cent. Sandy soils are more resistant than finer soils. Particles smaller than 20 microns indeed form the major part of the transported matter. But the very low gradient of the slope should be taken into account. Researches prior to Combeau (1960) have shown that erosion increases as the content of the humus decreases. But fines are much more abundant in forest than in savanna soils, therefore the clearing of the forest, which causes the destruction of the humus, has very grave consequences. The proportion of fines also varies according to the rocks; less common on granites, fines are more abundant on shales and micaschists, which explains the seriousness of erosion on a regolith derived from such rocks. Similar phenomena are found in subtropical forested regions, as Marques *et al.* (1951) have shown.

Absence or scarcity of humus also has repercussions on the permeability of the soil, as Segalen (1948) has shown in Madagascar. Water deflocculates the clay on the ground surface, after which desiccation causes the clay to bind the soil elements, to cement them into a hard, impermeable crust. Even after several days of rain the soil is often wet only to a depth of a few centimetres. Moreover, solutions of iron precipitated during dry spells contribute to this impermeability. The destruction of the humus indeed impedes the leaching of iron and prevents it from being removed to greater depths. That which is freed by plants consumed in brush fires remains close to the surface, which it hardens. Making certain allowances, a similar phenomenon occurs when chunks of soil are dug up with a spade and dried to form coherent blocks of 'laterite', which can be used as building stones, as d'Hoore (1954a) has reminded us. The iron content in savannas is always higher in the upper 50 cm (20 in) of soil, which makes such mechanisms function more readily even during a simple degradation of the vegetation and, *a fortiori*, during brush fires that precipitate all the solutions present in the soil. The iron adsorbed by clays, which are easily mobilised, progressively gives way to individual films of iron, which diminish the permeability of the soil. This evolution is facilitated by alternations of wetting and drying. Indeed, wetting enables the adsorbed iron to go back into solution, whereas desiccation precipitates the solutions in the form of films or coatings, especially if it is rapid. The laying bare of the soil, which considerably increases variations in humidity, and, especially, brush fires do much to accelerate this hardening process.

The agricultural development of tropical soils therefore constitutes a perilous undertaking, much more hazardous than that of midlatitude soils, even in 'mediterranean' regions. It explains the empirical practices adopted by African and Asian cultivators over the centuries. The irregular cleared patch, located at random, only permits a localised runoff, reduced downhill by natural or secondary vegetation. When extensive uninterrupted

[1] Index currently used by French pedologists.

clearings replace it, erosion increases catastrophically as runoff finds enough space to become concentrated into real torrents even on gentle slopes. The first clearings made in the Dabou savanna of lower Ivory Coast, to make room for rubber plantations, triggered off a furious wave of erosion on slopes of less than 5°. Now, before anything else is undertaken, terracettes are constructed. In a similar way traditional agriculture preserved a good part of the forest cover, notably all the large trees, which reduced splash erosion, produced shade, and reduced the amount of desiccation. Under their cover forest regrowth proceeded more favourably and more rapidly. The period of morphogenic imbalance was not as long. Introduction of the match and the axe, without any changes in cultivation methods, have considerably worsened the situation. Presentday clearing by fire is often so complete that nothing is spared. In the climax forest of western Ivory Coast, we have seen trees, 2 or 3 m (7 to 12 ft) in diameter, felled and superficially calcined, between which dry rice was being sowed. Such brutal interruptions in the balance of nature have consequences that are much more serious than those of the traditional methods of agriculture.

Nowhere else in the world are man-induced morphogenic processes so violent as in the tropics at the present time. An equal degree of intensity is only reached in some semiarid regions. The results are landforms that are to a certain extent unique.

Characteristics of tropical landforms caused by man

Besides soil erosion and gullying, which are in all cases induced by man, the humid intertropical zone has some peculiar landforms of its own: lavaka or voçoroka type ravines and large scale slumps. They are characteristic of regions whose climax vegetation would be forest, and which have a thick regolith.

LAVAKAS AND VOÇOROKAS

In Madagascar *lavakas* are abrupt ravines usually with crumbling walls, which, however, may be stabilised by a regrowth of vegetation. They are cut from the regolith, especially in a mantle of kaolinitic weathering. Their forms vary, so that it is preferable to give a genetic rather than a purely morphologic connotation to the term. *Voçorokas* are their Brazilian equivalent. Such landforms are not limited to these two countries; similar ones have been described in Hong Kong, for example (Panzer, 1954; Berry and Ruxton, 1960). Because the term lavaka is the better known in French geographical literature, we will use it here.

Lavakas may be as long as 1 km (0·6 mile) but usually are not more than 300 to 600 m (1 000 to 2 000 ft) because they are eroded from convex hills that are common in regions of forest climax where they ordinarily occur. Their volume may attain 500 000 cu m (18 million cu ft) (Brenon, 1952). They are several metres deep, sometimes 10 or 20 (35 to 70 ft). Because of their deeply incised and sharp forms, they contrast markedly with marshy, flat-bottomed, climactic vales, which, moreover, they fill and bury with their sandy fans.

The best type description is that of Brenon (1952). The overall shape is ovoid, with the wide end upstream, denting the upland. A narrow opening towards the lowland assures its outlet, which is prolonged by a fan of micaceous and argillaceous sands, angular rubble of not very coherent rocks, and especially quartz. The fan is very flat. It sometimes intersects the climactic bottomland, forcing its stream to undercut it and become loaded with sand. The height of the lavaka sidewalls increases upstream and generally reaches a peak at its head, which has the aspect of an abrupt niche. When live, the sidewalls are subvertical and crumbling. Lower down they may be less steep, with angles of 45°, and scree spread at the base. On their margins, uphill of the precipice, subparallel cracks, $\frac{1}{2}$ to 5 m (2 to 16 ft) wide, have torn open parallel to the sidewalls. The floor of the lavaka is often nearly flat although sloping and coincides more or less with the weathering front or with the contact of the disintegrated rock (Zone III) and the bedrock with weathered joints (Zone IV). It may be several metres, even several tens of metres wide (10 to 300 ft).

Seepages forming more or less well individualised springlets are found at the foot of the head wall. They are often buried by talus. When erosional activity slows down the lavaka becomes gradually stabilised. Vegetation appears on the slopes when they cease to crumble and acquire a minimum slope on which crumbling can occur, or about 45°. This evolution usually begins near the outlet and progressively proceeds towards the head. Plants finally also occupy the floor. From this moment on the waters of the springlets cut themselves a small channel, 1 to 2 m (3 to 6 ft) deep, and the fan in turn becomes covered with vegetation.

Brenon's type description is confirmed by Portères (1956) and by Riquier (1958). We have ourselves observed similar forms in Brazil. There are of course some variants. Often the lateral walls are not rectilinear but indented by short, very steep gullies, which give the whole the shape of an oak leaf. In this case there is no flat bottom. Indeed, as often happens in Brazil, the thickness of the regolith is such that the steep long-profile of a voçoroka (10 to 15° in its medial part) does not reach its base when the hills are steep (30 to 40°). The stream bed is then incised into the disintegrated rock of Zone III and there is a greater resemblance to an ordinary ravine. It seems that such forms characterise an early stage of evolution in Madagascar. According to Riquier (1958) differences in forms depend on the erosional origin of the lavaka. Those that are caused by a trail have a more linear form; those that result from sheet erosion are more digitated; finally, those that are derived from scars left by slumps or from a lateral undermining are more pear-shaped.

Average lavakas are 150 to 250 m (500 to 800 ft) long, 100 to 200 m (350 to 700 ft) wide, and span a total elevation of 30 to 40 m (100 to 130 ft). Giant ones in Madagascar reach a length of 2 km (1·2 miles) and a width of 300 m (1 000 ft). They are eroded out of a plateau rim, whereas average ones are found on hillsides, especially in the flanks of convex ridges. In Brazil the same holds true. The longest lavakas are usually the narrowest and least digitated, resembling ravines caused by fractures in dry regions. They owe this form to the predominance of underground drainage and tunnelling.

There are, then, various types of lavakas. But all are ravinelike forms forcefully incised in a thick tropical regolith and caused by runoff, subterranean drainage and tunnelling. The proportions of the three mechanisms differ from one case to another.

Lavakas begin with a linear incision that frequently originates in a desiccation crack. In spite of the kaolinitic nature of the clays, soil desiccation due to forest clearing and fires is sufficiently intense to cause cracks. Even when modest they play an important role because of the induration of the soil due to drought. Water draining into them reaches lower horizons, at a depth of $\frac{1}{2}$ to 3 m (1·6 to 10 ft), which are less indurated by iron oxides. The materials are watered-down, and a certain amount of entrainment results. Tunnelling begins. Barbier (1957) has described such initial forms, which we have also often observed ourselves. A subterranean streamlet develops in the regolith, normally in the zone of disintegrated rock, which is more friable and not as hard as the red soil or mottled clays. Small funnels, a few decimetres (one foot or more) or at the most one or two metres (3 to 6 ft) in diameter, mark its course. Then the roof caves in and a straight canyonlike gully appears. If the terrain is sufficiently flat this form persists: that is, a *linear lavaka*, which may attain a considerable length. On hillsides it widens and deepens until an *oak-leaf-shaped lavaka* develops. If conditions permit erosion to become stabilised on top of Zone IV, the result is a *pear-shaped lavaka* with flat bottom and abrupt sides, considered by Brenon to be the most typical, and in any case the most original. Hollowed out trails facilitating the concentration of runoff evolve in exactly the same way as the linear depressions caused by tunnelling. Scars caused by debris slides from the very start produce pear-shaped lavakas, which later, during a progressive stabilisation, may evolve into oak-leaf-shaped lavakas.

The abrupt walls of the lavakas and their rapid retreat are due to a combination of four factors:

1. The superposition of very hard, rather impermeable coherent red clays and the friable disintegrated rock, which undermines itself. Such a stratification of the regolith is typical of the forested tropics and subtropics. It is indeed in this kind of environment that Zone III is sufficiently thick, homogeneous and continuous to allow the development of lavakas. Such superposition controls the evolution of the lavaka throughout its development, from the initial tunnelling to the clearing on top of Zone IV and the development of a flat bottom.

2. The induration of the superficial clays due to the intervention of man,

which brings about an increase in contrast between the upper layers of the weathering profile and, especially, the beginning of tunnelling and the appearance of linear depressions, which never occur under forest. In savannas, on the other hand, even in empty regions like the north of Dahomey, similar linear incisions may form, but only in alluvial formations, as reported by Vogt (1961).

3. The existence of underground drainage in the regolith, feeding springlets and facilitating by undermining the retreat of the lavaka walls through debris falls. It presupposes the persistence of a certain amount of infiltration, indicating that degradation is not too far advanced.

4. The existence of a non-incisable floor and friable regolith at a sufficiently shallow depth to be removed.

The last two conditions are not always realised. This introduces a certain number of variants, thus assuring the transition from typical lavakas to ordinary ravines. In typical lavakas the retreat of the head wall is very rapid, especially during the rainy season when sliding occurs at the level saturated by springlets (at the contact between Zones III and IV). According to Brenon (1952) linear retreat may amount to $\frac{1}{2}$ m (1·6 ft) per year. When subterranean waters are deflected or dry up due to decreased infiltration caused by waterproofing of the superficial layer, the walls and then the whole lavaka become stabilised. For this reason, though the flanks of lavakas frequently intersect, their heads never go across ridge crests; because after a certain minimum interfluvial width has been reached the heads cease to retreat.

Like all forms produced by an interruption in the morphoclimatic equilibrium, lavakas evolve at a rate that increases like the curve of a bell. At first development is slow; it becomes rapid as soon as an aerial incision is realised, and, finally, it slows down when further evolution reduces water seepage and the rate of wall retreat. According to Brenon average lavakas have formed in 500 to 1 000 years, or slightly more. It seems, therefore, that they correspond to the period of colonisation of Madagascar by the Hovas whose pastoral habits were at the origin of a widespread increase in brush fires and the replacement of the forest by a repeatedly burnt over grassland, a condition essential to the development of lavakas. The interruption of the morphoclimatic equilibrium is more recent in Brazil, which in part explains the predominance of oak-leaf-shaped voçorokas. In the area of Rio de Janeiro, for example, the forest has been modified only a century and a half ago with the development of coffee plantations. Whereas it is only fifty years ago that the soils having become exhausted coffee was abandoned and replaced by mediocre pastures in which voçorokas have developed.

The lithology of the substratum influences the development of lavakas through the facies and thickness of the regolith. Jointed, acidic igneous rocks constitute the ideal environment because of the thickness of the regolith and the contrast in resistance between Zones I, II and III. In massive granites and gneisses a less thick regolith and the absence of Zone IV prevent the development of lavakas. In basic igneous rocks conditions are very unfavourable because Zone III is poorly developed. Sometimes, as Panzer (1954) has reported from Hong Kong, crushed zones in which weathering proceeds deeper and which drain the soil water favour the formation of lavakas, which extend into them. It seems that this is especially the case in regions of irregular weathering.

The formation of lavakas causes a geomorphic imbalance that progressively spreads to the whole drainage net, as Besairie (1948) has shown. In the Betsiboka Basin (Madagascar), covering 60 000 sq km (23 000 sq miles), where a very thick regolith on micaschists covers a vast area, lavakas formed between 1891 and 1946 have caused a considerable infilling of the estuary with clays derived from reworking products of kaolinitic weathering. The yearly supply amounts to an average of 15 106 cu m (542 000 cu ft). Higher up at the confluences of the Mahajamba, Kamoro and Ikopa rivers enormous sand accumulations form alluvial barriers that cause the appearance of distributaries and braided channels. The result has been the diversion of the Mahajamba towards the Kamoro and the Betsiboka, whereas otherwise it should have flowed towards the Tsimitondraka River. The consequences of the formation of lavakas are identical to the consequences of placer mining (gold, diamonds), which causes the dumping of vast quantities of sand in streams and the burial by sterile sands of lands that could have been developed into paddy fields and fed a dense population.

LANDSLIDES

Although less typical, landslides also play an important role in the formation of tropical landforms induced by man. They mainly occur in the regolith. But unlike lavakas they are also a natural phenomenon and the actions of man only multiply their numbers.

Desiccation cracks caused by clearing of the forest help to trigger them. They seem to be particularly numerous in recently cleared areas, perhaps because the induration of the soil that causes its waterproofing has not been completed. In these circumstances the infiltration of water is easier, as the soil permeability of the forest environment is still partly intact and desiccation cracks offer additional avenues of penetration. For this reason landslides are particularly frequent in the Brazilian Serro do Mar and, especially, the Val Paraíba, where their density surpasses that of lavakas. On the other hand few of them have been reported from Madagascar. The scars left by landslides are occasionally transformed into lavakas. It seems, therefore, that landslides occur more often during the initial phase of man-induced morphogenic processes.

Such landslides occur mainly during periods of heavy rains, for then the liquid limit is high in the rotten rocks that are normally affected. For example, in 1948, according to O'Reilly Sternberg (1949), between 400

241

and 500 mm (16 and 20 in) of rain fell in the Val Paraíba in twenty-four hours and were reinforced during the following days by totals of several tens millimetres (a few inches). It had also rained considerably during the three previous weeks, so that the regolith was saturated. The catastrophic land-slides took place in December, at the beginning of the rainy season.

The resultant forms are the same as those commonly associated with landslides. At the base of Zone III the water content must be sufficient to pass the liquid limit, so that pockets of liquid mud burst all of a sudden, causing the overlying regolith to be torn loose. Nichelike forms with clearly marked scarps predominate in the upper part. The human origin of the landslides has been emphasised by O'Reilly Sternberg, who has noted their exact locations. They have only affected cleared land transformed into mediocre pastures. In their vicinity patches of forest have been spared.

Occasionally, earthflows occur in addition to landslides. They are un-usually frequent as a result of man's interference in the area of Hong Kong, probably because of the violence of monsoonal downpours. Berry and Ruxton (1960) have described some of them, including 'fans', composed of boulders derived from Zone IV and irregularly set in a silt-clay matrix, the whole some 15 m (50 ft) thick. They suggest the term 'solifluval deposits' for them. Segalen (1948) mentions some small ones on micaschists in Mada-gascar.

For landslides and earthflows to form a thick regolith capable of water-logging is necessary. The most favourable conditions are found on acidic igneous rocks, especially micaschists whose regolith is sufficiently argillace-ous and in which the micas provide tiny sliding planes, according to Segalen.

Landslides and earthflows transport enormous masses of detrital mater-ials into lowlands in the form of fans and irregular mounds, which occasion-ally bar streams, as has happened in the Val Paraíba in December 1948. They tend to destroy the characteristic relief and to increase the solid load of streams.

Erosion resulting from the destruction of forests is particularly wide-spread in regions where the regolith is thick. A thick regolith forms slowly; it therefore exists only in regions that have endured a sufficient degree of climatic stability, which is the case of Atlantic Brazil and, probably, of Madagascar. It is not the case of West Africa, where climatic oscillations have been important during the Quaternary. They have caused other types of imbalance, which we will now study.

The influence of palaeoclimatic oscillations

Periodic variations in humidity of the intertropical zone have been common during the Cenozoic Era. They do not require important modifications in the climate of the earth as a whole, as occurred during the Quaternary. A modification in the configuration of land and sea, in atmospheric circula-tion or in relief is enough to cause a fluctuation in the humidity of the climate and may, if sufficiently pronounced, result in the substitution of forest by savanna or vice versa. We suspect such modifications to have occurred more than once during the Tertiary in the intertropical zone as well as in the midlatitudes, which then had warm climates. They, of course, have important repercussions on the evolution of landforms because of the great differences that exist between the morphogenic systems of warm and humid climates. We will now study them as revealed by certain deposits and landforms and by the evolution of cuirasses.

Palaeoclimatic oscillations as revealed by erosion and sedimentation

The number of proofs of palaeoclimatic oscillations in the tropics does not cease to multiply independently of geomorphic findings. For example, in regions such as Bas Congo and lower Ivory Coast, the forest presently tends to expand at the expense of the savanna (Meulenberg, 1949). Botanists consider these savannas to be palaeoclimatic relics which have maintained themselves on those soils that are most unfavourable to forest growth. The same occurs in Brazil, in the area of São Paulo. Inversely, in Senegal Maignien's (1961) pedologic studies have demonstrated an ancient northward extension of the forest, which is reflected in red (relict) forest soils on top of plateau remnants in a region that is presently characterised by beige savanna (zonal) soils.

Such facts should be brought together with divers geomorphic observations that have long suggested a palaeoclimatic origin for certain sediments and landforms.

RIVER TERRACES

It has long been an established tradition with armchair geographers to affirm the absence of terraces on tropical rivers. Nevertheless, since the beginning of exploration, mineral prospectors have seen them, gathered numerous empirical observations about them and created a complete technical vocabulary. They have done so because alluvial deposits in the tropics often hold economically valuable placers of cassiterite, diamonds or gold and, of more recent knowledge, radioactive materials. In Africa, southeast Asia and Guiana the floor of stream beds is often underlain by gravels, occasionally seen in low water pools, which permit a small amount of reworking. Usually they are covered by finer, sterile deposits of argillaceous, fine sands, which are presently deposited on the floodplain. For good reasons, therefore, miners call such gravels *graviers sous berges* or 'underbank gravels'. Such gravels have been recognised throughout a large part of West Africa, in the Congo Basin and in East Africa. Similar gravels are found in Malaya and Indonesia. But their study is simplest in West Africa where imbrications with formations due to tectonic deformations are reduced to a minimum.

243

Vogt (1959a) has reported a low gravel terrace along most of the rivers of Ivory Coast and along the Falémé; Michel (1959a) has reported a similar terrace in the upper valleys of the Gambia and Senegal basins. It can be found in the present forest environment as well as in savannas. In Ivory Coast it is normally quite wide, spread over more than a kilometre (0·6 mile), in such a way that the present river meanders are incised in it and only seldom undercut the valley sides. In the forest the gravels are usually buried under finer deposits, whereas in the savannas they are more often bare although they may also be covered by clays, as in the Precambrian schist area of northern Dahomey. The low terrace is commonly dissected, sometimes buried under colluvium at its inner margin.

Deposition of the terrace gravels implies a fluvial regime quite different from the present one, as is indicated by the nature of the material. A typical cross-section of the low terrace in the semi-evergreen seasonal forest of Ivory Coast, in the area of Ouellé and Béoumi, reveals, according to Vogt:

at the surface: cuirassed argillaceous sands with a scoriaceous aspect, in a beginning stage of induration but still friable, causing debris falls as the underlying gravels are undercut; thickness: $\frac{1}{2}$ to 3 m (1·5 to 10 ft);
fine, little worn gravels, in lenticular beds with cross-bedding; thickness: 2 to 3 m (6 to 10 ft);
coarser, more worn gravels alternating with lenses and beds of fine gravel and sand.

Siliceous materials are highly predominant. The petrographic composition varies rapidly, however, indicating an important lateral influx and a filling in of the valley bottom, which impedes downstream entrainment. The lateral influx explains the width of the alluvial deposits. The sedimentary structures indicate unstable channels in the midst of sand and gravel banks, or a regime similar to that of braided streams.

Such a stream bed implies a predominance of mechanical over chemical erosion and a reworking of weathered materials previously elaborated on the hillsides. Quartz pebbles are removed from veins that have imperfectly persisted in the regolith. Clays are rich in kaolinite but also contain traces of montmorillonite, which implies a deep dissection of profiles in certain places. Sands reveal traces of corrosion and ferrugination. They are mixed with ferruginous concretions indicating soil erosion (Le Bourdiec, 1958b).

In the low terrace Vogt (1959b) describes an alluvial formation embedded in a trench whose bottom is formed by 'underbank gravels' which, however, are usually hidden by more recent fine clastics. The 'underbank gravels' represent sinuous braided channels. The material is poorly sorted, slightly worn and of a highly changing nature, indicating sudden deposition by violent and successive floods. Potholes in the bedrock (at rapids) are frequently fossilised by the gravel and consolidated by iron oxide. Our own observations at Sotuba (near Bamako) have been confirmed by Vogt on the Bandama and the Bafing rivers. To account for the deposition of the underbank gravels, Vogt invokes a climate with violent showers falling on a soil poorly protected by vegetation. In short, a semiarid type of climate with, perhaps, a greater irregularity of precipitation than is common today, which would impede the development of vegetation. A high terrace composed of rounded siliceous pebbles and frequently capped by a cuirass indicates a similar history.

In other regions, at the foot of highlands such as those of Sierra Leone, there are alluvial aprons composed mainly of reworked cuirass fragments mixed with a few sandstone pebbles.[1] Streams are incised into them to a depth of about 10 m (33 ft). In the Siguiri Basin (Guinea) Pélissier and Rougerie (1953) have described transitions from cuirassed piedmont slopes to terraces. Further north at Kéniéba (Mali) Vogt (1956) has observed two stepped piedmont slopes, the lower partly covered by mudflow deposits 2 m thick (6 ft). At a later date this surface was entrenched and filled with gravels, which are coarse and slightly worn at the foot of the Ordovician cuesta. Close to the Falémé a filling of fine materials 5 to 10 m thick (16 to 33 ft), containing gravel lenses at the top, covers a large area. It was followed by a new entrenchment now filled with fine sediments. Presently an intense erosion caused by man is producing correlative gullies and mudflow deposits at the foot of intensely cultivated slopes. The field work of Michel (1959a and b) has demonstrated the widespread occurrence of such piedmont slopes, entrenchments and alluvial fills in a region extending from the Fouta Djallon to the middle Senegal.

The surroundings of Kéniéba, which have not been reached by Quaternary regressive erosion, are particularly interesting as their entire evolution depends exclusively on climatic oscillations. They show that several times during the Quaternary there have been alternations of periods with (a) predominant chemical weathering, of the type that is active today, with a liberation of fine sandy and clayey materials; and (b) violent mechanical erosion affecting previously weathered materials, causing their laceration by gullies and their spread into coarse textured aprons similar to the present phase of erosion triggered by man.

Similar alternations have also affected the forested tropics, particularly in Ivory Coast and in Ghana. It seems that only the southwestern extremity of Ivory Coast and Liberia have been able to preserve their forest cover more or less intact during the palaeoclimatic oscillations of the Quaternary. It appears to be the only region where there are few coarse materials in river terraces.

Climatic alternations of the same type have affected East Africa, notably the Great Lakes region. But essentially studied by prehistorians, they have been the object of interpretations that hardly conform with what we have learnt from the study of dynamic geomorphology. Whereas gravel deposits have been attributed to humid periods, fine materials have been associated with dry periods! However, it does seem that at least in Moçambique there

[1] Oral communication kindly contributed by F. Fournier.

is a similar disposition of phenomena as is found in West Africa. Indeed between the Zambezi and the Save River the alluvium of the Revué is composed, according to Amaro (1952), at the base of very coarse auriferous rubble, containing boulders the size of a metre (3 ft) and debris of very divers rocks: granites, greenstones, serpentines, conglomerates and, overlying them, sands and clays.

In many parts of the tropics, especially in Africa and in southeast Asia, certain periods of the Quaternary have caused a thinning of the plant cover as a result of a decrease in rainfall, which has permitted an intense sheet erosion of the hillsides. Considerable quantities of regolith have been swept away. The sands and gravels have been left behind in the valleys, eventually forming terraces, while the clays have reached the ocean. Such a morphoclimatic evolution has a practical aspect: gold, diamond or even cassiterite lodes too poor to be exploited *in situ* have undergone a certain amount of concentration, first through weathering, and later in valley bottoms under the effects of hydrodynamic sorting. They have given birth, in some cases, to economically valuable alluvial deposits. The study of them belongs to the field of applied geomorphology.

This evolution has also appreciably modified the relief of the interfluves.

The hillsides

In some areas of the wet tropics, hillsides have quite variable and irregular slopes with basal concavities and rectilinear sections, which are very different from the short, steep and convex sides of hills shaped like half oranges. Some exist in the centre and east of lower Ivory Coast, especially in areas underlain by schists. These hillsides, which resemble those of the savannas, always have a relatively thin regolith, a few metres (3 to 10 ft) thick, on the average three to five times thinner than on identical rocks on the coast of Liberia or in the state of Bahia, Brazil, which have reliefs that are typical of the wet tropics. Cross-sections reveal the explanation of the difference: the foot of these flaring hillsides is usually covered with colluvium, including rubble, quartz fragments and ferruginous concretions, sometimes abundant enough to serve for road construction. Inversely, the hill crests have a thin regolith with truncated soil horizons due to sheet erosion. Occasionally, the zone of rotten rock with red clays filling the joints is found at a depth of a metre (3 ft) only.

One may also observe significant facts on the horizontal plane. Reliefs sculptured from schists have numerous valleys whose streams have ceased to function and whose slopes continue to evolve by rillwash. Even the stream beds have vanished, resulting in a rolling topography with argillaceous colluvium occasionally aggrading the lower slopes. Some of the larger valleys debouch on the bottomlands by way of a flat, inactive fan. In regions of granitoid rocks with thick regolith, in a rather dissected environment, as in certain parts of the Serra do Mar, south of the state of Bahia, massive ridges exhibit scooped out, spoon-shaped depressions, 100 to 200 m

(330 to 660 ft) wide, which often coalesce laterally and look like flattened torrential catchment basins, on the floors of which seepages feed a certain amount of runoff (Tricart, 1958a).

All these forms may be explained in the light of palaeoclimatic oscillations as revealed by fluvial terraces. They are inherited from drier periods during which the regolith of the previous wet climate was partially swept away, mainly by slumps in the case of acidic igneous rocks, and by runoff in the case of schists. In southern Atlantic Brazil we have, moreover, several times observed the transition from the partly effaced outlines of past slumps to the sandy alluvial fans and portions of terraces containing rounded, sometimes coarse quartz fragments (10 to 15 cm: 4 to 6 in near Rio de Janeiro), and to the alluvial aprons veneering piedmont slopes (east of Niteroi).

Many hillsides of the humid tropics are therefore *polygenic*, which reduces their zonal characteristics. In present forest environments there is evidence that the hillsides once evolved under a savanna morphogenic system, as in Ivory Coast, and by countless catastrophic slumps, reflecting perhaps less severe and less prolonged climatic changes, as in the area of Rio de Janeiro and Santos. In the savannas some palaeoclimates were more humid than the present climates, such as the one that produced the red forest soils in central Senegal (Maignien, 1961) and that seems to have been responsible for the red soils on the ancient fixed dunes further north. Other palaeo-climates were drier; for example, near Bandiagara, Mali, Daveau (1959) has reported coarse slide-rock deposits covered by aeolian sands at the base of the enormous sandstone escarpments in a region where sandstone can presently weather into clay. In many savanna regions, as far south as the Fouta Djallon, there are stony slope deposits, often rich in cuirass fragments. They are live only in those places where normal degradation has reached a maximum; for example, on the faceslope of the Ordovician sandstone cuesta of Kéniéba or in certain parts of the Fouta Djallon (Vogt, 1956; Tricart, 1955).

Palaeoclimatic oscillations also permit a better understanding of the problem of the stone-line. Its origin is complex and there are certain convergences. As we have seen, the stone-line in some cases results from tropical solifluction, which causes the beheading of quartz veins and the displacement of rubble derived from slowly weathering rocks (certain schists, siltstones and cherts, for example), or even the reworking of beds of pebbles imbedded in terrace remnants at higher levels. In these cases of particular lithologic conditions the stone-line has only a genetic significance. In other cases, as noted by Ruhe (1956) or de Heinzelin (1952) or as we have ourselves observed in Brazil (1958a), the stone-line has a palaeoclimatic significance. It is an old, now buried, pavement produced during a period of more intense sheet erosion of the hillside above. In Brazil we have seen a stone-line in the process of formation due to man-induced erosion on the high, steep ridges, converted into pastures, in the south of the state of Minas

Gerais. The normal soil profile under semi-evergreen seasonal forest consists of a 1 to 2 m (3 to 6 ft) thick, leached, yellow sandy-clayey soil with a podzolic tendency, overlying reddish mottled clays and the other standard, very thick zones. On the denuded ridges, with thoroughly leached soils and a degraded vegetation, sheet erosion affects the upper slopes, and stony pavements mainly composed of vein quartz gradually form on the surface. Such an evolution may proceed along the entire slope down to its foot. Here the stone pavement may become interrupted if more violent erosion opens up gullies whose terminal fans of poorly sorted materials bury the pavement.

We have also been able to reconstruct a natural evolution in the same area of Brazil. During a short dry period, but with violent showers, which can be traced back to the Dunkirkian, a low sandy terrace passing gradually into colluvium and alluvial fans at the base of the hillsides was formed. The yellow soils of the hillsides were eroded except at the foot of the slopes, and stone pavements were emplaced. Later they were buried in a forest environment, on the one hand below a slow colluviation of fine material at the base of the slopes, and on the other hand below earth brought up by uprooted trees, by termites and other burrowing animals.

On the lowlands of the eastern margins of the Congo Cuvette, de Heinzelin (1952) has had recourse to the actions of termites to explain stone-lines buried under 0·2 to 1 m (8 to 40 in) of fine debris containing objects whose age has been estimated at a maximum of 10 000 years. There would, therefore, have been a period of drier climate in this region, permitting intense sheet erosion. It should be emphasised, in passing, that it came to an end right before the Holocene. It was followed by a more humid climate similar to the present one and during which a small amount of colluviation and, especially, termite action have brought about the burial of the stone-line pavements, which indicates a much greater stability from the geomorphological point of view.

Palaeoclimatic oscillations also have had their effects on the denudation of inselbergs and on the process of planation.

Inselbergs and savanna plains; the problem of planations

Alternations of wetter and drier climates have had different effects on the present forest and savanna environments. Indeed in the forested regions they have resulted in a profound modification of the morphogenic system, whereas in the savanna regions they have had mainly quantitative effects.

In the rainforest wet–dry alternations have caused sheet erosion. They have helped the exhumation of monolithic domes. Seasonal rainfall indeed facilitates landsliding. In the area of Rio de Janeiro and São Paulo, we have recognised the remains of important landslides at the foot of a number of monolithic domes: piles of sandy clay with boulders, which are sometimes quite large (Tricart, 1958a). Furthermore, in the mountains there were genuine earthflows, as, for example, in the Itatiaia massif. The boulders of these earthflows are presently washed by torrents that are incapable of carrying them away. Some earthflows tie in with the lower sandy terrace that dates back to the last dry period, probably Dunkirkian. In Barbados large earthflows have spread over the present littoral bench during the pre-Flandrian regression from the inface of the cuesta that forms the east coast of the island (see Fig. 2.29). Limestone blocks imbedded in them, sometimes 30 m (100 ft) in diameter, and now washed clean by the sea, show only two dissolution notches, one Dunkirkian, the other recent, which permits the conclusion that they were not in their present locations during the last high interglacial sea-level.

It appears, therefore, that it was during the wet–dry interludes that many monolithic domes were cleared of their regolith. Once bared, it is difficult for plants to reoccupy them. The chances are that they will remain bare, which completely modifies their evolution. The immunity that they presently experience on account of their steep slopes was heightened during wet–dry climatic intervals.

In savanna regions the climatic differences are less accentuated. Under open vegetation planation by rillwash and sheetfloods occurs as in drier lands. But the evolution of savanna plains is impeded by debris that are too coarse to be moved and form a pavement, such as of large ferruginous concretions. On the other hand weathering is still important enough to prepare material at a rapid rate, especially to produce rather easily mobilised deflocculated clays. Under a drier climate similar plains develop under a more brutal drainage endowed with a higher competence, but floods are less frequent. There is, however, less weathering, and the elaboration of debris is slower.

The succession of wet–dry periods of the savanna type and of semiarid periods must, therefore, be favourable to the development of planation surfaces or pediplains. This is verified in regions like the Sudan. During semiarid phases, but with torrential rains that cause the formation of alluvial aprons, planation surfaces are carved from the former regolith. The wet–dry phases, like the present one, are noted for their tendency of stream entrenchment. But as entrenchment remains linear, it does not prevent rillwash from perfecting the work of planation on the interfluves, already shaped as pediplains. As the latest climatic oscillations have gradually shortened during the late Quaternary, one occasionally encounters converging piedmont slopes with different gradients, as at Kéniéba in southwest Mali. Here the highest piedmont slope came into being after a long period of weathering while increased rillwash reworked considerable quantities of friable regolith on the hillsides, causing a steepening of the piedmont slope. The lower piedmont slope, on the contrary, developed after a slight lowering of the base-level during which the regolith was gradually and partially swept away and prevented from reaching its former thickness. As it was removed, the piedmont slope acquired a gentler gradient and became a true pediplain.

Just as in the rainforest, the exhumation of inselbergs seems to have been speeded up during semiarid periods when a swifter runoff removes more debris. But there is a want of precise studies on this point.

Dry and humid alternations of warm climates reigning over much longer periods of time, unlike those that occurred during the Quaternary, but rather like those of the Tertiary, are just as capable of influencing the evolution of the relief. Büdel (1957a and b) has drawn up an ingenious theory about this matter, that of the *double planation surface* (*doppelte Einebungsfläche*). According to him, it is necessary to distinguish two morphogenic surfaces in the humid tropics (especially the wet–dry and monsoonal tropics) : that of the unweathered bedrock below the regolith and that of the ground surface. This 'double planation surface' corresponds to our distinction between the ground surface and the weathering front. According to Büdel, because of the immunity of rockbars in the long-profile of tropical streams, aerial erosion is faster than linear erosion, and overland flow above the local or general base-level produces planations truncating the regolith. With moderate but intermittent crustal uplifts such pediplains would be preserved if weathering proceeds at about the same rate as the uplifts. However, at the margins of the pediplain, around the inselbergs, a series of stepped pediments, or piedmont benches, would eventually develop. With a more rapid uplift or as a result of an oscillation towards a drier climate, the regolith would be swept away and the unweathered rock exposed, producing a new, irregular, exhumed topography. Although much of Büdel's theory is too theoretical, and it is not clear what he exactly means by *feuchten Tropen*, the last statement is, in any case, true in the present wet–dry tropics.

Evolutions of this type have occurred several times during the Tertiary on the ancient massifs of France. On the Brazilian Arch in the state of Bahia the regolith of the humid Eogene was swept away during the semiarid climate of the Neogene. The Barreiras Series, composed of sands, quartz pebbles and kaolinitic clays, was deposited in the littoral zone, while coalescing erosional piedmont slopes (pediplains) developed inland around inselbergs, mainly at the level of the weathering front. This area was rather seriously warped, but when there is little or no warping semiarid phases may cause pediplains to be cut from the regolith itself, as in the Siguiri basin at Kéniéba, Mali, or in western Australia (Mabbutt, 1961). In the latter region a ferruginous cuirass developed on top of the deeply weathered kaolinitic clays of the zone of saturation probably during a wet–dry phase. Since then a slight uplift has caused some dissection. But at 150 km (90 miles) from the coast the resumed erosion of the early Quaternary only has produced a new level of pediplains carved from the pallid zone of the Tertiary kaolinitic regolith. Such a sandy material, lacking consistency, is indeed eminently apt to be eroded into piedmont slopes at the base of cuirassed residual mesas, as Mulcahy (1960) has shown.

Cuirasses and climatic oscillations

Cuirassing phenomena are also affected by climatic oscillations. The latter indeed control the intensity of the leaching process and the range of migration of the indurating solutions. But problems are different depending on the type of cuirass.

Climatic significance of ferruginous cuirasses

Ferruginous cuirasses evolve much more rapidly than bauxitic cuirasses. Some of them are Quaternary and in a number of places there are even several successive Quaternary cuirasses; for example, two of them on the terraced piedmont slopes of western Mali. Bauxitic cuirasses, on the contrary, develop slowly and the most recent go back to the end of the Tertiary. Whereas climatic oscillations are reflected in the ferruginous cuirassing process by the formation of successive cuirasses, in bauxitic cuirasses they are only recorded by successive epigenic modifications, sometimes in opposite directions, which complicates the problem. The study of climatic oscillations and their effects on the genesis and evolution of cuirasses must therefore start with the simplest case, that which is the most sensitive to climatic oscillations, i.e. the ferruginous cuirasses.

An important fact immediately appears to the geomorphologist who studies ferruginous cuirasses in the field: the discontinuity in time of their formation. Periods of cuirassing alternate with periods without cuirassing, marked either by erosion or by deposition of slope materials by mechanical means or, of course, by both at the same time, as they do not exclude one another.

In the Ferriferous Quadrilateral of Minas Gerais, for example, where iron is exceptionally abundant as the area is an ore deposit, the cuirassing processes have been discontinuous in time in such a manner that Tricart (1961b) has been able to distinguish two distinct cuirassed topographies. One, the older (Palaeocene), is found on ancient piedmont slopes at a high elevation, around 1 200 m (4 000 ft), occasionally placed in dominating positions by inversions of relief. The other, composed of hillside cuirasses fossilising late Tertiary lignites at Gandarela, seems to be of Pliocene age. It is found even on slopes as steep as 50° below the escarpments formed by the older cuirasses. Later erosion was not accompanied by any cuirassing, exactly as the dissection that separates the two cuirassing intervals. That the same rocks (Precambrian siliceous schists), exceptionally rich in iron, could give birth in the Pliocene to cuirasses on very steep slopes, whereas they did not permit any cuirassing at the beginning of the dissection of the 1 200 m level on slopes that were then much less abrupt demands an explanation. It cannot be but a dynamic one. One is forced to admit that climatic circumstances, through their variations, allow important cuirasses to form at certain times, even on very steep slopes, whereas at other times they do not, even on much gentler and more favourable reliefs. Now, our investigations on the Brazilian Atlantic Arch indicate the existence of a rather dry tropical climate during the Palaeocene, with phenomena of quartzification in the Cretaceous Basin of Bahia, followed by a long period of a rather humid climate with considerable chemical weathering during which the Eocene deposition of lacustrine limestones in small basins took place. Finally, a drier climate returns in the Neogene with the deposition of the Barreiras Series. As most authors concur on a Palaeocene age for the 1 200 m level, the correlation of periods of cuirass formation with wet–dry climatic intervals is well founded. On the contrary, the long intermediate period of a quite humid climate and the present climate, reflected by semi-evergreen seasonal forest, have not permitted the development of cuirasses and correspondingly facilitated erosion and made possible the relief inversions of the cuirassed piedmont slopes at the 1 200 m level.

West Africa has examples of similar alternations, but they are more rapid and in a different style. In the southeast of Ivory Coast, for example, the Sanwi area is characterised by two planation levels, according to Rougerie (1950). The highest, at about 200 m (660 ft), truncates steeply dipping Precambrian anticlinal structures later transformed into 'appalachian' ridges. The lowest, close to 130 m (430 ft), surrounds the ridges, which have become inselbergs. Both are covered by cuirasses formed during wet–dry climatic phases.

In the southwest of Ivory Coast Tricart (1962b) has described a cuirassed surface eroded into mesas. Iron oxides have penetrated the alluvial aprons, betraying a wet–dry climatic phase, which would be the equivalent of the *Continental Terminal*. They indurate sands containing beds of subangular gravels and mixtures of sand and clay that seem to have been produced by biotic actions. The topography of the thin but widely spread deposits and the nature of the materials denote a dry savanna type climate (i.e. semi-arid) with large, torrentially deposited alluvial aprons and easily flooded thickets at some distance from the main watercourses. The irregular ferruginous induration took place subsequently. At the present the cuirass is being dismantled under the rainforest. It would seem that the influx of iron here marks the passage from a dry savanna climate of the present Sahelian type with only occasional downpours to a perhumid climate that put a stop to the construction of alluvial aprons and the arrival of coarse clastics, and during which the entrenchment of the valleys took place after the cuirasses had ceased to form.

In the Siguiri Basin Pélissier and Rougerie (1953) have also described two stepped cuirassed piedmont slopes. Both are composed of alluvial aprons whose surfaces have been indurated and then dissected. Cuirassing therefore occurred in the same way as in southwest Ivory Coast (previous example), following the deposition of coarse clastics, which implies important mechanical actions, instability of the uplands and a torrential runoff during storms. On the other hand cuirassing preceded the subsequent dissection of the piedmont slopes, which, it seems, occurred under climatic conditions rather similar to those of the Holocene, which in this region are reflected by entrenchment of the watercourses. No cuirasses developed on the newly formed slopes of the down cutting phases in spite of the presence of a summital cuirass.

The south of Upper Volta, studied by Daveau *et al.* (1962), shows similar alternations although not quite as clearly. The main cuirassing periods affect the alluvial aprons as soon as they are formed, but, probably due to more favourable climatic conditions than in the Siguiri Basin, hillside cuirasses develop on the slopes resulting from the dissection of the piedmont slopes.

Similar climatic oscillations are recorded elsewhere on hillsides. In the Fouta Djallon, renowned for its cuirasses, de Chetelat (1938) and Maignien (1958) have reported hillside cuirasses cementing detrital materials, such as cuirass slabs slid down from *bowé* escarpments and other rubble. Tricart (1956b) has drawn attention to the palaeoclimatic significance of these slope facies. They imply alternations of periods with predominantly mechanical erosion and migration of an abundant waste on the not too densely vegetated slopes, and stable periods with predominantly chemical weathering and migration of solutions, which does not, however, exclude localised rockfalls from escarpments. Partially indurated slope deposits have varied facies, ranging from creeped materials to earthflows and from rockfalls to fans at gully mouths.

Highly significant cross-sections have been recorded by Brückner (1955) in Ghana. In an area located between 5° and 7° 30′ north latitude, the type-section is as follows:

(*a*) At the top, light, friable silty sands, with concentrations of loose limonite and calcareous matter, grading into silts at the surface. A metre (3·3 ft) thick on the hillsides, thicker at the base, this layer is disturbed by termites and reworked by colluviation. It contains Neolithic artifacts.

(*b*) A 1 to 3 m (3 to 10 ft) thick layer of quartz rubble and debris of ferruginous concretions with sandy matrix, sometimes reddish, filling holes and pockets, and with occasional limonite veins.

(*c*) A black, shiny scoriaceous cuirass with pisolites and tubular structures, indurating the subjacent layers to a depth of about half a metre (20 in).

(*d*) A layer of rubble in a sandy matrix, similar to the above one, with sandy and limonitic intercalations, grading downwards into a brecciaceous horizon with fragments of weathered rocks forming pockets in the zone of rotten rock with corestones. Its thickness may be as much as 4 m (13 ft).

Brückner, basing himself on the shiny aspect of certain loose rocks, appeals to wind action for an explanation. We think there is some confusion as iron oxides do not produce such aspects. He thus reconstitutes climatic oscillations swinging from arid to perhumid, which are not corroborated by other investigations, neither in the neighbouring areas, and which to us seem exaggerated. But, freed of its excesses and of purely hypothetic correlations with East Africa, the idea is right. Our own interpretation is as follows. There was first a period of intensive weathering resulting in the elaboration of a standard profile, which was truncated down to Zone IV during a semiarid phase. Intense slope wash then produced the lower rubble. The cuirass was produced during an oscillation towards a humid climate. A new semiarid phase accounts for the upper rubble and the present wet–dry climate beginning in the Neolithic, or slightly earlier, allowed the elaboration of silts and the mixing by termites. Resemblance of the last episode with the findings of de Heinzelin (1952) in central Africa will be noted. During certain periods silts may have formed, but they were swept away by later semiarid phases.

At the base of Mt Nimba (Guinea) Lamotte and Rougerie (1962) have made concordant observations but in an environment richer in iron. The cuirass at the Blandé cave gully shows the following cross-section:

	Thickness	
	m	ft
A homogeneous layer of coalescing iron concretions	1	3·3
A ferruginous breccia of large subangular crossbedded slabs	1·8	6
A homogeneous layer of coalescing iron concretions	0·8	2·6
A medium grained ferruginous conglomerate	1·2	4
A homogeneous ferruginous mass	0·8	2·6
A ferruginous conglomerate with large well rounded cobbles up to 40 cm (16 in) long	2	6·6
A fine grained ferruginous conglomerate	1	3·3
A coarse grained ferruginous conglomerate	2·5	8

In an area rich in iron and in the bottom of a small valley, everything is indurated. But the iron oxides have penetrated products that reveal climatic oscillations. Intervals of coarse detritus stand out against intervals of fine detritus and against essentially chemical deposits, producing either concretionary, conglomeratic or homogeneous layers.

When the dominating reliefs supply much iron, such as at the foot of Mt Nimba, cuirasses can therefore develop under rather varied climatic conditions, especially in valley bottoms that are always supplied with indurating solutions. But even in this case, as is shown by the example of the Ferriferous Quadrilateral, it is different on the hillsides, where cuirasses have a precise palaeoclimatic significance. And this is so *a fortiori* on alluvial plains and on piedmont slopes.

Maignien (1958a) has insisted on the fact that the cuirassing process is not an irreversible phenomenon and has shown that cuirasses are very slowly being dismantled in wet tropical climates, once more allowing the mobilisation of iron oxides. It may be observed in the rainforest of southwestern Ivory Coast where 1 m (3·3 ft) of fine greasy yellow clays cover a cuirass, penetrate its fissures, and sometimes surround completely isolated blocks.

In the wet tropics considerable quantities of iron oxides are dissolved and owing to organic compounds and bacteria are carried away. Iron leached out of an old cuirass is never redeposited in lower horizons but rather exported. There is not progressive induration of the cuirass as in the alternating wet–dry tropics, where migration is less easy and over shorter distances. The cuirass is thus progressively dismantled, its surface and fissures corroded, the latter gradually widening, especially as the weathering of the substratum causes movements that help to dissociate the mass.

The iron oxides that have gone into solution in perhumid climates are mostly exported to the sea. Only a small quantity remains in bottomlands, producing hydromorphic phenomena. If drainage is inland, as in Mali, the iron and silica solutions are transported towards drier interior basins where they are precipitated in flood plains. Climatic oscillations are thus reflected, on the middle Niger for example, by an alternation of semiarid phases and the construction of alluvial aprons with detrital materials derived from the regolith, and of semihumid (wet–dry) phases with development of ground water cuirasses and the influx of clays mixed with sands in large part removed from the deposits of riverbanks accumulated during the previous semiarid phase. The same evolution is found on less important streams, such as the marigots of Kéniéba and the Siguiri Basin, which have helped to form and cuirass the local piedmont slopes.

On interfluves, climates with a pronounced dry season, especially with a savanna vegetation, permit a partial mobilisation and limited migration of iron oxides, for they are very rapidly precipitated by evaporation. On *bowé*, the corrosion of the cuirass in its central part therefore nourishes a peripheral thickening and hardening. But under present conditions there is often no induration of slope deposits, except when iron is extremely abundant, as on Mt Nimba. When iron is not so abundant, the cuirassing process requires more humidity, less evaporation and more infiltration, conditions that were probably met before the vegetation was degraded by man. On the other hand, in drier climates, semiarid or subarid, infiltration no longer permits the mobilisation of iron over sufficient distances. There is only the formation of desert varnish. Mechanical processes erode the hillsides, which become covered with rubble and scree.

We have, then, schematically:

(a) In regions at present forested, cuirasses that belong to drier, semihumid palaeoclimates. They corrode slowly, which does not prevent but, on the contrary, helps inversions of relief, the cuirassed depressions being more resistant than the uplands that provided the iron (except in the case of ferruginous quartzites) and whose rate of weathering is increased in a perhumid climate.

(b) In regions that are now savannas, the cuirassing process is characteristic of a climate with a high seasonal variation of humidity, because under such conditions the iron oxides migrate only over short distances. The cuirassing process was interrupted by drier periods that decreased the volume and the amplitude of the migrations of iron. In semiarid regions they only permit the formation of desert varnish.

(*c*) Regions of inland drainage with dry climate located at the foot of humid highlands find themselves in ideal conditions to permit the development of cuirasses. It was the case, up to a recent date, of the middle Niger. But such cuirasses are also dependent on climatic oscillations that regulate the alluvial influx derived from the highlands and cause variations in the relative importance of solutions and porous detrital materials that are easily indurated.

Iron migrations are therefore closely integrated with the physical environment and play a role which for their amplitude goes far beyond pedologic dimensions. The considerable predominance of allogenic cuirasses, including stepped ones, can now be understood. It also explains the errors of pedologists; they will not be corrected until they have integrated the morphogenic factors into their analyses.

ORIGIN AND CLIMATIC SIGNIFICANCE OF BAUXITIC CUIRASSES

Much more restricted in extent, bauxitic cuirasses are not as significant as ferruginous cuirasses in the explanation of the relief, but they are of major importance to mining as the principal source of aluminium, which mankind consumes in ever increasing quantities. The problems are much more difficult to understand than in the case of ferruginous cuirasses because of a slower rate of evolution, which multiplies the polygenic aspects, and insufficiencies in the study of the problem, in which interdisciplinary cooperation has been little appealed to.

It may be said, schematically, that present discussions revolve around the respective importance of each of three series of factors.

1. *The initial weathering conditions* permitting the concentration of alumina. In the past alumina was considered to be derived from the dissociation of clays, and the ratio SiO_2/Al_2O_3 was considered to be the characteristic index of ferrallitic soils. As a matter of fact, this index leads to many frustrations, and most humid tropical soils do not fit into the typical categories, as the corresponding values of them are too high. At the present time the tendency is rather to consider that bauxitic cuirasses form in places where, because of the scarcity of silica and due to a basic or slightly acidic environment, the silica is leached because it does not combine with alumina in the form of clays, which are stable. In other words gibbsite is all the more easily formed as there is little or no development of kaolinite. Later, in order to be exploitable, the bauxite must be purified of its iron oxide. We have seen what is the importance of geomorphic conditions in this second phase. The climate that allows the formation of bauxite on basic igneous rocks is not a wet tropical one, for such a climate, as a result of the exuberant vegetation which it brings about, favours the acidification of the soil; this impedes

the leaching of silica, which in turn facilitates the formation of kaolinite, too stable to be later dissociated. It is, on the contrary, the climate of the semi-evergreen or deciduous seasonal forest which is the more apt to entertain the process of bauxitisation by restricting acidification while at the same time providing enough leaching. It is now clear why the ratio SiO_2/Al_2O_3 cannot correctly portray humid tropical weathering.

2. *The allogenic influxes*, as occur in ferruginous cuirasses. It is indeed impossible to understand the localisation of certain bauxitic ores if one tries to explain them by *in situ* weathering only, as for instance the bauxites of Surinam and Guyana located on the margin of an exhumed pre-Pliocene erosion surface truncating the Guiana Arch, here mainly composed of gneisses and granites. According to descriptions cited by Hose (1961), the bauxites are underlain by mottled clays, 15 to 30 m (50 to 100 ft) thick, becoming progressively more quartziferous and micaceous towards the base. On igneous rocks very rich in iron there is no bauxite here but a cuirass very rich in iron, nickel and titanium. For this reason Bakker (1959) thinks that bauxites do not result from the weathering of the substratum but from the desilication and dealkalinisation of littoral sediments. Bleakley (1961), discussing Hose's report, also recognises the weathering of kaolinitic sediments. In France Vaudour (1961) reports partially desilicified gravels and well sorted materials of the smaller dimensions in the bauxite pockets of Provence. He also recognises fluvial and aeolian (silts) deposits that, later, would have accumulated in karst sinks. The bauxite would have formed at a certain depth in the karst through the alteration of these elements. Whereas Roch suggests aeolian deposits as the original material of the bauxites, Butterlin (1958) proposes an autochthonous formation derived from decalcified products on pure or chalky limestone; crystalline, compact or coral limestones would, according to him, produce red ferrallitic soils. Butterlin (1961) also remarks that on the Ile de La Gonave (Haiti), which is entirely calcareous, bauxites can be derived only from red ferrallitic soils without possible addition of aeolian materials. The same is true in other parts of Haiti. But Zans *et al.* (1960) have not given up and attribute the raw material of bauxites to bottomland clays derived by runoff from the surrounding uplands.

3. *The subsequent epigenic accretions*, which stand out on the cross-sections provided by Hose (1961). In Guyana exploited bauxites are usually covered by a layer of kaolinitic clays, 0·3 to 1·2 m thick (1 to 4 ft), containing bauxitic nodules that are rather quartziferous near the top but purer at the base. Veins of kaolinitic clay, sometimes hardened, penetrate the bauxite. Reworked bauxitic pebbles forming alluvial aprons, 0·3 to 0·6 m (1 to 2 ft) thick, are spread at the foot of mesas capped by bauxite. In Johore, Malaya, it is on the contrary the bauxite that penetrates stalactite fashion into subjacent kaolinitic clays. Bauxite supposedly continues to form at the present time on the top of hills of a relative height of 45 to 60 m (150 to 200 ft). On the north flank of the Fouta Djallon, Michel (1959) reports blocks of bauxite

imbedded in ferruginous cuirasses on piedmont slopes at the foot of bauxit-ised *bowé*. In Guyana, according to Hose (1961), bauxites covering the pre-Pliocene erosion surface have subsequently been lowered below the watertable when the surface was lowered through tilting. They would then have been recemented.

The variety of facies and modes of occurrence of exploitable bauxites suggests that converging phenomena may explain their origin. One of these seems to be long continued *in situ* weathering of basic rocks, whether igneous or calcareous. As Hose (1961) has remarked (in part subsequent to Butter-lin's work) non-siliceous limestones produce a halloysite rich terra rossa, which may later bauxitise rather easily. On the contrary, siliceous lime-stones produce kaolinitic terra rossa, which does not bauxitise. At Koro, in the Chad Basin, an excellent study by Wacrenier (1961) describes mesas in which a bauxite overlies a sandy alluvium that becomes increasingly argillaceous near the top. The bauxite itself is overlain by a ferruginous cuirass containing quartz pebbles and blocks of bauxite, which indicates the bauxitisation was prior to the ferruginised detrital apron. The bauxite is ferruginous near the top, argillaceous and of oolitic texture at the base. It is almost incontestable, in this case, that it results from the weathering of probably lacustrine deposits, on account of the oolites. Similar oolites are found in the bauxites of Guyana.

It seems then that climatic oscillations, modifying the allogenic influxes, can intervene in the genesis of certain bauxites, in the same way that they intervene in that of ferruginous cuirasses. An original material poor in silica seems to be a necessary condition. The desilication of a kaolinitic clay pro-ducing a bauxite seems to be highly doubtful at the present, except perhaps in a basic lacustrine or lagoonal environment. In the Chad Basin, the Koro ore should be viewed in relation to the regional geomorphic evolution and the fluctuations of the lake. It seems likely that the increasingly argillaceous sedimentation ending in a bauxite indicates a lacustrine phase with little detrital sedimentation, whereas the overlying ferruginised rubble would correspond to a climatic desiccation followed by a new humid oscillation. Ferrugination of the bauxite would be secondary, connected to the evolu-tion of the overlying cuirass. A similar succession seems probable for the bauxites of Guyana and Surinam.

But the physical and chemical processes of bauxitisation are still insuffi-ciently known at the present to draw definite conclusions. They must be studied more and put into their morphogenic framework before the palaeo-climatic significance of divers types of bauxites can be defined.

Whatever the answers, we are now far removed from the concepts elabor-ated by armchair theoreticians and still current in the early 1950s: instan-taneous lateritic induration as soon as the forest is destroyed or instability of the clays in a tropical environment unleashing a characteristic freeing of alumina and automatic bauxitisation. The processes, although remaining essentially unique, are more complex and especially much slower. Sugarloaves do not dissolve under diluvial rains rich in acid as sugar does in coffee. A great deal of mistaken romanticism must be abandoned. Tropical weathering is slow; therefore it is capable of registering palaeoclimatic influences. Man-induced erosion is rapid when a slowly constituted capital that is the environment is deprived of the protective vegetative cover that is the forest. It is in the wet tropics that the human disturbance of the balance of nature unleashes the most violent catastrophes, as the erosive potential of the climate is large and the soils (regolith) fragile. All the more reason to understand better the dynamics of the natural environment, of which geomorphology is a part, in order to preserve this endowment and to utilise better the moisture and heat of cosmic origin to feed men, most of whom are hungry.

Bibliographic orientation

A large part of the literature used in this chapter has already appeared in the bibliographies of the preceding chapters.

The human factor in morphogenesis

One should again refer to Berry and Ruxton (1960) and to Birch and Friend (1956).

ALEXANDRE, J. and ALEXANDRE, S. (1960) 'Erosion en surface et érosion linéaire anthropiques dans une région tropicale à saison sèche', *XIXth Int. Geogr. Congr.*, Stockholm, Abstracts, p. 5.

AUBREVILLE, A. (1949) *Climats, forêts et désertification de l'Afrique tropicale*, Paris, Larose.
Basic work; a few romantic exaggerations.

AUBREVILLE, A. (1959) 'Erosion sous forêt et érosion en pays déforesté dans la zone tropicale humide', *Bois et Forêts Trop.* **68**, Nov.-Dec.
Good overall view.

BARBIER, R. (1957) 'Aménagements hydroélectriques dans le Sud du Brésil', *Bull. Soc. Géol. Fr.*, 7ᵉ sér. **7**, 877–92.

BELLOUARD. (1948) 'Erosion des sols du Sénégal oriental, du Soudan occidental et du Fouta Djalon', *Bull. Agric. Congo Belge*, **40**, 1299–1308.

BESAIRE, E. (1948) 'Deux exemples d'érosion accélérée à Madagascar', *Conf. Afr. Sols*, Goma, Comm. no. 11.

BHARUCHA, F. R. and SHANKARNARAYAN, K. A. (1952) 'Effects of overgrazing on the grasslands of the Western Ghats, India', *Ecology*, **39**, 152–3.

O'REILLY STERNBERG, H. (1949) 'Enchentes e movimentos coletivos do solo no vale do Paraíba em dezembro de 1948. Influéncia da explotação destructiva das terras', *Rev. Brasil. Geogr.* **11**, 223–61: English translation as 'Floods and landslides in the Paraiba Valley, December 1948. Influence of destructive exploitation of the land', *XVIth Int. Geogr. Congr.* (1949) Lisbon, vol. 3, 335–64.

PORTERES, R. (1956) 'Une forme spectaculaire d'érosion à Madagascar: le lavaka', *Science et Nature*, no. 14, 8–13.

QUANTIN, P. and COMBEAU, A. (1962) 'Erosion et stabilité structurale du sol', *Public. Int. Ass. Sci. Hydrol.* **59**, 124–30.

RIQUIER, J. (1958) 'Les "lavaka" de Madagascar', *Bull. Soc. Géogr.*, Marseille, **69**, 181–91.

SCHNELL, R. (1948) 'Sur quelques cas de dégradation de la végétation et du sol observés en Afrique Occidentale Française', *Bull. Agric. Congo Belge*, **40**, 1353–62.

SEGALEN, P. (1948) 'L'érosion des sols à Madagascar', *Conf. Afr. Sols*, Goma, 1127–37.
Good summation by a pedologist.

SETZER, J. (1956) 'A natureza e as possibilidades do solo no vale do Rio Pardo entre os municípios de Caconde, S.P. e Poços de Caldas, M.G.', *Rev. Brasil. Geogr.* **18**, 287–322.
Good granulometric analyses making it possible to recognise the effects of erosion.

SPIJKERMAN, A. C. (1960) 'Les problèmes de l'érosion en Afrique tropicale et les caractéristiques de quelques projets de conservation', *Int. Inst. Land Recl. Improv., Ann. Rept.*, 15–34.
Good summation.

TONDEUR, G. (1954) *Erosion du sol, spécialement au Congo Belge*, Brussels, Ministry of Colonies, 3rd edn, 240 p.
Abundant documentation.

TRICART, J. (1953) 'Erosion naturelle et érosion anthropogène à Madagascar', *Rev. Géomorph. Dyn.* **4**, 225–30.

TRICART, J. (1956) 'Dégradation du milieu naturel et problèmes d'aménagement au Fouta Djalon (Guinée)', *Rev. Géogr. Alpine*, **44**, 7–36.

Palaeoclimatic influences

Here again numerous regional works to which one should refer have already been cited. Below will be found references of studies concerning palaeoclimatic oscillations and the stoneline.

PALAEOCLIMATIC OSCILLATIONS

AMARO, L. (1952) 'Les alluvions aurifères du "Campo Mineiro de Macequece", Afrique Orientale Portugaise', *Proc. XIXth Int. Geogr. Congr.*, Algiers, **21**, 349–56.

BAKKER, J. P. (1957) 'Quelques aspects du problème des sédiments corrélatifs en climat tropical humide', *Z. Geomorph.*, n.s., **1**, 1–43.

BRAUN, W. A. (1961) 'Contribução ao estuo da erosão no Brasil e seu contrôle', *Rev. Brasil Geogr.* **23**, 591–642.

BRENON, P. (1952) 'Géomorphologie de l'Antsihanaka et de l'Antanosimboangy', *Erosion des sols à Madagascar*, Tananarive, 6–22.
Basic study on lavakas.

BÜDEL, J. (1957a) 'Die Flächenbildung in den feuchten Tropen und die Rolle fossiler solchen Flächen in anderen Klimazonen', *Dtscher Geogr.-Tag*, **31**, Würzburg, 89–121.

BÜDEL, J. (1957b) 'Die "doppelten Einebnungsflächen" in den feuchten Tropen', *Z. Geomorph.*, n.s., **1**, 201–28.

CARTER, J. (1958) 'Erosion and sedimentation from aerial photographs. A micro-study from Nigeria', *J. Trop. Geogr.* **11**, 100–6.
Good monograph of a much affected savanna region. Maps.

CHEVALIER, A. (1950a) 'Mesures urgentes à prendre pour entraver le déssèchement, l'ensablement et la décadence des sols et de la végétation en A.O.F. et spécialement au Soudan Français', *C.R. Acad. Sc.* **230**, no. 20, 1720–3.

CHEVALIER, A. (1950b) 'Régénération des sols et de la végétation en A.O.F.', *C.R. Acad. Sc.* **230**, no. 24, 2064–6.

COMBEAU, A. (1960) 'Quelques facteurs de la variation de l'indice d'instabilité structurale dans certains sols ferrallitiques', *C.R. Acad. Agric.* (France), **66**, 109–15.

CORMACK, R. M. (1951) 'Land erosion in Southern Rhodesia', *Int. Union Geod. Geophys., Int. Ass. Sci. Hydrol.*, Brussels Meeting, ii, 17–28.

COSTA, S. V. BOTELHO DA and AZEVEDO, A. LOBO (1952) 'Aspectos da erosão do solo em Angola', *Agros.* **33**, no. 1/2, 15–22.

FOURNIER, F. (1956) 'Les formes et types d'érosion du sol par l'eau en Afrique Occidentale Française', *C.R. Acad. Agric.* (France), **42**, 215–21.

FOURNIER, F. (1957a) 'Les valeurs d'érosion du sol dans les territoires français d'outre-mer', *Int. Union Geod. Geophys., Int. Ass. Sci. Hydrol.*, Toronto Meeting, i, 76–87.

Results of measurements on experimental parcels. Erosion: maximum in climates with the highest rainfall.

FOURNIER, F. (1957b) 'Les facteurs de l'érosion du sol en zone tropicale. Conclusions sur la conservation du sol', *Int. Union Geod. Geophys., Int. Ass. Sci. Hydrol.*, Toronto Meeting, i, 88–96.

GAUDY, M. (1958) 'La mesure du taux de ruissellement et de l'érosion en A.O.F.', *Serv. Cons. Sols*, Dakar, 21 p. mimeographed.

GROVE, A. T. (1951) 'Soil erosion and population problems in Southeastern Nigeria', *Geogr. J.* **117**, no. 3, 307–27.

GUENNELON, R. (1954) 'Erosion des sols et pédogenèse dans une île tropicale volcanique: la Réunion', *Bull. Assoc. Franç. Sols*, Feb., 12–19.

HAGGETT, P. (1961) 'Land use and sediment yield in an old plantation tract of the Serra do Mar, Brazil', *Geogr. J.* **127**, 50–62.

HARROY, J. P. (1949) *Afrique, terre qui meurt. La dégradation des sols africains sous l'influence de la colonisation*, 2nd edn, Brussels, M. Hayes, 557 p.

To compare with Aubreville (1949). Pessimistic outlook.

HUDSON, N. W. (1957) 'Erosion control and research. Progress report on experiments at Henderson research station, 1953–6', *Rhodesian Agric. J.* **54**, 297–323.

LE BOURDIEC, P. (1958) 'Contribution à l'étude géomorphologique du bassin sédimentaire et des régions littorales de Côte-d'Ivoire', *Etudes Eburnéennes*, **8**, 7–96.

LEDGER, D. C. (1961) 'Recent hydrographic change in the Rima basin, northern Nigeria', *Geogr. J.* **127**, 477–87.

MADAGASCAR, BUREAU GÉOLOGIQUE (1952) *Contribution à l'étude de l'érosion des sols à Madagascar*, Tananarive, 44 p.

MABBUTT, J. A. (1961) 'A stripped land surface in Western Australia', *Trans. Papers Inst. Brit. Geogr.* **29**, 101–14.

MABBUTT, J. and SCOTT, R. M. (1966) 'Periodicity of morphogenesis and soil formation in a savannah landscape near Port Moresby, Papua', *Z. Geomorph.*, n.s., **10**, 69–89.

MARQUES, J., GROMANN, F., BERTONI, J. and ALENCAR, F. AYRES DE (1951) 'Algumas conclusões gerais preliminares das determinações de perdas por erosã realizadas em São Paolo', *Recife, Inst. Agron.*, 44 p.

MEULENBERG, O. (1949) *Introduction à l'étude pédologique des sols du territoire du Bas-Fleuve*, Brussels, Inst. Roy. Col. Belge, 133 p.

NASCIMENTO, U. (1952) 'Estudo da regularização e protecção das barrocas de Luanda', *Min. Obras Publ.*, Lab. Nac. Engenharia Civil, Lisbon, no. 30, 39 p.

Excellent monograph.

PANZER, W. (1954) 'Verwitterungs- und Abtragungsformen im Granit von Hong-Kong', *Abh. Akad. Raumforschung und Landesplannung*, **28**, 41–60.

PLOEY, J. DE (1963) 'Quelques indices sur l'évolution morphologique et paléoclimatique des environs du Stanley-Pool (Congo)', *Studia Univ. Lovanium, Sc.*, no. 17, 16 p.

RIOU, G. (1965) 'Notes sur les sols complexes des savanes préforestières de Côte-d'Ivoire', *Ann. Univ. Abidjan, Lettres*, no. 1, 17–36.

TRICART, J., MICHEL, P. and VOGT, J. (1957) 'Oscillations climatiques quaternaires en Afrique occidentale', *Fifth Congr. INQUA*, Madrid-Barcelona.

VOGT, J. (1956) *Rapport provisoire de mission à Kéniéba*, Soudan, DFMG.

VOGT, J. (1959) 'Observations nouvelles sur les alluvions inactuelles de Côte-d'Ivoire et de Guinée', *Actes 84ᵉ Congr. Soc. Sav.*, Dijon, 205–10.

VOGT, J. (1959) 'Aspects de l'évolution morphologique récente de l'Ouest Africain', *Ann. Géogr.* **68**, 193–206.

THE STONE-LINE

BOURGEAT, F. and PETIT, M. (1966) 'Les "stone-lines" et les terrasses alluviales des hautes terres malgaches', *Cahiers Orstom, Pédologie*, **4**, no. 2, 3–19.

CAILLEUX, A. and TRICART, J. (1957) 'Termites et "stone-line"', *C.R. Somm. Soc. Biogéogr.* **34**, p. 12.

CRAENE, A. DE and SOROTCHINSKY, C. (1954) 'Essai d'interprétation nouvelle de la genèse de certains types de "stone-line"', *C.R. IIᵉ Conf. Interafr. Sols*, Leopoldville, vol. 1, 453–6.

CRAENE, A. DE (1954) 'Les sols de pédimentation ou les sols à "stone-line" du Nord-est du Congo Belge', *Proc. Vth Int. Congr. Soil Sc.*, Leopoldville, vol. 4, 451–60.

HEINZELIN, J. DE (1952) *Sols, paléosols et désertifications anciennes dans le secteur nord-oriental du Bassin du Congo*, Brussels, Publ. INEAC, 168 p., 8 pl.

HEINZELIN, J. DE (1955) *Observations sur la genèse des nappes de gravats dans les sols tropicaux*, Brussels Publ. INEAC, Sér. Sc., no. 64.

PARIZEK, E. J. and WOODRUFF, J. F. (1957) 'Description and origin of stone layers in soils of the southeastern States', *J. Geol.* **65**, 24–34.

RUHE, R. (1956) *Landscape Evolution in the High Ituri, Belgian Congo*. Brussels Publ. INEAC, Sér. Sc., no. 66, 91 p.

RUHE, R. (1959) 'Stone-line in soils', *Soil Sc.* **87**, 223–31.

SYS, C. (1958) *Projet de classification des sols congolais (3ᵉ approximation)*, Publ. INEAC, Yangambi, 91 p., mimeographed.

TRICART, J. (1958) 'Division morphoclimatique du Brésil atlantique central', *Rev. Géomorph. Dyn.* **9**, 1–22.

VOGT, J. and VINCENT, P. L. (1966) 'Terrains d'altération et de recouvrement en zone intertropicale', *BRGM* **4**, 1–111.

WAEGEMANS, G. (1953) 'Signification pédologique de la "stone-line" (note préliminaire)', *Bull. Agric. Congo Belge*, **44**, 521–31.

Climatic oscillations and ferruginous cuirasses

Here too one should refer back to references already cited in chapter 4, and completed below.

AUFRÈRE, L. (1932) 'La signification de la latérite dans l'évolution climatique de la Guinée', *Bull. Assoc. Géogr. Franç.*, June.

BRÜCKNER, W. (1955) 'The mantle-rock ("laterite") of the Gold Coast and its origin', *Geol. Rundschau*, **43**, 307–27.

BRÜCKNER, W. (1957) 'Laterite and bauxite profiles of West Africa as an index of rhythmical climate variations in the tropical belt', *Ecl. Geol. Helv.* **50**, 239–56.

ERHART, H. (1943) 'Les latérites du Moyen-Niger et leur signification paléoclimatique', *C.R. Acad. Sc.* **217**, 323–5.

ERHART, H. (1953) 'Sur les cuirasses termitiques fossiles dans la vallée du Niari et dans le massif du Chaillu (Moyen-Congo, A.E.F.)', *C.R. Acad. Sc.* **237**, 431–3.
    The termitaries indicate an interval of savanna climate in a presently forested region.

EYK, J. J. VAN DER (1957) *Reconnaissance Soil Survey in Northern Surinam*, Wageningen, 99 p., 9 pl.

PITOT, A. (1954) 'Végétation et sols et leurs problèmes en A.O.F.', *Ann. Ec. Sup. Sc.*, Dakar, **1**, 129–39.
    The problem of cuirasses is posed from a palaeoclimatic point of view.

ROUGERIE, G. (1950) 'Le pays du Sanwi. Esquisse morphologique dans le Sud-Est de la Côte-d'Ivoire', *Bull. Assoc. Géogr. Franç.*, no. 212–13, 138–45.

TRICART, J. (1962) 'Quelques éléments de l'évolution géomorphologique de l'Ouest de la Côte-d'Ivoire', *Recher. Africaines*, **1**, 30–9.

The problem of bauxitic cuirasses

This is not the place to list the enormous literature concerning bauxites. Only those publications are listed that may be useful to geomorphologists and which show the breadth of recent points of view. Brückner (1957) should be added to these.

ABOTT, A. T. (1958) 'Occurrence of gibbsite on the island of Kauai, Hawaiian Islands', *Econ. Geogr.*, **34**, 842–53.

BAKKER, J. P. (1959) 'Recherches néerlandaises de géomorphologie appliquée', *Rev. Géomorph. Dyn.* **10**, 67–84.

BERG, L. (1945) 'The origin of bauxites of the Ural', *Izv. Vsiess. Geogr. Obchtch.* **77**, p. 49.

BUTTERLIN, J. (1958) 'A propos de l'origine des bauxites des régions tropicales calcaires', *C.R. Somm. Soc. Géol. Fr.*, 121–3.

BUTTERLIN, J. (1961) 'Nouvelles précisions au sujet des sols rouges ferrallitiques trouvés sur les calcaires de la République d'Haïti (Grandes Antilles)', *C.R. Somm. Soc. Géol. Fr.*, 109–10.

HOSE, H. R. (1961) 'The origin of bauxites in British Guiana and Jamaica', *Proc. Vth inter-Guiana Geol. Conf.*, Georgetown, 1959, 185–98.
Symposium on bauxites, with discussions. Basic.

KERSTEN, J. F. VAN (1958) 'Bauxite deposits in Surinam and Demerara', *Leidse Geol. Med.* **21**, 247–375.

LAJOINIE, J. P. and BONIFAS, M. (1961) 'Les dolérites du Konkouré et leur altération latéritique', *Bull. Bur. Rech. Géol. Min.*, no. 2.

STRAKHOV, M. N. *et al.* (1958) 'Les bauxites, leur minéralogie et leur genèse', *Izv. Akad. Nauk. SSSR*, Moscow, 488 p., 155 fig., French trans. by BRGM, no. 2640.
Excellent summation compiling the views of Russian authors.

VAUDOUR, J. (1961) 'La vallée de l'Huyeaune', *Bull. Assoc. Géogr. Franç.* **298**, 59–78.

WACRENIER, P. (1961) 'Mission de recherche de bauxite au Logone et au Moyo-Kébi (Tchad)', *Bull. Inst. Equat. Rech. Et. Géol. Min.* **14**, 37–42.

WEBBER, B. N. (1959) 'Bauxitisation in the Poços de Caldas district, Brazil', *Mining Engineering*, Aug., 804–9.

ZANS, V. A. (1961) 'Classification and genetic types of bauxite deposits', *Proc. Vth inter-Guiana Geol. Conf.*, 1959, Georgetown, 205–11.

ZANS, V. A., LEMOINE, R. G. and ROCH, E. (1960) 'Les "calcaires latéritisés" des Grandes Antilles', *C.R. Somm. Soc. Géol. Fr.*, 177–8.

Consolidated bibliography

Abbreviations

AEF   Afrique Equatoriale Française
BRGM   Bureau de᾿ Recherches Géologiques et Minières—74, rue de la Fédération, Paris 15ᵉ.
CDU   Centre de Documentation Universitaire (Paris)
CNRS   Centre National de la Recherche Scientifique
DFMG   Direction Fédérale des Mines et de la Géologie (now SGPM—below)
IFAN   Institut Français d'Afrique Noire (now Institut Fondamental d'Afrique Noire, Dakar)
INEAC   Institut National pour l'Etude Agronomique du Congo (Yangambi, Congo-Kinshasa)
INQUA   International Union for Quaternary Research
ORSTOM   Office de la Recherche Scientifique et Technique pour les Pays d'Outre-Mer
SGPM   Service de Géologie et de Prospection Minière (Dakar)—formerly DFMG (see above)
AOF   Afrique Occidentale Française

ABECASSIS, F. M. (1958) 'Les flèches de sable de la côte d'Angola', *Lab. Nac. Engenharia Civil, Mem.* **140**, Lisboa, 9 p.

ABOTT, A. T. (1958) 'Occurrence of gibbsite on the island of Kauai, Hawaiian Islands', *Econ. Geogr.*, **34**, 842–53.

AB'SABER, A. NACIB (1953) 'Geomorfologia de uma linha de quesdas apalachiana típica do Estado de São Paulo', *An. Fac. Fil. Catol.*, São Paulo, 111–38, photographs.

ACKERMANN, E. (1936) 'Dambos in Nordrhodesien', *Wiss. Veröffentl. Dtschen Mus. Länderkunde zu Leipzig*, n.s., **4**, 149–57.

ALEXANDRE, J. (1966) 'L'action des animaux fouisseurs et des feux de brousse sur l'efficacité érosive du ruissellement dans une région de savane boisée', *Congr. and Coll. Univ. Liège*, **40**, 1967, 43–9.

ALEXANDRE, J. and ALEXANDRE, S. (1960) 'Erosion en surface et érosion linéaire anthropiques dans une région tropicale à saison sèche', *XIXth Int. Geogr. Congr.*, Stockholm, Abstracts, p. 5.

ALIA MEDINA (1951) 'Datos geomorfológicos de la Guinea continental española', *Consejo sup. Invest. Cient.*, 64 p., 9 pl.

AMARO, L. (1952) 'Les alluvions aurifères du "Campo Mineiro de Macequece", Afrique Orientale Portugaise', *Proc. XIXth Int. Geol. Congr.*, Algiers, **21**, 349–56.

ANDRADE, G. OSÓRIO DE (1955) *Itamaracá, contribuiçao para o estudio geomorfologico da costa pernambucana*, Recife, 84 p.

ANDRADE, G. OSÓRIO DE (1956) *A 'ria' do Rio Formoso na costa sul de Pernambuco*, Univ. de Recife, Fac. de Fil., Secçao E, no. 18, 13 p.

ANDRADE, G. OSÓRIO DE (1956) 'Furos, paranás e iguarapés. Anâlise genética de algumos elementos do sistema potamogrâfico amazônico', *Bol. Carioca de Geogr.* **9**, 15–50.

AUBERT, G. (1948) 'Observations sur le rôle de l'érosion dans la formation de la cuirasse latéritique', *Bull. Agron. Congo Belge*, **40**, no. 2, 1383–6.

AUBERT, G. (1953) 'Quelques problèmes pédologiques de mise en valeur des sols du Delta Central Nigérien (Soudan Français)', *Desert Res.*, Jerusalem, 392–9.

AUBERT, G. (1961) 'Influence des divers types de végétation sur les caractères et l'évolution des sols en régions équatoriales et subéquatoriales ainsi que leurs bordures tropicales semi-humides', UNESCO, *Humid Tropical Zone*, Abidjan Symposium, 1959, 41–7.

AUBERT DE LA RUE, E. (1954) *Reconnaissance géologique de la Guyane française méridionale*, Paris, Larose, 128 p., 22 pl., 5 maps.

AUBREVILLE, A. (1938) 'La forêt coloniale; les forêts de l'Afrique occidentale française', *Ann. Acad. Sc. Col.* **9**, 244 p.

AUBREVILLE, A. (1947) 'Erosion et bowalisation en Afrique noire française', *Agron. Trop.*, no. 7/8, 339–57.

AUBREVILLE, A. (1948) 'Ancienneté de la destruction de la couverture forestière primitive de l'Afrique tropicale', *Bull. Agron. Congo Belge*, **40**, no. 2, 1347–52.

AUBREVILLE, A. (1949) *Climats, forêts et désertification de l'Afrique tropicale*, Paris, Larose.

AUBREVILLE, A. (1949) *Contribution à la paléohistoire des forêts de l'Afrique tropicale*, Soc. Edit. Géogr. et Col.

AUBREVILLE, A. (1959) 'Erosion sous forêt et érosion en pays déforesté dans la zone tropicale humide', *Bois et Forêts Trop.* **68**, Nov.-Dec.

AUFRÈRE, L. (1932) 'La signification de la latérite dans l'évolution climatique de la Guinée', *Bull. Assoc. Géogr. Franç.*, no. 60, pp. 95–7.

AUFRÈRE, L. (1936) 'La géographie de la latérite', *C.R. Soc. Biogéogr.* **13**, 3–11.

BACHELIER, G. (1960) 'Sur l'orientation différente des processus d'humidification dans les sols bruns tempérés et les sols ferrallitiques des régions équatoriales', *Agron. Tropicale*, **15**, no. 3, 320–4.

BAKKER, J. P. (1951) 'Bodem en bodemprofielen van Suriname, in het bijzonder van de noordelijke savannenstrook', *Landbouwkundig Tijdschr.* **63**, 379–91.

BAKKER, J. P. (1957a) 'Die Flächenbildung in den Feuchten Tropen', *Dtscher Geogr.-Tag*, Würzburg, **31**, 86–8.

BAKKER, J. P. (1957b) 'Quelques aspects du problème des sédiments corrélatifs en climat tropical humide', *Z. Geomorph.*, n.s., **I**, 1–43.

BAKKER, J. P. (1959) 'Recherches néerlandaises de géomorphologie appliquée', *Rev. Gèomorph. Dyn.* **10**, 67–84.

BAKKER, J. P. and MULLER, H. (1957) 'Zweiphasige Flussanlagerungen und Zweiphasenverwitterung in der Tropen unter besonderer Berücksichtigung von Surinam', *Festschrift Lautensach*, 365–97.

BAKKER, J., MULLER, H., JUNGERIUS, P. and PORRENGA, H. (1957) 'Zur Granitverwitterung und Methodik der Inselbergforschung in Surinam', *Dtscher Geogr.-Tag*, Würzburg, **31**, 122–31.

BARAT, C. (1957) *Pluviologie et aquidimétrie dans la zone intertropicale*, Mém. IFAN, no. 49, 80 p., 15 pl.

BARAT, C. (1958) 'La Montagne d'Ambre (Nord de Madagascar)', *Rev. Géogr. Alpine*, **46**, 629–81.

BARBEAU, J. and GÈZE, B. (1957) 'Les coupoles granitiques et rhyolitiques de la région de Fort-Lamy (Tchad)', *Bull. Soc. Géol. Fr.*, 6ᵉ sér., **7**, 341–51.

BARBIER, R. (1957) 'Aménagements hydroélectriques dans le Sud du Brésil', *Bull. Soc. Géol. Fr.*, 7ᵉ sér., **7**, 877–92.

BASU, J. K. and PURANIK, N. B. (1951) 'Land erosion with particular reference to agricultural lands in the dry areas of the Bombay State', *Int. Union Geod. Geophys.*, *Int. Ass. Sci. Hydrol.*, *Brussels Meeting*, vol. 2, 57–63.

BATTISTINI, R. (1958) 'Structure et géomorphologie du littoral Karimbola (extreme south of Madagascar)', *Mém. Inst. Scient. Madagascar*, Sér. F, **2**, 1–77.

BATTISTINI, R. (1960) 'Quelques aspects de la morphologie du littoral Mikea (SW coast of Madagascar)', *Cahiers Océanogr.* **12**, 548–71.

BATTISTINI, R. and DOUMENGE, F. (1966) 'La morphologie de l'escarpement de l'Isalo et de son revers dans la région de Ranohira (Sud-Ouest de Madagascar)', *Madagascar, Rev. de Géogr.* **8**, 67–92.

BAYENS, J. (1938) *Les sols d'Afrique centrale, spécialement du Congo Belge*, Brussels, Publ. INEAC, hors-série.

BEADLE, N. (1940) 'Soil temperatures during forest fires and their effect on the survival of vegetation', *J. Ecology*, **28**, 180·92.

BEARD, J. S. (1953) 'The savanna vegetation of northern tropical America', *Ecol. Monogr.* **23**, 149–215.

BEHREND, F. (1919) 'Über die Entstehung der Inselberge und Steilstufen besonders in Afrika und die Erhaltung ihrer Formen', *Z. Dtschen Geol. Ges.* **70**, 154–67.

BELLOUARD (1948) 'Erosion des sols du Sénégal oriental, du Soudan occidental et du Fouta Djalon', *Bull. Agric. Congo Belge*, **40**, 1299–1308.

BERG, L. (1945) 'The origin of bauxites of the Ural', *Izv. Vsiess. Geogr. Obchtch.* **77**, 49.

BERGER, H. (1959/60) 'Beobachtungen zur Inselbergbildung in Nordost Uganda', *Geogr. Jahresver. aus Oesterr.* **28**, 72–9.

BERLIER, Y., DABIN, B. and LENEUF, N. (1956) 'Comparaison physique, chimique et micro-biologique entre les sols de forêt et de savane sur les sables tertiaires de la Basse Côte-d'Ivoire', *Rept. VIth Int. Congr. Soil Sc.*, Paris, vol. 5, 499–502.

BERNARD, E. (1945) *Le Climat écologique de la cuvette congolaise*, Brussels, Publ. INEAC, 240 p.

BERRY, L. and RUXTON, B. (1960) 'The evolution of Hong-Kong harbour basin', *Z. für Geom.* **4**, 97–115.

BERTHOIS, L. and GUILCHER, A. (1956) 'La plaine d'Ambilobé (Madagascar), étude mor-phologique et sédimentologique', *Rev. Géom. Dynamique*, **7**, 35–52.

BESAIRIE, E. (1948) 'Deux exemples d'érosion accélérée à Madagascar', *Conf. Afr. Sols*, Goma, Comm. no. 11.

BETREMIEUX, R. (1951) 'Etude expérimentale de l'evolution du fer et du manganèse dans le sol', *Ann. Agron.*, sér. A, **2**, no. 3, 193–295.

BHARUCHA, F. R. and SHANKARNARAYAN, K. A. (1952) 'Effects of overgrazing on the grass-lands of the Western Ghats, India', *Ecology*, **39**, 152–3.

BIRCH, H. F. and FRIEND, M. T. (1956) 'Humus decomposition in East African soils', *Nature, Lond.* **178**, 500–1.

BIROT, P. (1955) *Les Méthodes de la morphologie*, Paris, PUF, Coll. Orbis, 177 p.

BIROT, P. (1957) 'Esquisse morphologique de la région littorale de l'état de Rio de Janeiro', *Ann. de Géogr.* **66**, 80–91.

BIROT, P. (1959) *Géographie physique générale de la zone intertropicale*, Paris, CDU, 'Cours de Sorbonne', mimeographed, 244 p., 15 fig.

BIROT, P. (1965) *Les Formations végétales du globe*, Paris, SEDES, 508 p.

BIROT, P. (1968) *The Cycle of Erosion in different Climates*, trans. from the French edn (1960) by C. Jan Jackson and Keith M. Clayton, Univ. Calif. Press, 144 p.

BLEAKLEY, D. (1961) 'Status of investigations into the bauxite deposits of British Guiana', *Proc. Fifth Inter-Guiana Geol. Conf.*, Georgetown, 1959.

BLEAKLEY, D. (1964) 'Bauxite and laterites of British Guiana', *Geol. Surv. Brit. Guiana*, Bull. no. 34, 156 p., 21, fig. 1, 21 pl.

BLUMENSTOCK, D., FOSBERG, F. R. and JOHNSON, C. G. (1961) 'The resurvey of typhoon effect on Jaluit atoll in the Marshall Islands', *Nature, Lond.* **199**, 618–20.

BONNET, P., VIDAL, P. and VEROT, P. (1958) *Premiers résultats des parcelles expérimentales d'étude de l'érosion de Sérédou en Guinée forestière*, Sérédou, station expérimentale du quinquina. Mimeographed, 38 p.

BORNHARDT, W. (1900) *Zur Oberflächengestaltung und Geologie Deutsch Ostafrikas*, collection *Deutsch Ostafrika*, Berlin, vol. 7, 595 p., 28 pl., atlas.

BOURGEAT, F. and PETIT, M. (1966) 'Les "stone-lines" et les terrasses alluviales des hautes terres malgaches', *Cahiers Orstom, Pédologie*, **4**, no. 2, 3–19.

BOYÉ, M. (1960) 'Morphométrie des galets de quartz en Guyane française', *Rev. Géom. Dyn.* **11**, 13–27.

BOYER, P. (1955) 'Premières études pédologiques et bactériologiques des termitières', *C.R. Acad. Sc.* **240**, 569–71.

BOYER, P. (1959) 'De l'influence des termites de la zone intertropicale sur la configuration de certain sols', *Rev. Géomorph. Dyn.* **10**, 41–4.

BRAJNIKOV, B. (1953) 'Les "pains de sucre" du Brésil sont-ils enracinés?' *C.R. Somm. Soc. Géol. Fr.*, 267–9.

BRAMAO, L. and BLACK, C. (1955) 'Nota preliminar sobre o estudo solo-vegetação de Barreiras (Bahia)', *Bol. Serv. Nac. Pesquisas Agron.*, Rio de Janeiro, no. 9.

BRANNER, J. C. (1896) 'Decomposition of rocks in Brazil', *Bull. Geol. Soc. Amer.* **7**, 255–314.

BRANNER, J. C. (1913) 'The fluting and pitting of granites in the tropics', *Proc. Amer. Phil. Soc.* **52**, 163–74.

BRANNER, J. C. (1948) 'Decomposição das rochas do Brasil', *Bol. Geográfico*, Rio de Janeiro, **59**, 1266–1300.

BRAUN, W. A. (1961) 'Contribução ao estuo da erosão no Brasil e seu contróle', *Rev. Brasil. Geogr.* **23**, 591–642.

BRENON, P. (1952) 'Géomorphologie de l'Antsihanaka et de l'Antanosimboangy', *Erosion des sols à Madagascar*, Tananarive, 6–22.

BRÜCKNER, W. (1955) 'The mantle-rock ("laterite") of the Gold Coast and its origin', *Geol. Rundschau*, **43**, 307–27.

BRÜCKNER, W. (1957) 'Laterite and bauxite profiles of West Africa as an index of rhythmical climatic variations in the tropical belt', *Ecl. Geol. Helv.* **50**, 239–56.

BUCHANAN, F. (1807) *A Journey from Madras through the Countries of Mysore, Canara and Malabar*, London, 2 vols.

BÜDEL, J. (1957a) 'Die "doppelten Einebnungsflächen" in den feuchten Tropen', *Z. Geomorph.*, n.s., **1**, 201–28.

BÜDEL, J. (1957b) 'Die Flächenbildung in den feuchten Tropen und die Rolle fossiler solcher Flächen in anderen Klimazonen', *Dtscher. Geogr.-Tag*, **31**, Würzburg, 89–121.

BÜDEL, J. (1965) 'Die Relieftypen der Flächenspülzone Süd-Indiens am Ostabfall Dekans gegen Madras', *Colloquium geographicum*, Bd. 8, Bonn Univ., F. Dümmler, 100 p.

BUNTING, B. T. (1965) *The Geography of Soil*, London, Hutchinson, 213 p.

BUTTERLIN, J. (1953) 'Données nouvelles sur la géologie de la République d'Haïti', *Bul. Soc. Géol.*, 6ᵉ sér., **3**, 283–91.

BUTTERLIN, J. (1958) 'A propos de l'origine des bauxites des régions tropicales calcaires', *C.R. Somm. Soc. Géol. Fr.*, 121–3.

BUTTERLIN, J. (1961) 'Nouvelles précisions au sujet des sols rouges ferrallitiques trouvés sur les calcaires de la République d'Haïti (Grandes Antilles)', *C.R. Somm. Soc. Géol. Fr.*, 109–10.

CACHAN, P. (1960) *L'étude des microclimats et de l'écologie de la forêt sempervirente en Côte d'Ivoire*, IDERT-UNESCO, Adiopodioumé, mimeographed, 10 p.

CAHEN, L. and LEPERSONNE, J. (1948) 'Notes sur la géomorphologie du Congo occidental', *Ann. Musée Congo Belge*, Tervuren, Sér. in-8°, Sc. Géol., **1**, 95 p.

CAILLERE, S. and HENIN, S. (1951) 'Etude de l'altération de quelques roches en Guyane', *Ann. Inst. Agron.*, Série A, **2**, no. 4, July-August.

CAILLEUX, A. (1959) 'Etudes sur l'érosion et la sédimentation en Guyane', *Mém. Serv. Carte Géol. Fr.*, 49–73.

CAILLEUX, A. and TRICART, J. (1957) 'Termites et "stone-line"', *C.R. Somm. Soc. Biogéogr.* **34**, 12.

CARTER, D. B. (1954) *Climates of Africa and India according to Thornthwaite's 1948 classification*, Johns Hopkins Univ., Publ. in climatology, vol. 7, no. 4, 25 p., 3 fig., 2 pl.

CARTER, J. (1958) 'Erosion and sedimentation from aerial photographs. A micro-study from Nigeria', *J. Trop. Geogr.* **11**, 100–6.

CASTRO SOARES, L. DE (1949) 'Observacões sobre a morfologia das margens do Baixo-Amazonas e Baixo-Tapajós, Pará, Brasil', *Proc. Int. Geogr. Congress*, Lisbon, **2**, 748–61.

CARVALHO, G. SOARES DE (1961) *Nota sobre as formações superficiais da pediplanicie de região da Quibala-Catofe (Angola)*, Lisbon, Garcia de Orta, **9**, 779–90.

CHETELAT, E. DE (1938) 'Le modelé latéritique de l'ouest de la Guinée française', *Rev. Géogr. Phys. Géol. Dyn.* **11**, 5–120.

CHEVALIER, A. (1929) 'Sur la dégradation des sols tropicaux causée par les feux de brousse et sur les formations végétales régressives qui en sont la conséquence', *C.R. Acad. Sc.* **188**, 84–6.

CHEVALIER, A. (1950a) 'Mesures urgentes à prendre pour entraver le déssèchement, l'ensablement et la décadence des sols et de la végétation en A.O.F. et spécialement au Soudan Français', *C.R. Acad. Sc.* **230**, no. 20, 1720–3.

CHEVALIER, A. (1950b) 'Régénération des sols et de la végétation en A.O.F.', *C.R. Acad. Sc.* **230**, no. 24, 2064–6.

CHOUBERT, B. and BOYE, M. (1959) 'Envasements et dévasements du littoral en Guyane française', *C.R. Acad. Sc.* **249**, 145–47.

CLAYTON, R. W. (1956) 'Linear depressions (Bergfussniederungen) in savannah landscapes', *Geogr. Studies*, **3**, 102–26.

COLLIGNON, J. and ROUX, C. (1955) 'Indications concernant les caractères physico-chimiques de quelques eaux douces du Moyen-Congo', *Bull. Et. Centrafricaines*, **9**, 5–14.

COLLOQUIUM: 'Le problème des savanes et campos dans les régions tropicales', *XVIIIth Int. Geogr. Congr.*, Rio de Janeiro, 1956, vol. 1, 299–348.

COMBEAU, A. (1960) 'Quelques facteurs de la variation de l'indice d'instabilité structurale dans certains sols ferrallitiques', *C.R. Acad. Agric.* (France), **46**, 109–15.

COORAY, P. G. (1958) 'Earthslips and related phenomena in the Kandy district, Ceylon', *The Ceylon Geogr.* **12**, 75–90.

CORMACK, R. M. (1951) 'Land erosion in Southern Rhodesia', *Int. Union Geod. Geophys., Int. Ass. Sci. Hydrol.*, Brussels Meeting, vol. 2, 17–28.

CORNET, J. (1896) 'Les dépots superficiels et l'érosion continentale dans le bassin du Congo', *Bull. Soc. Belge Géol.* **10**, M44–116.

CORRENS, C. W. (1943) 'Die Stoffwanderungen in der Erdrinde', *Naturwissenschaften*, **21**, no. 3-4.

COSTA, S. V. BOTELHO DA and AZEVEDO, A. LOBO (1952) 'Aspectos da erosão do solo em Angola', *Agros.* **33**, no. 1/2, 15–22.

COTTON, C. A. (1961) 'The theory of savanna planation', *Geography*, **46**, 89–101.

CRAENE, A. DE (1954a) 'De la néoformation du quartz dans le sol', *Sols Afr.* **3**, 298–300.

CRAENE, A. DE (1954b) 'Les sols de pédimentation ou les sols à "stone-line" du Nord-est du Congo Belge', *Proc. Vth Int. Congr. Soil Sc.*, Leopoldville, vol. 4, 451–60.

CRAENE, A. DE and LARUELLE, J. (1955) 'Genèse et altération des latosols tropicaux et équatoriaux', *Bull. Agron. Congo Belge*, **46**, no. 5, 1113–1243.

CRAENE, A. DE and SOROTCHINSKY, C. (1954) 'Essai d'interprétation nouvelle de la genèse de certains types de "stone-line"', *C.R. IIᵉ Conf. Interafr. Sols*, Leopoldville, vol. 1, 453–6.

CZAJKA, W. (1957) 'Das Inselbergproblem auf Grund von Beobachtungen in Nordost-brasilien', *Matchatschek Festschrift*, 321–33.

DABIN, B. (1957) *Note sur le fonctionnement des parcelles expérimentales pour l'étude de l'érosion à la station d'Adiopodoumé (Côte d'Ivoire)*, Dakar, Decrét. Permanent Bureau Sols AOF, mimeographed, 16 p.

DABIN, B., LENEUF, N. and RIOU, G. (1960) *Carte pédologique de la Côte-d'Ivoire*, scale 1 : 2 000 000. Explanatory text. Adiopodioumé. Office de la Recherche Scientifique et Technique pour les Pays d'Outre-Mer, 31 p.

DAVEAU, S. (1959) *Recherches morphologiques sur la région de Bandiagara*, Mem. IFAN, no. 56, 120 p., 19 fig., 30 pl.

DAVEAU, S. (1960) 'Les plateaux du Sud-Ouest de la Haute-Volta, étude géomorphologique', *Trav. Dépt. Géogr. Univ. Dakar*, no. 7, 65 p., 11 fig.

DAVEAU, S. (1962) 'Principaux types de paysages morphologiques des plaines et plateaux soudanais dans l'Afrique de l'Ouest', *Inf. Géogr.* **26**, 61–72.

DAVEAU, S., LAMOTTE, M. and ROUGERIE, G. (1962) 'Cuirasses et chaînes birrimiennes en Haute-Volta', *Ann. Géogr.* **71**, 460–82.

DELHAYE, F. and BORGNIEZ, G. (1947) 'Contribution à la connaissance de la géographie et de la géologie de la région de la Lukénie et de la Tshuapa supérieures', *Bull. Soc. Belge Géol. Paléont. Hydrol.* **56**, 349–71.

DELHAYE, F. and BORGNIEZ, G. (1948) 'Contribution à la connaissance de la géographie et de la géologie de la région de la Lukénie et de la Tshuapa supérieures', *Ann. Musée Congo Belge*, Sér., in-8°, Sc. Géol., **3**, 155 p.

DENISOFF, I. (1957a) 'Un type particulier de concrétionnement en cuvette centrale congolaise', *Pédologie* (Ghent), **7**, 119–23.

DENISOFF, I. (1957b) 'Contribution à l'étude des formations latéritiques du Parc National de la Garamaba (Congo Belge)', *Pédologie* (Ghent), **7**, 124–32.

DIXEY, F. (1943) 'The morphology of the Congo-Zambezi watershed', *South Afr. Geogr. J.* **25**, 20–41.

DOMINGUES, A. PORTO and KELLER, E. (1956) *Guidebook* no. 6, Bahia, XVIIIth Int. Geogr. Congr., Rio de Janeiro, 254 p., 35 phot.

DOMMERGUES, Y. (1956) 'Etude sur la biologie des sols des forêts tropicales sèches et de leur évolution après défrichement', *Rept. IVth Int. Congr. Soil Sc.*, Paris, vol. 5, 605–10.

DOUGLAS, I. (1967) 'Erosion of granite terrains under tropical rain forest in Australia, Malaysia and Singapore', *Int. Ass. Sci. Hydrol.*, Bern, *Symp. River Morph.*, 31–9.

DRESCH, J. (1952a) 'Dépots de couverture et relief en Afrique Occidentale française', *Proc. XVIIth Int. Geogr. Congr.*, Washington, 323–6.

DRESCH, J. (1952b) 'Dépots superficiels et relief du sol au Dahomey septentrional', *C.R. Acad. Sc.* **234**, 1566–8.

DRESCH, J. (1953) 'Plaines soudanaises', *Rev. Géomorph. Dyn.* **4**, 39–44.

DUCHAUFOUR, P. (1960, 1965) *Précis de pédologie*, Paris, Masson, 481 p.

DUVIGNEAUD, P. (1955) 'Etudes écologiques de la végétation en Afrique tropicale', *Année Biol.*, 3rd ser., **31**, 375–92.

EDEN, T. (1964) *Elements of Tropical Soil Science*, 2nd edn, London, Macmillan, 164 p.

EMBERGER, L., MANGENOT, G. and MIEGE, J. (1950) 'Existence d'associations végétales typiques dans la forêt dense équatoriale', *C.R. Acad. Sc.* **231**, 640–2.

ERHART, H. (1935, 1938) *Traité de pédologie*, Strasbourg, Edit. Institut pédologique du Bas-Rhin, 2 vols.

ERHART, H. (1943) 'Les latérites du Moyen-Niger et leur signification paléoclimatique', *C.R. Acad. Sc.* **217**, 323–5.

ERHART, H. (1947) 'Les caractéristiques des sols tropicaux et leur vocation pour la culture des plantes oléagineuses', *Oléagineux*, **2**, no. 6–7.

ERHART, H. (1951a) 'Sur l'importance des phénomènes biologiques dans la formation des cuirasses ferrugineuses en zone tropicale', *C.R. Acad. Sc.* **233**, 804–6.

ERHART, H. (1951b) 'Sur le rôle des cuirasses termitiques dans la géographie des régions tropicales', *C.R. Acad. Sc.* **233**, 966–8.

ERHART, H. (1953a) 'Sur la nature minéralogique et la genèse des sédiments dans la cuvette tchadienne', *C.R. Acad. Sc.* **237**, 401–3.

ERHART, H. (1953b) 'Sur les cuirasses termitiques fossiles dans la vallée du Niari et dans le massif du Chaillu (Moyen Congo, A.E.F.)', *C.R. Acad. Sc.* **237**, 431–4.

ERHART, H. (1954) 'Sur les phénomènes d'altération pédogénétiques des roches silicatées et alumineuses en Malaisie britannique et à Sumatra', *C.R. Acad. Sc.* **238**, 2012–14.

EYK, J. J. VAN DER (1957) *Reconnaissance Soil Survey in Northern Surinam*, Wageningen, 99 p., 9 pl.

EYRE, S. R. (1968) *Vegetation and Soils*, 2nd edn, Chicago, Aldine; London, Arnold, 328 p.

FAIR, T. J. (1944) 'The geomorphology of the Ladysmith basin, Natal', *South Afr. Geogr. J.* **26**, 35–43.

FAUCK, F. (1954) 'Les facteurs et les intensités de l'érosion en Moyenne Casamance', *Proc. & Trans. Vth Int. Congr. Soil Sc.*, Leopoldville, vol. 3, 376–9.

FLORENCANO and AB'SABER (1950) 'A Serra do Mar e a mata atlântica em São Paulo', *Bol. Paulista de Geogr.*, no. 4, 61–70.

FOURNIER, F. (1955) 'Les facteurs de l'érosion du sol par l'eau en Afrique Occidentale Française', *C.R. Acad. Agric.* (France), **41**, no. 15, 660–5.

FOURNIER, F. (1956) 'Les formes et types d'érosion du sol par l'eau en Afrique Occidentale Française', *C.R. Acad. Agric.* (France), **42**, 215–21.

FOURNIER, F. (1957) 'Les valeurs d'érosion du sol dans les territoires français d'outre-mer', *Int. Union Geod. Geophys., Int. Ass. Sci. Hydrol.*, Toronto Meeting, vol. 1, 76–87.

FOURNIER, F. (1957) 'Les facteurs de l'érosion du sol en zone tropicale. Conclusions sur la conservation du sol', *Int. Union Geod. Geophys., Int. Ass. Sci. Hydrol.*, Toronto Meeting, vol. 1, 88–96.

FRANÇA, A. (1956) 'La route du café et les fronts pionniers', XVIII Int. Geogr. Congr., Rio de Janeiro, *Excursion Guidebook*, no. 3.

FREISE, F. W. (1933) 'Beobachtungen über Erosion an Urwaldgebirgsflüssen des brasilianischen Staates Rio de Janeiro', *Z. für Geom.* **7**, 1–9.

FREISE, F. W. (1933/35) 'Brasilianische Zuckerhutberge', *Z. für Geom.* **8**, 49–66.

FREISE, F. W. (1935) 'Erscheinungen des Erdfliessens in Tropenwälder; Beobachtungen aus brasilianischen Küstenwälder', *Z. für Geom.* **9**, 88–98.

FREISE, F. W. (1936) 'Bodenverkrüstung in Brasilien', *Z. Geomorph.* **9**, 233–48.

FREISE, F. W. (1936/38) 'Inselberge und Inselberg-Landschaften im Granit- und Gneissgebiete Brasiliens', *Z. für Geom.* **10**, 137–68.

FREISE, F. W. (1939–43) 'Der Ursprung der brasilianischen Zuckerhutberge', *Z. für Geom.* **11**, 93–112.

GALVAO, R. (1955) 'Introdução ao conhecimento da área maranhense abrangida pelo plano de valorização econômica da Amazônia', *Rev. Brasil. de Geogr.* **17**, 241–99.

GARNER, H. F. (1959) 'Stratigraphic-sedimentary significance of contemporary climate and relief in four regions of the Andes Mountains', *GSA Bull.* **70**, 1327–68.

GARNIER, B. J. (1953) 'The incidence and intensity of rainfall at Ibadan, Nigeria', *C.R. IVᵉ Réunion Conf. Int. Africanistes de l'Ouest*, Abidjan, p. 87.

GARNIER, B. J. (1956) 'A method of computing potential evapo-transpiration in West Africa', *Bull. IFAN*, sér. A, **18**, 665–76.

GAUDY, M. (1958) 'La mesure du taux de ruissellement et de l'érosion en A.O.F.', *Serv. Cons. Sols*, Dakar, 21 p., mimeographed.

GAUSSEN, H. (1952) 'L'indice xérothermique', *Bull. Assoc. Géogr. Franç.*, no. 222–3, 10–16.

GEIJSKES, D. C. (1952) 'On the structure and origin of the sandy ridges in the coastal zone of Suriname', *Tijdschr. Kon. Nederl. Aardrijkskundig Gen.* **69**, no. 2, 225–37.

GIERLOFF-EMDEN, H. (1958a) 'Analyse de l'évolution et des conditions de développement actuel du littorel du Salvador', *Bull. Assoc. Géogr. Franç.*, no. 278–9, 2–23.

GIERLOFF-EMDEN, H. (1958b) 'Erhebungen und Beiträge zu den physikalisch-geographischen Grundlagen von El Salvador', *Mitt. Geogr. Ges. Hamburg*, **53**, 7–140.

GIERLOFF-EMDEN, H. (1959) 'Die Küste von El Salvador. Eine morphologisch-ozeanographische Monographie', *Acta Humboldtiana*, Geogr.-Ethonol. Ser., no. 2, 183 p., 38 fig., 24 maps, 22 pl.

GIERLOFF-EMDEN, H. (1959) 'Die Küste von El Salvador', *Dtsch. Hydr. Zeitschr.* **12**, 14–24.

GIGOUT, M. (1949) 'Définition d'un étage ouljien', *C.R. Acad. Sc.* **229**, 551–2.

GLANGEAUD, L. (1941) 'Evolution des minéraux résiduels et notamment du quartz dans les sols autochtones en Afrique Occidentale Française', *C.R. Acad. Sc.* **212**, 862–4.

GOUROU, P. (1956) 'Milieu local et colonisation réunionaise sur les plateaux de la Sakay (Centre-Ouest de Madagascar)', *Cahiers d'Outre-Mer*, **9**, 36–57.

GRASSÉ, P. and NOIROT, C. (1957) 'La genèse et l'évolution des termitières géantes en Afrique Equatoriale Française', *C.R. Acad. Sc.* **244**, 974–9.

GRASSÉ, P. and NOIROT, C. (1959) 'Rapports des termitières avec les sols tropicaux', *Rev. Géomorph. Dyn.* **10**, 35–40.

GROVE, A. T. (1951) 'Soil erosion and population problems in Southeastern Nigeria', *Geogr. J.* **117**, no. 3, 307–27.

GUENNELON, R. (1954) 'Erosion des sols et pédogenèse dans une île tropicale volcanique: la Réunion', *Bull. Assoc. Franç. Sols*, Feb., 12–19.

GUILCHER, A. (1954a) *Morphologie littorale et sous-marine*, Paris, Orbis, PUF, 216 p., 40 fig., 8 phot. pl. Translated by B. W. Sparks and R. H. W. Kneese as *Coastal and Submarine Morphology*, London, Methuen, 1958, 274 p.

GUILCHER, A. (1954b) *Rapport sur les causes de l'envasement du Rio Kapatchez (Guinée Française)*, Gouv. Gén. AOF, Serv. de l'Hydraulique, 29 p., mimeographed.

GUILCHER, A. (1956a) 'Aspects morpho-végétaux de côtes alluviales tropicales (Suriname et Nigeria)', *Norois*, **9**, 95–8.

GUILCHER, A. (1956b) 'Etude géomorphologique des récifs coralliens du Nord-Ouest de Madagascar', *Ann. Inst. Océanogr.* **33**, 65–136.

GUILCHER, A. (1957) 'Formes de corrosion littorale du calcaire sur les côtes du Portugal', *Tijdschr. Kon. Nederl. Aardr. Gen.* **74**, 263–9.

GUILCHER, A. (1959) 'Coastal ridges and marshes and their continental environment near Grand Popo and Ouidah, Dahomey', *2nd Coastal Geogr. Conf.*, Baton-Rouge, 189–212.

GUILCHER, A. (1961) 'Le "beach-rock" ou grès de plage', *Ann. de Géogr.* **70**, 113–25.

GUILCHER, A., BERTHOIS, L., BATTISTINI, R. and FOURMANOIR, P. (1958) 'Les récifs coralliens des îles Radama et de la Baie Ramanetaka', *Mém. Inst. Scient. Madagascar*, Sér. F, **2**, 117–99.

GUILCHER, A., BERTHOIS, L., and BATTISTINI, R. (1962) 'Formes de corrosion littorale dans les roches volcaniques particulièrement à Madagascar et au Cap Vert (Sénégal)', *Cahiers Océanogr.* **14**, 208–40.

GUILCHER, A. and JOLY, F. (1954) 'Recherches sur la morphologie de la côte atlantique du Maroc', *Trav. Inst. Sc. Chérifien*, Sér. Géol. et Géogr. Phys., no. 2, 140 p.

GUILCHER, A. and PONT, P. (1957) 'Etude expérimentale de la corrosion littorale des calcaires', *Bull. Assoc. Géogr. Franç.*, no. 265–6, 48–62.

GUILLOTEAU, J. (1958) 'Le problème des feux de brousse et des brûlis dans la mise en valeur et la conservation des sols en Afrique au Sud du Sahara' (part I), *Sols Africains*, **4**, no. 2, 65–104.

HAGGETT, P. (1961) 'Land use and sediment yield in an old plantation tract of the Serra do Mar, Brazil', *Geogr. J.* **127**, 50–62.

HANDLEY, J. R. (1952) 'The geomorphology of the Nzega area, Tanganyika, with special reference to the formation of granite tors', *Proc. XIXth Int. Geol. Congr.*, Algiers, 201–10.

HANNEMAN, M. (1951/52) 'Eine Inselberglandschaft in Zentraltexas', *Erde*, 354–65.

HARMS, J. E. and MORGAN, B. D. (1964) 'Pisolitic limonite deposits in North-West Australia', *Proc. Australasian Inst. Mining Metall.* **212**, 91–124, 2 maps, 2 cross-sections, 5 phot., 2 tables.

HARRASSOVITZ, H. (1930) 'Boden der tropischen Regionen', *Handbuch der Bodenlehre*, **2**, Berlin, 362–432.

HARROY, J. P. (1949) *Afrique, terre qui meurt. La dégradation des sols africains sous l'influence de la colonisation*, 2nd edn, Brussels, M. Hayes, 557 p.

HAY, R. L. (1960) 'Rate of clay formation and mineral alteration in a 4 000 year old volcanic ash soil on St Vincent, B.W.I.', *Amer. J. Sc.* **258**, 354–78.

HEINZELIN, J. DE (1952) *Sols, paléosols et désertifications anciennes dans le secteur nord-oriental du Bassin du Congo*, Brussels, Publ. INEAC, 168 p., 8 pl.

HEINZELIN, J. DE (1955) *Observations sur la genèse des nappes de gravats dans les sols tropicaux*, Brussels Publ. INEAC, Sér. Scient., no. 64.

HENIN, S. (1952) 'Quelques remarques concernant l'infiltration de l'eau dans les sols', *Rev. Gén. Hydr.*, Mar. Apr., 77–82.

HIERNAUX, C. (1955) 'Sur un nouvel indice d'humidité proposé pour l'Afrique occidentale', *Bull. IFAN*, sér. A, **17**, 1–6.

HINGSTON, F. J. and MULCAHY, M. J. (1961) 'Laboratory examination of the soils of the York-Quairading area, W.A.', *Commonw. Scientif. Industr. Res. Org., Div. Soils Rept.*, Adelaide, 9/61, 76 p.

HOORE, J. D' (1954a) *L'Accumulation des sesquioxydes libres dans les sols tropicaux*, Brussels, Publ. INEAC, Sér. Scient., no. 62, 132 p., 23 fig., 37 photographs.

HOORE, J. D' (1954b) 'Essai de classification des zones d'accumulation de sesquioxydes libres sur des bases génétiques', *Sols Afr.* **3**, 66–81.

HOORE, J. D' (1961) 'Influence de la mise en culture sur l'évolution des sols dans la zone de forêt dense de basse et moyenne altitude', UNESCO, *Humid Tropical Zone*, Abidjan Symposium, 1959, 49–58.

HOORE, J. D' and FRIPIAT, J. (1948) *Recherches sur les variations des structures du sol à Yangambi (Congo Belge)*, Publ. INEAC, Sér. Scient., no. 38, 60 p.

HOPKINS, B. (1960) 'Rainfall interception by a tropical forest in Uganda', *East Afr. Agric. J.* **25**, 255–8.

HOSE, H. R. (1961) 'The origin of bauxites in British Guiana and Jamaica', *Proc. Vth inter-Guiana Geol. Conf.*, Georgetown, 1959, 185–98.

HUDSON, N. W. (1957) 'Erosion control and research. Progress report on experiments at Henderson research station, 1953–6', *Rhodesian Agric. J.* **54**, 297–323.

HUMBERT, H. (1952) 'Le problème du recours aux feux courants', *Int. Union Prot. Nat.*, 3rd meeting, Caracas, mimeographed, 9 p.

ISNARD, H. (1955) *Madagascar*, Paris, Colin, 219 p.

JAEGER, P. (1952) 'Mission au Soudan occidental', *Encycl. Mens. d'Outre-Mer*, Suppl. to no. 24, August, 7 p.

JAEGER, P. (1953) 'Contribution à l'étude du modelé de la dorsalre guinéenne: les Monts Loma (Sierra Leone)', *Rev. Géomorph. Dyn.* **3**, 105–13.

JAEGER, P. (1956) 'Contribution à l'étude des forêts reliques du Soudan occidental', *Bull. IFAN*, sér. A, **18**, 993–1053.

JAEGER, P. and JAVAROY, M. (1952) 'Les grès de Kita (Soudan occidental); leur influence sur la répartition du peuplement végétal', *Bull. IFAN*, sér. A, **14**, 1–18.

JAHN, A. (1954) 'Balance de dénudation du versant', *Czasopismo Geogr.* **25**, 38–64.

JAHN, A. (1968) 'Denudational balance of slopes', *Geographia Polonica* **13**, 9–29.

JENNY, H., GESSEL, S. P. and BINGHAM, F. T. (1949) 'Comparative study of decomposition rates of organic matter in temperate and tropical regions', *Soil. Sc.* **68**, 419–32.

JESSEN, O. (1936) *Reisen und Forschungen in Angola*, Berlin, Reimer.

KAISIN, F. (1949) 'L'érosion et la stabilité des tranchées en climat tropical', *Bull. Soc. Belge Géol.*, 292–7.

KELLOGG, C. G. (1941) 'Climate and soil', *Yearbook of Agriculture 1941*, US Dept. Agric., 265–91.

KERSTEN, J. F. VAN (1958) 'Bauxite deposits in Surinam and Demerara', *Leidse Geol. Med.* **21**, 247–375.

KING, L. C. (1948) 'A theory of bornhardts', *Geogr. J.* **112**, 83–7.

KING, L. C. and FAIR, T. (1944) 'Hillslopes and dongas', *Trans. Geol. Soc. S. Afr.* **47**, 4 p.

KITTREDGE, J. (1948) *Forest Influences*, New York, McGraw-Hill, 394 p.

KRAUS, R. W. and GALLOWAY, R. A. (1960) 'The role of algae in the formation of beach-rock in certain islands of the Caribbean', *Coastal St. Inst., Tech. Rep.* II E, 55 p., numerous tables and measurements.

KREBS, N. (1933) 'Morphologische Beobachtungen in Südindien', *Sitz. Ber. Preuss. Akad. Wiss., Phys.-Math. Kl.*, no. 23.

KREBS, N. (1932) 'Morphologische Beobachtungen in Central-India und Rajputana', *Z. Ges. Erdk.* Berlin.

KREBS, N. (1942) 'Über Wesen und Verbreitung der tropischen Inselberge', *Abh. Preuss. Akad. Wiss., Math.-Nat. Kl.*, no. 6.

LACROIX, A. (1913) 'Les latérites de la Guinée et les produits d'altération qui leur sont associés', *Nouv. Arch. Mus.*, série 5, vol. 5, 255–358.

LADD, H. S. (1961) 'Reef building', *Science*, **134**, 703–15.

LAFOND, L. R. (1957) 'Aperçu sur la sédimentologie de l'estuaire de la Betsiboka', *Rev. Inst. Fr. Pétrole-Ann. Combustibles liq.*, no. 4, 425–31.

LAJOINIE, J. P. and BONIFAS, M. (1961) 'Les dolérites du Konkouré et leur altération latéritique', BRGM, no. 2.

LAMEGO, A. RIBEIRO (1955) 'Geologia das quadrículas de Campos: São Tomé, Feia e Xéxé', *Bol. Div. Geol. e Miner.*, no. 154, 60 p.

LAMOTTE, M. and ROUGERIE, G. (1952) 'Coexistence de trois types de modelé dans les chaînes quartzitiques du Nimba et du Simandou (Haute Guinée Française)', *Ann. Géogr.* **61**, 432–42.

LAMOTTE, M. and ROUGERIE, G. (1953) 'Les cuirasses ferrugineuses allochtones. Significa- tion paléoclimatique et rapports avec la végétation', *C.R. IVᵉ Réunion Afric. de l'Ouest*, Abidjan, 89–90.

LAMOTTE, M. and ROUGERIE, G. (1962) 'Les apports allochtones dans la genèse des cuirasses ferrugineuses', *Rev. Géomorph. Dyn.* **13**, 145–60.

LAMPRECHT, H. (1961) 'Tropenwälder und tropische Waldwirtschaft', *Schweiz. Z. Forst- wesen*, **32**, 1–110.

LAPLANTE, A. and BACHELIER, G. (1954) 'Un processus pédologique de la formation des cuirasses latéritiques dans l'Adamaoua (Nord Cameroun)', *Rev. Géomorph. Dyn.* **6**, 214–19.

LAPLANTE, A. and ROUGERIE, G. (1950) 'Etude pédologique du bassin français de la Bia', *Bull. IFAN*, **12**, Sér. A, 883–904.

LAPPARENT, J. DE (1939) 'La décomposition latéritique du granite dans la région de Macenta (Guinée Français). L'arénisation prétropicale et prédésertique en Afrique Occidentale Française et au Sahara', *C.R. Acad. Sc.* **208**, 1767–9; **209**, 7–9.

LASSERRE, G. (1961) *La Guadaloupe, étude géographique*, Bordeaux, Union Française d'Impres- sion, vol. 1, 448 p.

LAUER, W. (1951) 'Hygrische Klimate und Vegetationszonen der Tropen mit besonderer Berücksichtigung Ostafrikas', *Erdkunde*, **5**, 284–93.

LE BOURDIEC, P. (1958a) 'Aspects de la morphogénèse plio-quaternaire en Basse Côte d'Ivoire (A.O.F.)', *Rev. Géom. Dynamique*, **9**, 33–42.

LE BOURDIEC, P. (1958b) 'Contribution à l'étude géomorphologique du bassin sédimentaire et des régions littorales de Côte-d'Ivoire', *Etudes Eburnéennes*, **7**, 7–96.

LEDGER, D. C. (1961) 'Recent hydrographic change in the Rima basin, northern Nigeria', *Geogr. J.* **127**, 477–87.

LEE CHOW (1956) 'The silt problem in Taiwan', UNO, *Proc. Reg. Conf. on Water Res. Dev. in Asia and the Far East*, Bangkok, 265–70.

LEENHEER, L. DE, HOORE, J. D' and SYS, K. (1952) *Cartographie et caractérisation pédologique de la catena de Yangambi*, Publ. INEAC, sér. scient., no. 55, 62 p.

LEFEVRE, M. (1953) 'Notes sur la morphologie du Katanga', *Bull. Soc. Belge Et. Géogr.* **22**, 407–31.

LEHMANN, H. (1936) 'Morphologische Studien auf Java', *Geogr. Abh.* **3**.

LEINZ, V. and VIEIRA DE CARVALHO, A. M. (1957) 'Contribução à geologia da Bacia de S. Paulo', *Bol. Fac. Fil. Univ. S. Paulo*, no. 205, 61 p., 3 maps.

LEMEE, G. (1961) 'Effets des caractères du sol sur la localisation de la végétation en zones équatoriale et tropicale humide', UNESCO, *Humid Trop. Zone, Abidjan Symposium*, 1959, 25–39.

LEMEE, G. (1967) *Précis de biogéographie*, Paris, Masson, 358 p.

LENEUF, N. (1959) *L'Altération des granites calco-alcalins et des grano-diorites en Côte d'Ivoire forestière et les sols qui en sont dérivés*, Paris, ORSTOM, 210 p., 15 phot. pl.

LOUIS, H. (1959) 'Beobachtungen über die Inselberge bei Hua-Hin am Golf von Siam', *Erdkunde*, **13**, 314–19.

LOUIS, H. (1964) 'Über Rumpflächen- und Talbildung in den wechselfeuchten Tropen besonders nach Studien in Tanganyika', *Z. Geomorph.*, n.s., **8**, Sonderheft, 43–70.

LOVERING, T. S. (1959) 'Significance of accumulator plants in rock weathering', *Bull. Geol. Soc. Amer.* **70**, 781–800, reviewed by A. Cailleux (1960) in *Rev. Géom. Dyn.* **11**, 173–4.

MAACK, R. (1950) 'Notas preliminares sôbre clima, solos e vegetação do Estado do Paraná', *Bol. Geogr.*, Rio de Janeiro, **7**, no. 84, 1401–87.

MABBUTT, J. A. (1961) 'A stripped land surface in Western Australia', *Trans. Papers Inst. Brit. Geogr.* **29**, 101–14.

MABBUTT, J. and SCOTT, R. M. (1966) 'Periodicity of morphogenesis and soil formation in a savannah landscape near Port Moresby, Papua', *Z. Geomorph.*, n.s., **10**, 69–89.

MACAR, P. (1947) 'Les chutes de l'Inkisi (Congo occidental) et leurs divers modes d'érosion', *Ann. Soc. Géol. de Belgique*, **82**, 38–51.

MADAGASCAR, BUREAU GÉOLOGIQUE (1952) *Contribution à l'étude de l'érosion des sols à Madagascar*, Tananarive, 44 p.

MAIGNIEN, R. (1954a) 'Formation de cuirasses de plateau, région de Labé (Guinée Française)', *Proc. Vth Int. Congr. Soil Sc.*, vol. 4, 13–18.

MAIGNIEN, R. (1954b) 'Cuirassement de sols de plaine, Bally (Guinée Française)', *Proc. Vth Int. Congr. Soil Sc.*, vol. 4, 19–22.

MAIGNIEN, R. (1954c) 'Différents processus de cuirassement en A.O.F.', *C.R. IIe Conf. Interafr. Sols*, doc. 116, 1469–86.

MAIGNIEN, R. (1956) 'De l'importance du lessivage oblique dans le cuirassement des sols en A.O.F.', *Rept. VIth Int. Congr. Soil Sc.*, Paris, vol. 5, 463–7.

MAIGNIEN, R. (1958a) *Le cuirassement des sols en Guinée (Afrique Occidentale)*, Mem. Serv. Carte Geol. Alsace-Lorraine, no. 16, 239 p.

MAIGNIEN, R. (1958b) 'Le cuirassement des sols en Afrique tropicale de l'Ouest', *Sols Afr.* **4**, no. 4, 5–41.

MAIGNIEN, R. (1961) 'Le passage des sols ferrugineux tropicaux aux sols ferrallitiques dans les régions Sud-Ouest du Sénégal', *Sols Afr.* **6**, 113–72.

MAIGNIEN, R. (1966) *Review of Research on Laterites*, UNESCO, ser. Research on natural resources, no. 4, 148 p.

MANGENOT, G. (1951) 'Une formule simple permettant de caractériser les climats de l'Afrique tropicale dans leur rapports avec la végétation', *Rev. Gén. Botanique*, 353–69.

MARQUES, J., GROMANN, F., BERTONI, J. and ALENCAR, F. AYRES DE (1951) 'Algumas conclusões gerais preliminares das determinações de perdas por erosõa realizadas em São Paolo', *Recife, Inst. Agron.*, 44 p.

MARTONNE, E. DE (1926) 'Une nouvelle fonction climatologique, l'indice d'aridité', *La Météorologie*.

MARTONNE, E. DE (1939) 'Sur la formation des pains de sucre au Brésil', *C.R. Acad. Sc.* **208**, 1163–5.

MARTONNE, E. DE (1940) 'Problèmes morphologiques au Brésil tropical atlantique', *Ann. Géogr.* **49**, 1–27, 106–29.

MARTONNE, E. DE (1946) 'Géographie zonale: la zone tropicale', *Ann. Géogr.* **55**, 1–18.

MARTONNE, E. DE and BIROT, P. (1944) 'Sur l'évolution des versants en climat tropical humide', *C.R. Acad. Sc.* **218**, 529–32.

MASSIP, S. and MASSIP, S. DE (1949) 'La cordillera de los Organos en la porción occidental de Cuba', *Proc. Int. Geogr. Congr.*, Lisbon, vol. 2, 734–47.

MATHIEU, F. (1931) 'Notes sur la géographie physique des bassins de la Dungu et de la Duru (Haut Uele)', *Ann. Soc. Géol. Belgique*, **53**, 663–8.

MATON, G. (1960) *Introduction à un important programme de construction de barrages en Haute-Volta*, Ouagadougou, Min. of Agric., mimeographed, 136 p.

MEULENBERG, O. (1949) *Introduction à l'étude pédologique des sols du territoire du Bas-Fleuve*, Brussels, Inst. Roy. Col. Belge, 133 p.

MICHEL, P. (1958) *Rapport de mission dans le Nord du Fouta-Djalon et dans le pays bassaris (Guinée)*, Dakar, SGPM, 50 p., mimeographed, 31 phot., 20 pl. Reviewed in *Rev. Géomorph. Dyn.* **11** (1960), 72–4.

MICHEL, P. (1959a) 'L'évolution géomorphologique des bassins du Sénégal et de la Haute-Gambie, ses rapports avec la prospection minière', *Rev. Géomorph. Dyn.* **10**, 117–43.

MICHEL, P. (1959b) 'Rapport de mission au Soudan occidental et dans le Sud-Est du Sénégal, II: dépôts alluviaux et dynamique fluviale', *SGPM*, Dakar, 9/59, 57 p., mimeographed.

MICHEL, P. (1960) *Notes sur les formations cuirassées de la région de Kédougou* (March–April 1960), BRGM, Paris, mimeographed, 23 p.

MICHEL, P. (1962a) *Etude géomorphologique des sondages dans les formations cuirasses de la région de Kédougou (Sénégal)*, BRGM, Dakar, A5, mimeographed, 56 p., 20 pl.

MICHEL, P. (1962b) 'Observations sur la géomorphologie et les dépôts alluviaux des cours moyens du Bafing et du Bakoy (Rép. du Mali)', *BRGM*, Dakar, 39 p., 16 phot., 11 pl., mimeographed.

MIÈGE, J. (1953) 'Relations entre savanes et forêts en Basse Côte-d'Ivoire', *C.R. Vᵉ Réunion Afric. Ouest*, Abidjan, 27–9.

MIÈGE, J. (1955) 'Les savanes et forêts claires de Côte d'Ivoire', *Etudes Eburnéennes*, **4**, 62–83.

MILLIES-LACROIX, A. (1960) 'La presqu'île tombolisée de Vung-Tau, étude des sables marins et de leur contexte physique', *Univ. Saigon, Fac. Sc. Annales*, 425–52.

MILLOT, G. and BONIFAS, M. (1955) 'Transformations isovolumétriques dans les phéno-mènes de latéritisation et de bauxitisation', *Bull. Serv. Carte Géol. Alsace-Lorraine*, **8**, 1–20.

MOHR, E. C. J. (1944) *The Soils of Equatorial Regions, with special reference to the Netherlands East Indies*, trans. R. L. Pendleton, Ann Arbor, Michigan, 766 p.

MOHR, E. C. J. and VAN BAREN, F. A. (1954) *Tropical Soils: a critical study of soil genesis as related to climate, rock and vegetation*, The Hague, van Hoeve, 498 p.

MORETTI, A. (1951) 'Fenomeni d'erosione marina nei pressi di Porto Torres (Sardegna)', *Riv. Geogr. Ital.* **58**, 181–97.

MORTELMANS, G. (1951) 'Observations sur la morphologie de la région Mituaka-Haute Kalumengongo (Monts Kilara, Katanga)', *Bull. Soc. Belge Géol. Paléont. Hydrol.* **59**, 383–99.

MORTENSEN, H. (1929) 'Inselberglandschaft im Nordchile', *Z. Geomorph.* **4**.

MULCAHY, M. (1960) 'Laterites and lateritic soils in south-western Australia', *J. Soil Sc.* **11**, no. 2, 206–26.

MULCAHY, M. and HINGSTON, F. (1961) 'The development and distribution of soils of the York-Quairading area, Western Australia, in relation to landscape evolution', *Commonw. Scientif. Industr. Res. Org., Soil Publ.* no. 17, 43 p., 3 pl., 1 map.

NASCIMENTO, U. (1952) 'Estudo da regularização e protecção das barrocas de Luanda', *Min. Obras Publ.*, Lab. Nac. Engenharia Civil, Lisbon, no. 30, 39 p.

NEIVA, J. M. COTELO and NEVES, J. M. CORREIRA (1957) 'Latérites de l'Ile du Prince', *Mem. e Noticias, Mus. e Lab. Miner. e Geol. Univ. Coimbra*, no. 44, 1–9.

NICOLAS, J. P. (1956) 'Essais sur deux nouveaux indices climatiques saisonniers pour la zone intertropicale', *Bull. IFAN*, sér. A, **18**, 653–64.

NIZERY, A. and CRESPY, S. (1949) 'Un laboratoire pour l'étude de la résistance des maté-riaux en climats tropical', *Rev. Gén. Electr.* **58**, 455–99.

NOWACK, E. (1936) 'Zur Erklärung der Inselberglandschaften Ostafrikas', *Z. Ges. Erdkunde*, Berlin, no. 7/8.

275

NYE, P. H. (1954) 'Some soil forming processes in the humid tropics: I. A field study of a catena in West African forest', *J. Soil Sc.* **5**, 7–21.

OLLIER, C. D. (1960) 'The Inselbergs of Uganda', *Z. Geomorph.*, n.s., **4**, 43–52.

OLLIER, C. D. and TUDDENHAM, W. G. (1961) 'Inselbergs of Central Australia', *Z. Geomorph.*, n.s., **5**, 257–76.

ORSTOM (1957) *Etudes hydrologiques des petits bassins-versants d'Afrique Occidentale Française. Rapport préliminaire pour la campagne de 1957*, mimeographed.

O'REILLY STERNBERG, H. (1949) 'Enchentes e movimentos coletivos do solo no vale do Paraíba em dezembro de 1948. Influência da explotação destrutiva das terras', *Rev. Brasil. Geogr.* **11**, 223–61: English translation as 'Floods and landslides in the Paraiba Valley, December 1948. Influence of destructive exploitation of the land', *XVIth Int. Geogr. Congr.*, Lisbon (1959), vol. 3, 335–64.

OTTMANN, F. and OTTMANN, J. M. (1959) 'Les sédiments de l'embouchure du Capibaribe, Recife, Brésil', *Trab. Inst. Biol. Marît. e Oceanogr.*, Recife, **1**, 51–69.

OTTMANN, F., NOBREGA, R., COUTINHO, P. and OLIVEIRA, S. F. (1959) 'Estudo topográfico e sedimentológico de um perfil da praia de Piedade', *Trab. Inst. Biol. Marît. e Oceanogr.*, Recife, **1**, 19–38.

PAGNEY, P. (1958) 'Mouvements marins d'origine cyclonique. Leurs manifestations dans la mer Caraïbe, spécialement sur les côtes de la Martinique', *Bull. Assoc. Géogr. Fr.*, no. 276–7, 61–72.

PALLISTA, J. W. (1953) 'Erosion levels and laterite in Buganda province, Uganda', *Proc. XIXth Int. Geol. Congr.*, Algiers, **21**, 193–9.

PALMER, H. S. (1927) 'Karrenbildung in den Basaltgesteinen der Hawaiischen Inseln', *Mitt. Geogr. Ges.* Wien, **70**, 89–94.

PANZER, W. (1954) 'Verwitterungs- und Abtragungsformen im Granit von Hong-Kong', *Abh. Akad. Raumforschung und Landesplanung*, **28**, 41–60.

PARIZEK, E. J. and WOODRUFF, J. F. (1957) 'Description and origin of stone layers in soils of the southeastern States', *J. Geol.* **65**, 24–34.

PASSARGE, S. (1904a) 'Die Inselberglandschaften im tropischen Afrika', *Natur. Wochenschr.*, Iena, **19**, 657–65.

PASSARGE, S. (1904b) 'Rumpfflächen und Inselberge', *Z. Dtschen Geol. Ges.* **56**, 193–215.

PASSARGE, S. (1912) 'Inselberglandschaften der Massaisteppen', *Pet. Geogr. Mitt.*

PASSARGE, S. (1924) 'Das Problem der Skulptur Inselberglandschaften', *Petermanns Geogr. Mitt.* 66–70, 117–20.

PEGUY, C. P. (1961) *Précis de climatologie*, Paris, Masson (new edn in the press).

PELISSIER, P. (1960) 'Un point de vocabulaire relatif à la morphologie tropicale', *Inf. Geogr.* **24**, 113–14.

PELISSIER, P. and ROUGERIE, G. (1953) 'Problèmes morphologiques dans le bassin de Siguiri (Haut Niger)', *Bull. IFAN*, sér. A, **15**, 1–47.

PICHLER, E. (1958) 'Aspectos geológicos dos escorregamentos de Santos', *Noticia Geom.* no. 2, 40–4.

PIMIENTA, J. (1956) *Evolution du delta lagunaire du Rio Tubarão à Laguna (Brésil méridional)*, XVIII Int. Geogr. Congr., Rio de Janeiro, mimeographed.

PITOT, A. (1953) 'Feux sauvages, végétation et sols en A.O.F.', *Bull. IFAN*, sér. A, **15**, 1369–83.

PITOT, A. (1954) 'Végétation et sols et leurs problèmes en A.O.F.', *Ann. Ec. Sup. Sc.*, Dakar, **1**, 129–39.

PITOT, A. and MASSON, H. (1951) 'Quelques données sur la température au cours des feux de brousse aux environs de Dakar', *Bull. IFAN*, sér. A, **13**, no. 3.

PLAYFORD, P. E. (1954) 'Observations on laterite in Western Australia', *Aust. J. Sc.* **17**, no. 1, 11–14.

PLOEY, J. DE (1963) 'Quelques indices sur l'évolution morphologique et paléoclimatique des environs du Stanley-Pool (Congo)', *Studia Univ. Lovanium, Sc.*, no. 17, 16 p.

PLOEY, J. DE (1964) 'Nappes de gravats et couvertures argilo-sableuses au Bas-Congo. Leur genèse et l'action des termites', in A. Bouillon, *Etudes sur les termites africains*, Leopoldville, Edit. Univ., 399–415, 6 fig., 2 phot., bibl.

POLINARD, E. (1948) 'Les grands traits de la géographie physique et les particularités des formations du plateau dans le Nord-est de la Lunda (Angola). Interprétations des observations des premières missions de recherches', *Bull. Soc. Belge Géol. Paléont. Hydrol.* **57**, 541–53.

PORTERES, R. (1934) 'Sur un indice de sécheresse dans les régions tropicales forestières. Indices en Côte d'Ivoire', *Bull. Comm. Et. Hist. Sc. A.O.F.*, 417–27.

PORTERES, R. (1956) 'Une forme spectaculaire d'érosion à Madagascar: le lavaka', *Science et Nature*, no. 14, 8–13.

PORTO DOMINGUES, A. and KELLER, E. (1955) *Guidebook*, no. 6, *Bahia*, XVIII Int. Geogr. Congr., Rio de Janeiro, 254 p.

POUQUET, J. (1954) 'Altérations de dolérites de la presqu'île du Cap Vert (Sénégal) et du plateau du Labé (Fouta-Djalon, Guinée Française)', *Bull. Assoc. Geogr. Franç.*, 173–82.

POUQUET, J. (1956a) 'Aspects morphologiques du Fouta Djalon, régions de Kindia et de Labé, Guinée Française, A.O.F. Aspects alarmants des phénomènes d'érosion des sols déclenchés par les activités humaines', *Rev. Géogr. Alpine*, 231–46.

POUQUET, J. (1956b) 'Methodes d'études des versants et principaux résultats obtenus sur le Labé, Guinée Française, A.O.F.', *1st* Report Comm. Evolution of Slopes, I.G.U., Amsterdam, 85–95.

PREEZ, J. W. DU (1954) *Notes sur la présence d'oolithes et de pisolithes dans les latérites de Nigeria*, Bur. Interafricain Sols, 7 p., mimeographed.

PRESCOTT, S. A. and PENDLETON, R. L. (1952) 'Laterite and lateritic soils', *Commonwealth Bur. Soil Sc. Tech. Comm.* no. 47, 51 p.

QUANTIN, P. and COMBEAU, A. (1962) 'Erosion et stabilité structurale du sol', *Publ. Int. Ass. Sci. Hydrol.*, no. 59, 124–30.

RADWANSKI, S. A. and OLLIER, C. D. (1959) 'A study of an East African catena', *J. Soil Sc.* **10**, no. 2, pp. 149–68.

RANSON, G. (1955) 'Observations sur les îles coralliennes de l'archipel des Tuamotu (Océanie française)', *C.R. Somm. Soc. Géol. de Fr.*, 47–9.

RATHJENS, C. (1957) 'Physische-geographische Beobachtungen im nordwestindischen Trockengebiet (Ein erster Forschungsbericht)', *Erdkunde*, **11**, 49–58.

REFORMATSKY, N. (1935) 'Quelques observations sur les latérites et les roches ferrugineuses de l'Ouest de la colonie du Niger', *Bull. Soc. Géol. Fr.* **5**, 575–89.

REVELLE, R. and EMERY, K. O. (1957) 'Chemical erosion of beach-rock and exposed reef rock', *Geol. Surv.*, Prof. Paper 260 T, Washington.

RIBEIRO, O. (1954) 'Contribution à l'étude géographique des pays de l'Océan Indien', *Pan-Indian Oc. Sc. Ass.*, March 8, mimeographed.

RICH, J. L. (1953) 'Problems in Brazilian geology and geomorphology suggested by reconnaissance in summer of 1951', *Univ. S. Paulo, Fac. Fil., Ciencias e Letras, Bol.* 146, 80 p.

RICHARDS, P. W. (1952) *The Tropical Rainforest*, Cambridge University Press, 450 p., 45 fig., 15 phot. pl.

RICHARDS, P. W. (1961) 'The types of vegetation in the humid tropics in relation to the soil', Unesco, *Humid Trop. Zone, Abidjan Symposium*, Jan. 1959, 15–20.

RIOU, G. (1965) 'Notes sur les sols complexes des savanes préforestières de Côte-d'Ivoire', *Ann. Univ. Abidjan, Lettres*, no. 1, 17–36.

RIQUIER, J. (1954) 'Formation d'une cuirasse ferrugineuse et manganésifère ou latéritique', *Proc. and Trans. Vth Int. Congr. Soil Sc.*, vol. 4, 229–36.

RIQUIER, J. (1958) 'Les "lavaka" de Madagascar', *Bull. Soc. Géogr.*, Marseille, **69**, 181–91.

ROBERTY, G. (1942) *Contribution à l'étude phytogéographique de l'Afrique occidentale*, doct. diss., Geneva, 150 p., 1 separate map.

ROBYNS, W. (1952) 'Les feux courants et la végétation', *Int. Union Prot. Nat.*, 3rd meeting, Caracas, mimeographed, 4 p.

ROCH, E. (1950) 'La genèse de certains sables rouges en Afrique Equatoriale Française', *C.R. Acad. Sc.* **230**, 670–1.

ROCH, E. (1952) 'Les reliefs résiduels ou Inselberge du bassin de la Bénoué (Nord Cameroun)', *C.R. Acad. Sc.* **234**, 117–18.

ROCH, E. (1953) 'Itinéraires géologiques dans le Nord du Cameroun et le Sud-Ouest du Territoire du Tchad', *Bull. Serv. Mines Cameroun*, no. 1, 110 p.

ROCHEFORT, M. (1958) 'Rapports entre la pluviosité et l'écoulement dans le Brésil subtropical et le Brésil tropical atlantique (Etude comparée des bassins du Guaíba et du Paraíba do Sul)', *Trav. et Mém. Inst. Hautes Etudes Amérique Latine*, vol. 2, 279 p.

RODIER, J. (1961) 'Transports solides en Afrique Noire à l'Ouest du Congo', *Bull. Int. Ass. Sci. Hydrol.* **6**, 32–4.

RODRIGUES DA SILVA, R., LIMA ALMEIDA, G. C. and SOUTO MAIOR, R. (1957) 'Identificação microscópica dos componentes minerais dos solos', *Inst. Pesquisas Agron. Pernambuco*, n.s., no. 3, 51 p.

ROUGERIE, G. (1950) 'Le pays du Sanwi. Esquisse morphologique dans le Sud-Est de la Côte d'Ivoire', *Bull. Assoc. Géogr. Fr.*, no. 212–13, 138–45.

ROUGERIE, G. (1951) 'Etude morphologique du bassin français de la Bia et des régions littorales de la lagune Aby', *Et. Eburnéennes*, no. 2, 108 p.

ROUGERIE, G. (1955) 'Un mode de dégagement probable de certains dômes granitiques', *C.R. Acad. Sc.* **240**, 327–9.

ROUGERIE, G. (1956) 'Etude des modes d'érosion et du façonnement des versants en Côte-d'Ivoire équatoriale', *1st Rep. Comm. Evolution of Slopes, I.G.U.*, Amsterdam, 136–41.

ROUGERIE, G. (1958) 'Existence et modalités du ruissellement sous forêt dense de Côte d'Ivoire', *C.R. Acad. Sc.* **246**, 290–2.

ROUGERIE, G. (1959) 'Latéritisation et pédogenèse intertropicales', *Inf. Géogr.* **23**, 199–206.

ROUGERIE, G. (1960a) 'Le façonnement actuel des modelés en Côte d'Ivoire forestière', *Mém. Institut Français d'Afrique Noire*, no. 58, 542 p., 134 fig., 92 phot.

ROUGERIE, G. (1960b) 'Sur les versants en milieux tropicaux humides', *Z. für Geom.*, Suppl. 1, 12–18.

ROUGERIE, G. (1961) 'Modelés et dynamiques de savene en Guinée orientale', *Etudes Africaines* (Conakry), no. 4, 24–50.

ROUGERIE, J. and LAMOTTE, M. (1952) 'Le Mont Nimba (Guinée, Côte-d'Ivoire). Etude de morphologie tropicale', *Bull. Assoc. Géogr. Franç.*, no. 226–8, 113–20.

RUELLAN, F. (1945) 'Evolução geomorfológica da baía de Guanabara e das regiões vizinhas', *Rev. Brasil. Geogr.* **6**, 445–508.

RUELLAN, F. (1953) 'O papel das enxurradas no modelado do relévo brasileiro', *Bol. Paulista de Geogr.*, no. 13, 5–18 and no. 14, 3–25.

RUHE, R. (1956) *Landscape Evolution in the High Ituri, Belgian Congo*, Brussels Publ. INEAC, Sér. Scient. no. 66, 91 p.

RUHE, R. (1959) 'Stone-line in soils', *Soil Sc.* **87**, 223–31.

RUSSELL, R. J. (1959) 'Caribbean beach-rock observations', *Z. für Geom.* **3**, 227–36.

RUSSELL, R. J. (1962) 'Origin of beach-rock', *Z. für Geom.* **6**, 1–16.

RUXTON, B. P. (1958) 'Weathering and subsurface erosion in granite at the piedmont angle, Balos, Sudan', *Geol. Mag.* **95**, 353–77.

RUXTON, B. (1967) 'Slopewash under mature primary rain forest in Northern Papua', in J. N. Jennings and J. A. Mabbutt, eds, *Landform Studies from Australia and New Guinea*, Cambridge University Press, 85–94.

RUXTON, B. and BERRY, L. (1957) 'Weathering of granite and associated erosional features in Hong-Kong', *Bull. Geol. Soc. Amer.* **68**, 1263–92.

RUXTON, B. and BERRY, L. (1961) 'Weathering profiles and geomorphic position on granite in two tropical regions', *Rev. Géom. Dyn.* **12**, 16–31.

SABOT, J. (1952) 'Les latérites', *Proc. XIXth Int. Geol. Congr.*, Algiers, **21**, 181–92.

SALVADOR, O. (1959) 'La température et le flux de la chaleur dans le sol à Dakar', *Ann. Fac. Sc. Univ. Dakar*, **4**, 47–54.

278

SAPPER, K. (1935) *Geomorphologie der feuchten Tropen*, Teubner, Leipzig, 150 p.

SAURIN, E. and CARBONNEL, J. P. (1964) 'Les latérites sédimentaires du Cambodge oriental', *Rev. Géogr. Phys. Geol. Dyn.*, n.s., **16**, 241–56.

SAUTTER, G. (1951) 'Note sur l'érosion en cirque des sables au Nord de Brazzaville', *Bull. Inst. Et. Centrafr.*, n.s., no. 2, 49–61.

SAVIGEAR, R. (1960) 'Slopes and hills in West Africa', *Z. Geomorph.*, Suppl. i, 156–71.

SCAETTA, H. (1934) *Le climat écologique de la dorsale Nil-Congo*, Inst. Royal Col. Belge, Sect. Sc. Nat. et Méd., Mém. in 4°, no. 3.

SCAETTA, H. (1938) 'Sur la genèse et l'évolution des cuirasses latéritiques. Rôle des cuirasses latéritiques dans l'évolution ultérieure des sols sous-jacents', *C.R. Soc. Biogéogr.* **15**, no. 125, 14–18; no. 126, 26–7.

SCAETTA, H. (1941) 'Limites boréales de la latéritisation actuelle en Afrique occidentale. L'évolution des sols et de la végétation dans la zone des latérites en Afrique occidentale', *C.R. Acad. Sc.* **212**, 129–30, 169–71.

SCHAUFELBERGER, P. (1955) 'Eignen sich die Regenfaktoren langs zur exakten Klassifikation der tropischen Böden', *Petermanns Geogr. Mitt.* **99**, 99–106.

SCHNELL, R. (1945) 'Structure et évolution de la végétation des Monts Nimba (A.O.F.) en fonction du modelé et du sol', *Bull. IFAN*, sér. A, **7**, 80–100.

SCHNELL, R. (1948a) 'Observations sur l'instabilité de certaines forêts de la Haute Guinée Française en rapport avec le modelé et la nature du sol', *Bull. Agron. Congo Belge*, **40**, no. 1, 671–6.

SCHNELL, R. (1948b) 'Sur quelques cas de dégradation de la végétation et du sol observés en Afrique Occidentale Française', *Bull. Agric. Congo Belge*, **40**, 1353–63.

SCHNELL, R. (1950) *La Forêt dense; introduction à l'étude botanique de la région forestière d'Afrique occidentale*, Paris, Lechevalier, 330 p., 13 fig., 22 phot. pl.

SCHNELL, R. (1960) 'Notes sur la végétation et la flore des plateaux gréseux de la Moyenne Guinée et de leurs abords', *Rev. Gén. Botanique*, **67**, 325–99.

SCHOFFIELD, A. N. (1957) 'Nyasaland laterites and their indication on aerial photographs', *Overseas Bull.*, Road. Res. Lab., 5 p.

SCHOKALSKAYA, S. (1953) *Die Böden Afrikas. Die Bedingungen der Bodenbildung. Die Böden und ihre Klassifikation*, Berlin, Akademie Verlag, 403 p., 18 figs., 1 folded map. German translation of a Russian work.

SECK, A. (1955) 'La Moyenne Casamance, étude de géographie physique', *Rev. Géogr. Alpine*, **43**, 707–55.

SEGALEN, P. (1948) 'L'érosion des sols à Madagascar', *Conf. Afr. Sols*, Goma, 1127–37.

SETZER, J. (1955/56) 'Os solos do municipio de São Paulo', *Bol. Paulista de Geogr.*, no. 20, 2–30; no. 22, 26–54; no. 24, 35–56.

SETZER, J. (1956) 'A natureza e as possibilidades do solo no vale do Rio Pardo entre os municipios de Caconde S. P. e Poços de Caldas M.G.', *Rev. Brasil. de Geogr.* **18**, 287–322.

SHERMAN, G. and KANEHIRO, Y. (1954) 'Origin and development of ferruginous concretions in Hawaiian latosols', *Soil Sc.* **77**, 1–8.

SPIJKERMAN, A. C. (1960) 'Les problèmes de l'érosion en Afrique tropicale et les caractéristiques de quelques projets de conservation', *Int. Inst. Land Recl. Improv., Ann. Rept.*, 15–34.

STRAKHOV, M. N. *et al.* (1958) 'Les bauxites, leur minéralogie et leur genèse', *Izv. Akad. Nauk. SSSR*, Moscow, 488 p., 155 fig. (French trans. by BRGM, no. 2640).

STRAUCH, N. (1955) 'A bacia do Rio Doce', *Cons. Nac. Geogr.*, Rio de Janeiro, 195 p.

SYS, C. (1955) 'L'importance des termites sur la constitution des latosols de la région d'Elizabethville', *Sols Afr.* **3**, 392–5.

SYS, C. (1958) *Projet de classification des sols congolais (3ᵉ approximation)*, Publ. INEAC, Yangambi, 91 p., mimeographed.

TAILLEFER, F. (1957) 'Les rivages des Bermudes et les formes de dissolution littorale du calcaire', *Cahiers Géogr.*, Québec, no. 2, 115–38.

TALTASSE, P. (1957) 'Les cabeças de jacaré et le rôle des termites', *Rev. Géomorph. Dyn.* **6**, 166–70.

TEIXEIRA, D. (1960) 'Relêvo e padrões de drenagem na soleira cristalina de Queluz (São Paulo), notas prévias', *Bol. Paulista Geogr.*, no. 36, 3–10.

TEIXEIRA GUERRA, A. (1951a) 'Algunos aspectos geomofológicos do litoral Amapaense', *Bol. Geogr.*, Rio de Janeiro, **9**, 167–78.

TEIXEIRA GUERRA, A. (1951b) 'Reflexões em torno de uma geografia da laterização', *Bol. Geogr.*, Rio de Janeiro, **9**, 669–72.

TEIXEIRA GUERRA, A. (1952) 'Contribuição ao estudo da geologia do território federal Amapá', *Rev. Brasil. Geogr.* **14**, 3–26.

TEIXEIRA GUERRA, A. (1952) 'Importáncia da alteração superficial das rochas', *Bol. Geogr.* Rio de Janeiro **10**, 42–7.

TEIXEIRA GUERRA, A. (1953) 'Formação de lateritas sol a floresta equatorial amazônica (Territorio Federal do Guaporé)', *Rev. Brasil. Geogr.* **14**, 407–26.

TEIXEIRA GUERRA, A. (1954) *Estudio geografico do territorio do Amapá*, IBGE, 366 p., 246 fig.

TEIXEIRA GUERRA, A. (1955a) 'Ocorréencia de lateritas na bacia do Alto Purús', *Rev. Brasil. Geogr.* **17**, 107–14.

TEIXEIRA GUERRA, A. (1955b) 'Os lateritos dos campos do Rio Branco e sua importáncia para a geomorfologia', *Rev. Brasil. Geogr.* **17**, 220–4.

TEIXEIRA GUERRA, A. (1957) 'Estudio geográfico do territorio do Rio Branco', *Cons. Nac. Geogr.*, Rio de Janeiro, 252 p., 150 fig.

TESSIER, F. (1954) 'Oolithes ferrigineuses et fausses latérites dans l'Est de l'Afrique Occidentale Française', *Ann. Ec. Sup. Sc.*, Dakar, vol. 1, 113–28.

TESSIER, F. (1959a) 'La latérite du Cap Manuel à Dakar et ses termitières fossiles', *C.R. Acad. Sc.* **248**, 3320–2.

TESSIER, F. (1959b) 'Termitières fossiles dans la latérite de Dakar (Sénégal). Remarques sur les structures latéritiques', *Ann. Fac. Sc.*, Univ. Dakar, vol. 4, 91–132.

THORBECKE, F. (1921) 'Die Inselberglandschaft von Nord-Tikar', *Festschr. A. Hettner*, Breslau, 215–42.

THORP, M. (1967) 'Closed basins in younger granite massifs, northern Nigeria', *Z. Geomorph.*, n.s., **11**, 459–80.

TONDEUR, G. (1954) *Erosion du sol, spécialement au Congo Belge*, 3rd edn, Brussels, Ministry of Colonies, 240 p.

TOTHILL, J. D. (1952) 'A note on the origin of the soils of the Sudan from the point of view of the man in the field', *Agriculture in the Sudan*, 2nd edn, Oxford University Press, 129–43.

TRICART, J. (1953) 'Erosion naturelle et érosion anthropogène à Madagascar', *Rev. Géomorph. Dyn.* **4**, 225–30.

TRICART, J. (1955) 'Types de fleuves et systèmes morphogénétiques en Afrique occidentale', *Bull. Sect. Géogr. Comité Trav. Hist. et Sc.* **68**, 303–45, 21 pl.

TRICART, J. (1956a) 'Comparaison entre les conditions de façonnement des lits fluviaux en zone tempérée et zone intertropicale', *C.R. Acad. Sc.* **245**, 555–7.

TRICART, J. (1956b) 'Dégradation du milieu naturel et problèmes d'aménagement au Fouta Djalon (Guinée)', *Rev. Géogr. Alpine*, **44**, 7–36.

TRICART, J. (1956c) 'Aspects géomorphologiques du delta du Sénégal', *Rev. Géom. Dyn.* May–June, pp. 65–85.

TRICART, J. (1957a) 'Aspects et problèmes géomorphologiques du littoral occidental de la Côte d'Ivoire', *Bull. IFAN*, sér. A, **19**, 1–20.

TRICART, J. (1957b) 'L'évolution des versants', *Inf. Géogr.* **21**, no. 3 (May–June), 108–16.

TRICART, J. (1957c) 'Observations sur le rôle ameublisseur des termites', *Rev. Géomorph. Dyn.* **8**, 170–2, 179.

TRICART, J. (1958) 'Division morphoclimatique du Brésil atlantique central', *Rev. Géom. Dyn.* **9**, 1–22.

TRICART, J. (1958) 'Les variations quaternaires du niveau marin', *Information Géographique*, **20**, no. 3, 100–4.

TRICART, J. (1959a) 'Géomorphologie dynamique de la moyenne vallée du Niger, Soudan', *Ann. Géogr.* **68**, 333–43.

TRICART, J. (1959b) 'Observations sur le façonnement des rapides des rivières inter-tropicales', *Bull. Sect. Géogr. Comité Trav. Hist. et Sc., 1958*, 289–313.

TRICART, J. (1959c) 'Problèmes géomorphologiques du littoral oriental du Brésil', *Cahiers océanogr.* **11**, 278–308. Photographs.

TRICART, J. (1961a) 'Les caractéristiques fondamentales du système morphogénétique des pays tropicaux humides', *Inform. Géogr.* **25**, 155–69.

TRICART, J. (1961b) 'Le modelé du Quadrilatero Ferrifero, au Sud de Belo Horizonte (Brésil)', *Ann. Géogr.* **70**, 255–72.

TRICART, J. (1961c) 'Note explicative de la carte géomorphologique du delta du Sénégal', *Mém. Bur. Rech. Géol. Min.*, no. 8, 137 p., 9 plates, 3 coloured maps, scale 1 : 100,000.

TRICART, J. (1962a) 'Etude générale de la desserte portuaire de la "SASCA", I : conditions morphodynamiques générales du littoral occidental de Côte d'Ivoire; II : les sites portuaires', *Cahiers océanogr.* **14**, 88–97, 146–61.

TRICART, J. (1962b) 'Quelques éléments de l'évolution géomorphologique de l'Ouest de la Côte-d'Ivoire', *Recher. Africaines*, no. 7, 30–9.

TRICART, J. and CAILLEUX, A. (1965) *Introduction à la géomorphologie climatique*, Paris, SEDES, 306 p., English translation by KIEWIET DE JUNGE, C. J. : *Introduction to Climatic Geomorphology*, London, Longman, 1972.

TRICART, J. and CAILLEUX, A. (1969) *Le Modelé des régions sèches*, Paris, SEDES, 472 p.

TRICART, J. and CARDOSO DA SILVA, T. (1958) 'Algumas observações concernentes as possi-bilidades do planejamento hidráulico no estado da Bahia', *Est. de Geogr. da Bahia*, Salvador, 51–110.

TRICART, J. and SILVA, T. CARDOSO DA (1958) 'Observações de geomorfologia litoral do Rio Vermelho (Salvador)', *Est. de Geogr. da Bahia* (Salvador), 225–43.

TRICART, J., CAILLEUX, A. and RAYNAL, R. (1962) *Les Particularités de la morphogenèse dans les régions de montagnes*, Paris, CDU, 'Cours de l'Université de Strasbourg', mimeographed, 136 p.

TRICART, J., MICHEL, P. and VOGT, J. (1957) 'Oscillations climatiques quaternaires en Afrique occidentale', *Fifth Congr. INQUA*, Madrid-Barcelona.

TROCHAIN, J. (1952) 'Les territoires phytogéographiques de l'Afrique noire française d'après leur pluviométrie', *Recueil Trav. Sc. Montpellier*, Sér. Botanique, no. 5.

TROLL, C. (1958) 'Zur Physiognomik der Tropengewächse', *Jahresber. Ges. Freunden. Univ. zu Bonn*, 75 p., 68 phot.

TROLL, C. (1959) 'Der Physiognomik der Gewächse als Ausdruck der ökologischen Lebens-bedingungen', *Deutscher Geographentag*, **32**, Berlin, 97–122.

TSCHANG TSI LIN (1957) 'Potholes in the river beds of North Taiwan', *Erdkunde*, **11**, 296–303.

TURTON, G. A., MARSH, N. L., MCKENZIE, R. M. and MULCAHY, M. J. (1962) 'The chemistry and mineralogy of laterite soils in the southwest of Western Australia', *Commonw. Scient. Industr. Res. Organ., Soil Publ.* no. 20, 40 p., 5 fig.

US Dept. Agr., Washington, *Reconnaissance Soil Survey of Liberia*, 1951, 108 p.

VAN DER EYCK, J. J. (1957) *Reconnaissance Soil Survey in Northern Surinam*, 99 p., 9 pl.

VANN, J. H. (1959) 'The geomorphology of the Guiana Coast', *2nd Coastal Geogr. Conf.*, Baton Rouge, 153–87.

VANN, J. H. (1963) 'Developmental processes in laterite terrain in Amapá', *Geogr. Rev.* **53**, 406–17.

VAUDOUR, J. (1961) 'La vallée de l'Huyeaune', *Bull. Assoc. Géogr. Franç.*, no. 298, 59–78.

VERSTAPPEN, H. TH. (1953) *Djakarta Bay. A Geomorphological Study on Shoreline Development*, The Hague, Trio, 101 p., 10 fig., 26 phot.

VERSTAPPEN, H. TH. (1954) 'Het kustgebied van noordelijk West-Java op de luchtfoto', *Tijdschr. Kon. Nederl. Aardrijks. Gen.* **71**, 146–52, aerial photos.

VERSTAPPEN, H. (1955) 'Geomorphologischen Notizen aus Indonesien', *Erdkunde*, **9**, 134–44.

VERSTAPPEN, H. TH. (1957) 'Short note on the dunes near Parangritis (Java)', *Tijdschr. Kon. Nederl. Aardrijks. Gen.* **74**, 1–6.

VOGT, J. (1956) *Rapport provisoire de mission à Kéniéba*, Soudan, DFMG.

VOGT, J. (1959a) 'Aspects de l'évolution morphologique récente de l'Ouest Africain', *Ann. Géogr.* **68**, 193–206.

VOGT, J. (1959b) 'Observations nouvelles sur les alluvions inactuelles de Côte-d'Ivoire et de Guinée', *Actes 84ᵉ Congr. Soc. Sav.*, Dijon, 205–10.

VOGT, J. (1959c) 'Note sur la Lobo (Côte d'Ivoire)', Dakar, SGPM, 14 p., mimeographed.

VOGT, J. (1961) 'Badlands du Nord-Dahomey', *Actes 85ᵉ Congr. Soc. Sav.*, 1960, Geogr., 227–39.

VOGT, J. and VINCENT, P. L. (1966) 'Terrains d'altération et de recouvrement en zone intertropicale', *Bull. Bur. Rech. Géol. Min.*, no. 4, 1–111.

WACRENIER, P. (1961) 'Mission de recherche de bauxite au Logone et au Moyo-Kébi (Tchad)', *Bull. Inst. Equat. Rech. Et. Géol. Min.* **14**, 37–42.

WAEGEMANS, G. (1951) 'Introduction à l'étude de la latéritisation et des latérites du Centre Africain', *Bull. Agron. Congo Belge*, **42**, no. 1, 13–56.

WAEGEMANS, G. (1953) 'Signification pédologique de la "stone-line" (note préliminaire)', *Bull. Agric. Congo Belge*, **44**, 521–31.

WAEGEMANS, G. (1954) *Les latérites de Gimbi (Bas Congo)*, Brussels Publ. INEAC, sér. sc., no. 60, 27 p.

WAIBEL, L. (1925) 'Gebirgsbau und Oberflächengestalt der Karrasberge in Südwestafrika', *Mitt. Dtschen Schutz.* **33**.

WALTER, H. (1939) 'Grasland, Savanna und Busch der ariden Teile Afrikas in ihren oekologischen Bedingtheit', *Jahrb. für wiss. Bot.* **87**, no. 750.

WEBBER, B. N. (1959) 'Bauxitisation in the Poços de Caldas district, Brazil', *Mining Engineering*, Aug., 804–9.

WECK, J. (1959) 'Regenwälder, eine vergleichende Studie forstlichen Produktionspotential', *Die Erde*, **90**, 10–37.

WEISSE, G. DE (1952) 'Note sur quelques types de latérites de la Guinée Portugaise', *Proc. XIXth Int. Geol. Congr.*, Algiers, **21**, 171–9.

WENTWORTH, C. K. (1943) 'Soil avalanches in Oahu, Hawaii', *Bull. Geol. Soc. Amer.* **54**, 53–64.

WHITE, L. S. (1949) 'Process of erosion on steep slopes on Oahu, Hawaii', *Amer. J. Sc.* **247**, 168–86.

WILHELMY, H. (1958) *Klimamorphologie der Massengesteine*, Westermann, Braunschweig, 238 p., 137 fig.

WILHELMY, H. (1958) 'Umlaufseen und Dammuferseen tropischer Tieflandflüsse', *Z. für Geom.* **2**, 27–54.

ZANS, V. A. (1961) 'Classification and genetic types of bauxite deposits', *Proc. Vth interGuiana Geol. Conf.*, 1959, Georgetown, 205–11.

ZANS, V. A., LEMOINE, R. G. and ROCH, E. (1960) 'Les "calcaires latéritisés" des Grandes Antilles', *C.R. Somm. Soc. Géol. Fr.*, 177–8.

Biographical sketch of the author

Jean Tricart was born in Montmorency, France, in 1920. In 1943 he finished his university studies and obtained the title of *agrégé*. He received his geography education at the Sorbonne under the direction of the great masters of French geography, Emm. de Martonne, Albert Demangeon, and Max Sorre. From 1945 to 1948 he was the assistant of A. Cholley, under whose direction he wrote his dissertation, *La Partie orientale du Bassin de Paris*, a geomorphological study in two volumes (1949, 1952). At the same time he studied geology and sedimentology under the direction of L. Lutaud and J. Bourcart.

After the retirement of H. Baulig from the University of Strasbourg, Tricart took over the latter's classes and later became director of the Institut de Géographie of that university. In 1949 he created the Laboratoire de Géographie Physique, and in 1956 the Centre de Géographie Appliquée of which he became the director. Also, in 1948, he joined the service charged with the geologic mapping of France and later became principal collaborator, entrusted to direct the mapping of Quaternary formations.

Within the framework of the Laboratoire and later of the Centre de Géographie Appliquée, Tricart developed several new approaches to the study of landforms, including methods of sedimentological investigation in connection with morphogenic processes, the use of palaeoclimatic data in geomorphic cartography (as exemplified in the map of the Senegal delta at the scale of 1 : 50 000, finished in 1956), the study of the Quaternary and of climatic oscillations, particularly in France, Africa, and South America. His main effort, however, went into applied geography which was thereby considerably stimulated. At the date of its foundation in 1956 he became secretary of the Commission of Applied Geography of the International Geographical Union and in 1960 its president. He helped to develop an international geomorphic cartography within the framework of the Commission and later founded a team charged with the establishment of a detailed geomorphic map of France with the help of the Centre National de la Recherche Scientifique. Lastly, in conjunction with the development of the Centre de Géographie Appliquée, he increasingly occupied himself with the study of the interactions between the various elements of the physical geographical environment (geology, geomorphology, climate, vegetation, and soils) in order to understand better the laws which control them and thus to provide a surer foundation to the solution of problems of land improvement and economic development.

As consultant for UNESCO, FAO, the Council of Europe, the Ministerio de Obras Públicas of Venezuela, the Institut de Recherches Agronomiques Tropicales and with teams of the Center of Applied Geography, Professor Tricart has conducted various low cost projects concerned with problems of land classification, agricultural development and management, including soil and hydrologic surveys with detailed maps, whether in France, Senegal, Mali, Argentina, Chile, Peru, Venezuela or Algeria. The Center of Applied Geography of the University of Strasbourg therefore not only provides a unique service among institutes of geography but ensures part of its own income, which is equally unique and may well serve as a model.

Books by J. Tricart

La Partie Orientale du Bassin de Paris: Etude morphologique, 2 vols., Paris, SEDES, 1949, 1952, 474 p.

Le Modelé des chaînes plissées, Paris, CDU, 'Cours de l'Université de Strasbourg', 1954, 1963, 330 p.

Les Massifs anciens, Paris, CDU, 'Cours de l'Université de Strasbourg', 1957, 1963, 252 p.

Les Particularités de la morphogenèse dans les régions de montagnes (with A. Cailleux and R. Raynal), Paris, CDU, 'Cours de l'Université de Strasbourg', 1962, 136 p.

L'Epiderme de la terre. Esquisse d'une géomorphologie appliquée, Paris, Masson, 1962, 167 p.

Le Modelé glaciaire et nival (with A. Cailleux), Paris, SEDES, 1962, 508 p.

Géomorphologie des régions froides, Paris, P.U.F., Collection 'Orbis' directed by A. Cholley, 1963, 289 p. English translation by Edward Watson: *Geomorphology of Cold Environments*, Macmillan and St. Martin's Press, 1969, 320 p.

Le Relief des côtes (cuestas), Paris, CDU, 'Cours de l'Université de Strasbourg', 1963, 137 p.

Le Type de bordures de massifs anciens, Paris, CDU, 'Cours de l'Université de Strasbourg', 1963, 118 p.

Introduction à la géomorphologie climatique (with A. Cailleux), Paris, SEDES, 1965, 306 p. English translation by Conrad J. Kiewiet de Jonge: *Introduction to Climatic Geomorphology*, London, Longman, 1972.

Le Modelé des régions chaudes, forêts et savanes, Paris, SEDES, 1965, 322 p. English translation by Conrad J. Kiewiet de Jonge: *The Landforms of the Tropics, Forests and Savannas*, London, Longman, 1972.

284

Principes et méthodes de la géomorphologie, Paris, Masson, 1965, 496 p.

Le Modelé des régions périglaciaires, Paris, SEDES, 1967, 512 p.

Précis de géomorphologie, vol. I, *Géomorphologie structurale,* Paris, SEDES, 1968, 322 p.

Le Modelé des régions sèches (with A. Cailleux), Paris, SEDES, 1969, 472 p.

Introduction à l'utilisation des photographies aériennes (with S. Rimbert and G. Lutz), vol. I, Paris, SEDES, 1970, 247 p.

Précis de géomorphologie, vol. II, Paris, SEDES, 1972.

Books co-authored by J. Tricart

P. GEORGE and J. TRICART, *L'Europe centrale,* Paris, P.U.F., Collection 'Orbis' directed by A. Cholley, 1963, 2 vols., 753 p.

A. CAILLEUX and J. TRICART, *Initiation à l'étude des sables et des galets,* Paris, CDU, 1963, 3 vols.

Conrad J. Kiewiet de Jonge
(California State University, San Diego)

Index of authors

Index of place names

Subject index

A

acacias, 45, 51

accelerated erosion, 151, 153

acidic: environment, 177; rocks, 27, 28, 30, 31, 32, 134, 143; igneous rocks, 41, 135, 159, 200, 240, 242, 247

actions of man, 51, 53, 142, 151, 168, 170, 194; *see also* intervention; impact of anthropic

adsorption, 34, 132, 179, 180, 236

aeolian: deflation, 97, 98, 176, 194, 196; sands, 190, 247, 257 (deposits)

aerial photographs, XIV, 154

aerophytes, 158

agreste, 45, 167

agricultural development, 236

algae, 25, 27, 58, 89, 95, 173; blue lithophytic, 115; encrusting, 89, 112, 113, 114, 115, 116 (lithothamnion), 119

algal: crusts, 86; ridge, 117, 119

alkaline: earths, 132; minerals, 31, 32, 35; rocks, 28

alluvial: aprons, 199, 245, 247, 249, 252, 253, 255, 257; fans, 29, 209, 210 (cuirassed), 238, 242, 247, 248, 253; fills, 245; grade sizes, 63, 64; plains, 60, 67, 254

alluvium (alluvial deposits), 28, 32, 58, 60, 61, 62, 64, 65, 66, 98, 100, 101, 103, 104, 175, 189, 217, 219, 241, 243, 246, 258

alluviation of fluvio-marine plains, 98–103

alumina, 33, 34, 35, 36, 37, 136, 137, 178, 180, 204, 205, 206, 256

aluminium (oxides, hydroxides), XV, 176, 256

alternations of wetting and drying, 78, 115, 130, 195, 198, 217, 218, 235, 236

amphiboles, 33, 176

amphitheatres, 148, 153, 157, 195, 216

ancient massifs, 141 (European), 250 (France)

andesites, 155

andesitic glass, 140–1

anorthite, 141

anthropic degradations, influences, 12, 50, 52, 54

aplite dikes, 157

'appalachian' relief, 207, 214, 252

applied tropical geomorphology, 145, 246

aragonite, 94

araucarias, 47

arches, 79

arid: climates, 118; lands, regions, 44, 118; zone, 2, 40

aridity index, 9

Atlantic period, 87

atolls, 117

augite, 141

auriferous rubble, 246

avicennia, 91

B

bacteria, 17, 18 (thermal optimum), 31, 47, 95, 176, 177, 178, 255

badlands, 150

bajadas, 199

baobab, 172

bar, 99

Barreiras Series, 141, 250, 252

barrier beaches: Dunkirkian, 17, 97; Eemian, 17; *see also under* offshore bars

barrier reefs, 117, 118

Barriguda, 172

basalts, 19, 30, 39, 65, 103, 147, 151, 155, 159, 160, 200

base-level, local, 83, 86

basement, crystalline, 86, 88

bases, 32, 33, 34, 35, 37, 137

open pits, 43
Ordovician cuesta of West Africa, 211, 212, 218, 245, 247
organic acids, 32, 41, 177, 179, 181; compounds, 255; filling, 103; matter, 216, 235; solutions, 183
oriçangas, 22, 27
orographic rain, 127, 147
Oujlian, 90
overland flow, XV, 16, 17, 19, 29, 40, 42, 53, 61, **124**, 127, 128, 129, 131, 132, 142, 146, 147, **148–52**, 153, 154, 155, 159, 167, 173, 174, 181, 195, 196, 197, 199, 217, 218, 219, 220
oxalic acids, 31, 177
oxidation, 177, 179

P

Pack ice, 58
Palaeocene, 251, 252
paleoclimates, 50, 54, 160, 175
paleoclimatic: legacies, 222; reconstructions, 209
paleosols, 183, 194
pallid zone, 31, 34, 134, 135
palmettos, spiny, 104
parallel retreat of slopes, 147, 148, 153, 155, 195, 211, 222
passes, 117
pastures, 134, 145, 170, 233, 240, 242, 247
patina, 179, 180, 198
peat, peaty soils, 17, 102
pebbles, 29, 58, 59, 62, 64, 65, 66, 69, 73, 75, 80, 90, 94, 97, 103, 110, 116, 140, 146, 152, 184, 185, 186, 215, 218, 244, 245, 247, 250, 257, 258
pediments, **199**, 218, 250 (stepped)
pediplains, 86, 167, **199**, 215, **218–19**, 223, 233, 249, 250
pedogenesis, 16, 151, 187, 188, 194–6, 204
pedologic studies, XIII, 243
pedologists, 189, 193, 198, 206, 256
pellets, 194, 195, 208; pisolitic, 181
perhumid: forested regions, 146; tropical climate, 155, 222, 252, 255; tropics, 118
permeability of soils, 236
petrographic: petrochemical composition, 39, 40, 41, 158; methods, 189; studies, 206, 221
pH, 17, 18, 25, 26, 27, 28, 115
phosphorous, 182
photosynthesis, 14–17, 115, 125
phreatic zone, 138

phyllites, 219
piedmont slopes, 171, 195, **199**, 200, 206, 207, 212, 213, 218, 219, 220, 222, 223, 245, 247, 251, 254, 255, 258; alluvial, 208, 209; antithetic cuirassed, 214; benches, 250; converging, 249; cuirassed, 210, 213, 245, 252; erosional, 218, 250; stepped, 209, 211, 218, 245, 252; terraced, 251
pingen, 22
piping, 200
pisolites, 67, 184, 185, 215, 219, 253
pittings, 106, 107, 108, 109
placer mining, 241
placers, 243
planations, 218, 219, 248–50, 252; stepped microplanations, 147
plank buttresses, 132
plant: formations, 46, 132; formation-types, 19, 49; geographers, 173; screen, 170–4, 235
plants, 14, 177, 234; spiny, 172
plateaus, 170, 196, 199, 206, 209, 218, 239, 243; cuirassed, 132, 190, 197, 201, 206, 209, 211, 213
platform regions, 79, 143
Pleistocene, 77, 97, 111, 117, 118, 119, 132, 140, 153, 169, 186, 192, 209
Pliocene, 83, 207, 251
plurizonal, 118
plutons, 158, 159
podzolic soils, tropical, 134, 148
podzolisation, 31, 32, 34
podzols, tropical, 17, 134
polizonal, 58, 118
polygenic landforms, 247
pool covered (rock) benches, 108, 109, 110
pools, 116 (on reef flat), 117; stepped, 112, 114, 115
porous formations, 29
post-Hercynean surface, 141
potholes, 22, 72–4, 77, 78, 80, 84, 244
pre-Eemian regression, 101
pre-Flandrian regression, 29, 101, 111, 169, 249
prehistorians, 245
prehumic acids, 17
promontories, rocky, 86, 87, 97, 98, 101, 104, 106, 109, 113
pseudo-sands, 133, 136
puddling, 148
pyroclasts, 62
pyroseres, 53
pyroxenes, 33, 176